T0201765

WiMAX SECURITY AND QUALITY OF SERVICE

WiMAX SECURITY AND QUALITY OF SERVICE

AN END-TO-END PERSPECTIVE

Edited by

Seok-Yee Tang
Think Wireless Tech Pte. Ltd., Singapore

Peter Müller
IBM Zurich Research Laboratory, Switzerland

Hamid R. Sharif
University of Nebraska-Lincoln, USA

A John Wiley and Sons, Ltd., Publication

Library of Congress Cataloging-in-Publication Data

WiMAX security and quality of service : an end-to-end perspective / edited by Seok-Yee Tang,
  Peter Müller, and Hamid Sharif.
    p. cm.
  Includes bibliographical references and index.
  ISBN 978-0-470-72197-1 (cloth)
  1. Wireless metropolitan area networks–Security measures. 2. IEEE 802.16 (Standard)
  I. Tang, Seok-Yee, 1968- II. Müller, Peter, 1961 July 8- III. Sharif, Hamid R. (Hamid Reza), 1958-
    TK5105.85.W55 2010
    621.382′1 – dc22

                                                                    2010003319

A catalogue record for this book is available from the British Library.

ISBN 978-0-470-72197-1 (H/B)

Typeset in 10/12 Times by Laserwords Private Limited, Chennai, India
Printed and Bound in Singapore by Markono Print Media Pte Ltd

Contents

Preface

The rapid increase in demand for high-speed broadband wireless networks has spurred the development of new technologies in recent years. Worldwide Interoperability for Microwave Access, known as WiMAX, is one of these technologies. WiMAX is based on the IEEE 802.16 family of standards and offers flexible fixed and mobile wireless solutions along with high-bandwidth services for extended distance coverage and a variety of applications including support of an array of multimedia functions.

IEEE 802.16e is the most popular implementation of this standard; it defines a path of evolution to support high throughput wireless technology for mobile systems. The WiMAX mobile wireless standard, which was defined originally by the IEEE 802.16e-2005 amendment, is now being deployed in more than 140 countries by more than 475 operators.

The 802.16 Medium Access Control (MAC) is designed to support high data transfer for uplink and downlink communications between a base station and a large number of clients for continuous and bursty traffic. WiMAX also supports significant flexible operations across a wide range of spectrum allocation including both licensed and license-exempt frequencies of 2 to 11 GHz. It provides an access system which is based on a request-grant mechanism designed to support service requirements, scalability and efficiency. Along with the bandwidth allocation task, the IEEE 802.16 access mechanism provides a sublayer designed to support privacy and authentication for network access and establishment of connection.

Quality of Service (QoS) is an important factor in WiMAX technologies. WiMAX can provide QoS for wireless broadband communications over an extended coverage area for real-time delay-sensitive applications such as Voice over IP and real-time streaming in stationary or mobile environments. It offers different access methods for different classes of traffic. The 802.16e protocol is a connection-oriented medium access control with service flows as well as a grant-based system which allows centralized control and eliminates overheads and delay of acknowledgements. This in turn provides an effective QoS handling which is fundamentally different from connectionless wireless protocols such as IEEE 802.11. The IEEE 802.16 grant-based MAC can react to QoS requests in real time which reduces the workload of the base stations and produces lower overheads since connections are aggregated.

Additionally, in order to guarantee the QoS of competing services, the fragmentation of the 802.16 Protocol Data Units allows for very large Service Data Units to be sent across frame boundaries. OFDM and OFDMA also provide error correction and interleaving in order to improve QoS. Furthermore, the adaptive modulation techniques used in WiMAX technology result in extended wireless distance coverage areas.

Security is also an important feature of WiMAX and was included in the 802.16 protocol after the failures that restricted the early IEEE 802.11 networks. Security is handled by a privacy sublayer within the WiMAX MAC. WiMAX provides a flexible means for authenticating subscriber stations and users in order to prevent unauthorized use. The 802.16 protocol provides several mechanism designed to protect the service provider and the customer from unauthorized information disclosure.

'WiMAX Security and Quality of Service: An End-to-End Perspective' is a collection of carefully selected articles by researchers with extensive experience with WiMAX. Determining how to provide QoS and security for different applications is a significant issue and the aim of this book is to provide readers with an in-depth discussion of security and QoS considerations in WiMAX based communications. Many books and articles have addressed WiMAX and the IEEE 802.16e protocol, but an end-to-end prospective on security and QoS has been missing. This book is split into four parts. Part A introduces an overview of the end-to-end WiMAX architecture, its protocols and system requirements. Three chapters in Part B discuss security issues in WiMAX, while in Part C five chapters examine QoS in detail. Advanced topics on WiMAX architecture, resource allocation, mobility management and interfacing WiFi and WiMAX are discussed in Part D.

Part A: Introduction

Chapter 1 provides an overview of end-to-end WiMAX network architecture. The objective of this chapter is to discuss the detail of different wireless communications technologies, mobile WiMAX, radio interface specifications for WiMAX, different interface specifications and various interoperability issues of WiMAX networks, as well as interoperability among the different WiMAX network vendors.

Part B: Security

Chapter 2 analyzes WiMAX security as defined in the different released versions of the IEEE 802.16 standards. It provides an overview of the WiMAX 802.16 networks and discusses the main security requirements to be met by a standard for broadband access. It then describes the security mechanisms that are to be guaranteed by the security sublayer and describes the weaknesses revealed in the initial versions, namely those related to fixed WiMAX. In this chapter, the security amendments made in the recent versions of mobile WiMAX are described and analyzed.

Key management in 802.16e is an important security issue and is discussed in Chapter 3. This chapter focuses specifically on the key management scheme of 802.16. Key derivation procedures and the key hierarchy of PKM version 2 are examined and discussed thoroughly. The weaknesses and countermeasures are identified and analyzed. Some comparisons with IEEE 802.11i and Third Generation (3G) mobile networks standards are also provided.

In Chapter 4, WiMAX network security is examined. The analysis is based on WiMAX Forum specification 1.2 and focuses on the standards, technical challenges the solutions for the issues of; 1) integration of authentication techniques and management of AAA (Authorization, Authentication, Accounting); 2) IP addressing and networking issues; and

3) distribution of the QoS parameters. These topics are analyzed from the perspective of the network manager and the interaction between the access network and the back-end.

Part C: Quality of Service

Chapter 5 focuses on cross-layer QoS architecture, highlighting both PMP and mesh topology aspects and the differences between them. Each type of topology presents a different means of obtaining QoS; however other important elements such as bandwidth allocation scheduling and call admission control algorithms are left to vendor implementation. This deficiency with reference to the MAC and PHY layers as well as other important issues are discussed in this chapter. The challenges for WiMAX QoS are also discussed, focusing the future of QoS in the IP world for multimedia applications.

QoS in Mobile WiMAX is addressed in Chapter 6. Here, QoS management in WiMAX networks is discussed. The analysis focuses on demonstrating how mobile WiMAX technology offers continuity of services while providing enhanced QoS guarantees in order to meet subscribers' demands. The architectural QoS requirements that have to be fulfilled during subscribers' mobility and the mechanisms constructed by the Mobile WiMAX network to provide QoS are discussed in this chapter. Service flow, the 'connection-oriented' nature of the MAC layer, the bandwidth request, and allocation procedures and the scheduling service are also examined.

Mobility Management in WiMAX Networks is addressed in Chapter 7. The authors discuss the amendment of the IEEE 802.16d-2004 standard which provides improvements related mainly to mobility management. This chapter also examines the logical architecture of a mobile WiMAX network defined by the Network Working Group1 (NWG) of the WiMAX Forum. Other topics discussed in this chapter include horizontal and vertical handover mechanisms and means for their improvement, as well as analysis of co-existence with other access technologies in networks in the future.

Chapter 8 discusses the challenges facing QoS in the handover process. This chapter describes the challenges that the handover process represents for the QoS performance indicators in full mobility scenarios. It also describes the application of QoS requirements for full mobility and the requirements relating to end-to-end performance. Timing and performance considerations in the handover process and the Media Independent Handover Initiative (MIH or IEEE802.21) are also discussed. The efficient scheduling of the handover process and its influence on handover performance, end-to-end quality of service and a handover performance analysis are the other topics presented in this chapter.

Resource Allocation in Mobile Networks is discussed in Chapter 9. Here, a technical overview is presented of the emerging Mobile WiMAX solution for broadband wireless and important issues related to QoS in Mobile WiMAX are discussed. Additionally, resource allocation in Mobile WiMAX is examined in this chapter. Issues related to scheduling and method of channel access for different Service Flows in MAC layer and burst profiles based on the AMC slot structure in OFDMA frame are examined. Multiuser resource allocation, which involves OFDMA, AMC and multiuser diversity, is presented for downlink mobile WiMAX networks. Furthermore, the Channel Aware Class Based Queue (CACBQ), which is an adaptive cross-layer for scheduling and slot allocation, is introduced.

Part D: Advanced Topics

Chapter 10 provides a discussion of QoS issues and challenges in WiMAX and WiMAX MMR networks. MAC-level QoS scheduling algorithms in WiMAX networks for multimedia traffic are also provided. This includes scheduling algorithms designed for a WiMAX mobile multi-hop relay (MMR) network. This chapter also discusses the characteristics of real-time traffic and the different codecs used for voice and video. A description of a few algorithms on uplink scheduling for real-time traffic inWiMAX networks is also provided. Additionally, MMR based WiMAX networks and downlink scheduling schemes for MMR based WiMAX networks are examined.

The Integration of WiFi and WiMAX Networks is an important issue and is discussed in Chapter 11. The deployment of an architecture that allows users to switch seamlessly between WiFi and WiMAX networks would afford several advantages to both users and service providers. However, WiMAX and WiFi networks have different protocol architectures and QoS support mechanisms; therefore an adaptation of protocol is required for their internetworking. This chapter outlines the design tenets for an interworking architecture between both WiFi and WiMAX technologies. The authors also define the various functional entities and their interconnections as well as end-to-end protocol layering in the interworking architecture, network selection and discovery and IP address allocation. Additionally, details are provided for the functional architecture and processes associated with security, QoS and mobility management.

QoS simulation and an enhanced solution for cell selection for WiMAX networks is discussed in Chapter 12. In this chapter, the authors examine the major WiMAX network simulation tools. A detailed system model for a cell selection algorithm is presented in this chapter. The authors have also performed simulation for QoS in a WiMAX network for several scenarios. An analysis of their simulation results are also provided.

The editors believe that this book is unique and significant in that it provides a complete end-to-end perspective on QoS and security issues in WiMAX and that it can be of great assistance to a large group of scientists, engineers and the wireless community with regard to the fast growing era of multimedia applications over wireless networks.

Seok-Yee Tang
Hamid R. Sharif
Peter Müller

Acknowledgement

To Ursula, Samira, Francis and Alena.

Peter Müller

To my three boys and the love of my life for her encouragement,
inspiration and support.

Hamid space

In memory of my mother.
To my husband Chong Ming, my best friend Bibi, and my sister Seok Hun.

Seok-Yee Tang

The editors would like to thank and acknowledge all authors for their contribution to the book content and their cooperation during this book's preparation process.

We would also like to thank the John Wiley & Sons Ltd team for their assistance and encouragement in making of this book.

List of Contributors

Editors

Peter Müller
IBM Zurich Research Laboratory, Switzerland;
Formerly with Siemens R&D, Switzerland

Hamid R. Sharif
University of Nebraska-Lincoln, USA

Seok Yee Tang
Think Wireless Tech Pte. Ltd., Singapore

Authors

Luca Adamo
Department of Electronics and Telecommunications
University of Florence, Italy

Marina Aguado
ETSI, Departamento de Electrónica y Telecomunicaciones
University of the Basque Country, Spain

Marion Berbineau
INRETS (Institut National de recherche sur les Transports et leur Sécurité),
Université Lille Nord de France,
Villeneuve d'Ascq, France

Debika Bhattacharyya
Head, Department of CSE
Institute of Engineering & Management, Salt Lake,
Kolkata, India

Noureddine Boudriga
Communication Networks and Security Research Laboratory (CNAS),
University of the 7th November at Carthage, Tunisia

Daniel Câmara
EURECOM
Mobile Communications Department
Sophia-Antipolis Cedex, France

Mohuya Chakraborty
Head, Department of Information Technology
Institute of Engineering & Management, Salt Lake,
Kolkata, India

Hakima Chaouchi
Telecom and Management Sud Paris
Evry cedex, France

Floriano De Rango
DEIS Department
University of Calabria, Italy

Romano Fantacci
Head of LaRT Laboratory
Department of Electronics and Telecommunications
University of Florence, Italy

Fethi Filali
QU Wireless Innovations Center
Doha, Qatar

Stefanos Gritzalis
Laboratory of Information and Communication Systems Security
Department of Information and Communication Systems Engineering
University of the Aegean, Karlovassi, Greece

Shen Gu
Department of Electronic Engineering
Shanghai Jiaotong University
Shanghai, China

Eduardo Jacob
ETSI, Departamento de Electrónica y Telecomunicaciones
University of the Basque Country, Spain

Georgios Kambourakis
Laboratory of Information and Communication Systems Security
Department of Information and Communication Systems Engineering
University of the Aegean, Karlovassi, Greece

Neila Krichene
Communication Networks and Security Research Laboratory (CNAS),
University of the 7th November at Carthage, Tunisia

Kiran Kumari
Indian Institute of Technology Madras
Chennai, India

Leonardo Maccari
Department of Electronics and Telecommunications
University of Florence, Italy

Andrea Malfitano
DEIS Department
University of Calabria, Italy

Salvatore Marano
DEIS Department
University of Calabria, Italy

Ikbal Chammakhi Msadaa
EURECOM
Mobile Communications Department
Sophia-Antipolis Cedex, France

Srinath Narasimha
Indian Institute of Technology Madras
Chennai, India

Slim Rekhis
Communication Networks and Security Research Laboratory (CNAS),
University of the 7th November at Carthage, Tunisia

Ivan Lledo Samper
Bournemouth University, UK

Krishna M. Sivalingam
Indian Institute of Technology Madras, Chennai, India;
Formerly with University of Maryland Baltimore County,
Baltimore, USA

Jiajing Wang
Department of Electronic Engineering
Shanghai Jiaotong University
Shanghai, China

Xinbing Wang
Department of Electronic Engineering
Shanghai Jiaotong University
Shanghai, China

Yuan Wu
Department of Electronic Engineering
Shanghai Jiaotong University
Shanghai, China

Tara Ali Yahiya
Computer Science Laboratory, Paris-Sud 11 University, France

List of Acronyms

2G	Second Generation mobile networks
3G	Third Generation mobile networks
3GPP	Third Generation Partnership Project
3GPP2	Third Generation Partnership Project 2
4G	Fourth Generation mobile networks
AAA	Authorization, Authentication and Accounting
AAS	Adaptive Antenna System
AAT	Advanced Antenna Technology
AC	Access Category
ACK	Acknowledge
ACM	Adaptive Coding and Modulation
ACs	Access Categories
AES	Advanced Encryption Standard
AIFS	Arbitration Interframe Space
AK	Authorization Key
AKA	Authentication and Key Agreement
AKID	Authentication Key Identifier
AMC	Adaptive Modulation and Coding
AMR	Adaptive Multi Rate
AP	Access Point
AR	Access Router
ARQ	Automatic Repeat Request
AS	Authentication Server
ASN	Access Service Network
ASN	Abstract Syntax Notation
ASN-GW	Access Service Network Gateway
ASP	Application Service Provider
ATM	Asynchronous Transfer Mode
AUTN	Authentication Token
AV	Authentication Vector
AWGN	Additive White Gaussian Noise
BCID	Basic Connection Identity
BE	Best Effort

BER	Bit Error Rate
BLER	Block Error Rate
BPSK	Binary Phase Shift Keying
BR	Bandwidth Request
BRAS	Broadband Access Server
BS	Base Station
BSID	Base Station Identity
BW	Bandwidth
BWA	Broadband Wireless Access
CA	Certification Authority
CAC	Call Admission Control
CACBQ	Channel Aware Class Based Queue
CAPF	Cost Adjusted Proportional Fair
CBC	Cipher Block Chaining
CBR	Constant Bit Rate
CCM	Counter with CBC-MAC
CDMA	Code Division Multiple Access
CELP	Code Excited Linear Prediction
CID	Connection Identifier
CINR	Carrier to Interference plus Noise Ratio
CK	Cipher key
CMAC	Cipher Message Authentication Code
CMIP	Client-MIP
COA	Care-of-Address
COTS	Commercial Off-The-Shelf
CPE	Consumer Premises Equipment
CPS	Common Part Sublayer
CQI	Channel Quality Indicator
CQICH	Channel Quality Indicator Channel
CRC	Cyclic Redundancy Check
CRL	Certificate Revocation List
CS	Convergence Sublayer
CSC	Connectivity Service Controllers
CSCl	Convegence Sublayer Classifiers
CSMA CA	Carrier Sense Multiple Access with Collision Avoidance
CSN	Connectivity Service Network
CSP	Common Part Sub-layer
CSs	Service Classes
CW	Contention Window
DAD	Duplicate Address Detection
DCD	Downlink Channel Descriptor
DCF	Distributed Coordination Function
DER	Distinguished Encoding Rule
DES	Data Encryption Standard
DFR	Decode and Forward Relay
DFS	Dynamic Frequency Selection

DHCP	Dynamic Host Configuration Protocol
DHMM	Dynamical Hierarchical Mobility Management
DIAMETER	Protocol extending RADIUS
DiffServ	Differentiated Service
DL	Downlink
DOCSIS	Data Over Cable Service Interface Specification
DoD	Department of Defence
DoS	Denial of Service
DSA-REQ	Dynamic Service Addition request
DSA-RSP	Dynamic Service Addition response
DSL	Digital Subscriber Line
DSSS	Direct Sequence Spread Spectrum
EAP	Extensible Authentication Protocol
EAP-AKA	EAP-Authentication and Key Agreement
EAPOL	EAP over LAN
EAP-TTLS	EAP-Tunneled Transport Layer Security
EC	Encryption Control
EDCA	Enhanced Distributed Channel Access
EDCF	Enhanced Distributed Coordination Function
EDF	Earliest Deadline First
EFR	Enhanced Full Rate
EIK	EAP Integrity Key
EKS	Encryption Key Sequence
ertPS	Extended Real Time Polling Service
ETSI	European Telecommunications Standards Institute
E-UTRAN	Evolved UMTS Terrestrial Radio Access Network
FA	Foreign Agent
FBack	Fast Binding Acknowledgment
FBSS	Fast Base Station Switching handover
FBU	Fast Binding Update
FCH	Frame Control Header
FDD	Frequency Division Duplex
FDMA	Frequency Division Multiple Access
FEC	Forward Error Correction
FFT	Fast Fourier Transform
FHSS	Frequency Hopping Spread Spectrum
FIFO	First In First Out
FPC	Fast Power Control
FTP	File Transfer Protocol
FUSC	Full Usage of Subchannels
GKDA	Group-based Key Distribution Algorithm
GKEK	Group Key Encryption Key
GKMP	Group Key Management Protocol
GMH	Generic MAC Frame Header
GPC	Grant Per Connection
GPRS	General Packet Radio Service

GSA	Group Security Association
GSAID	Group SAID
GSM FR	GSM Full rate
GSM	Global System for Mobile Communications
GTEK	Group Traffic Encryption Key
GTK	Group Transient Key
HA	Home Agent
HAck	Handover Acknowledgment
HAP	High Altitude Platform
HARQ	Hybrid Automatic Repeat Request
HCCA	HCF Controlled Channel Access
HCF	Hybrid Coordination Function
HCS	Header Check Sequence
HDR	High Data Rate
HDTV	High-definition TV
HHO	Hard Handover
HI	Handover Initiation
HIPERMAN	High Performance Radio Metropolitan Area Network
HMAC	Hash Message Authentication Code
HNSP	Home Network Service Provider
HO	Handover
HOA	Home-of-Address
HOKEY	Handover Keying (Group)
HoL	Head of Line
HSPA	High-Speed Packet Access
HSPA+	Evolved HSPA
HT	Header Type
HUF	Highest Urgency First
ICV	Integrity Checking Value
ID	Identifier
IE	Information Element
IEEE	Institute of Electrical & Electronics Engineers, Inc.
IETF	Internet Engineering Task Force
IK	Integrity Key
IKE	Internet Key Exchange (protocol)
ILBC	Internet Low Bit rate Codec
IP	Internet Protocol
IPv6	Internet Protocol version 6
ISI	Intersymbol Interference
ISO	International Standard Organization
ISP	Internet Service Provider
ITU	International Telecommunication Union
IV	Initialization Vector
KDF	Key Derivation Function
KEK	Key Encryption Key
L2	Layer 2

L3	Layer 3
LAN	Local Area Network
LDPC	Low Density Parity Check
Link ID	Link Identifier
LOS	Line of Sight
LRC	Low Runtime Complexity
LTE	Long Term Evolution
M3	Mesh Mobility Management
MAC	Media Access Control
MAC	Message Authentication Code
MAN	Metropolitan Area Network
MAP	Media Access Protocol
MAP	Mesh Access Point
MBRA	Multicast and Broadcast Rekeying Algorithm
MBS	Multicast and Broadcast Service
MCS	Modulation and Coding Scheme
MDHO	Macro Diversity Handover
MIB	Management Information Base
MIC	Message Integrity Code
MICS	Media-Independent Command Service
MIES	Media-Independent Event Service
MIH	Media-Independent Handover
MIHF	Media-Independent Handover Function
MIHU	Media-Independent Handover User
MIIS	Media-Independent Information Service
MIM	Man In the Middle
MIMO	Multiple Input Multiple Output
MIP	Mobile IP
MMR	Mobile Multi-hop Relay
MMS	Multimedia Messaging Service
MN	Mobile Node
MOS	Mean Opinion Score
MP	Mesh Point
MPDU	MAC Protocol Data Unit
MPEG	Moving Picture Expert Group
MPP	Mesh Portal Point
MRR	Minimum Reserved Rate
MS	Mobile Station
MS	Mobile Subscriber Station
MSB	Most Significant Bit
MSCHAPv2	Microsoft Challenge-Handshake Authentication Protocol
mSCTP	Mobile Stream Control Transmission Protocol
MSDU	MAC Service Data Unit
MSE	Mean Square Error
MSID	Mobile Station Identifier
MSK	Master Session Key

MSO	Multi-Services Operator
MSR	Maximum Sustained Rate
MSS	Mobile Subscriber Station
MTK	MBS Traffic Key
MVNO	Mobile Virtual Network Operator
NAP	Network Access Provider
NAP	Network Access Point
NAR	New Access Router
NBR	Neighbor
NCoA	New Care of Address
NGWS	Next Generation Wireless System
NLOS	Non Line-of-Sight
NMS	Network Management System
Node ID	Node Identifier
NRM	Network Reference Model
nrtPS	Non-Real-Time Polling Service
NSP	Network Service Provider
NSSK	Needham Schroeder Secret Key Protocol
NTSC	National television System Committee
NWG	Network Working Group
OCSP	Online Certificate Status Protocol
O-DRR	Opportunistic- Deficit Round Robin
OFDM	Orthogonal Frequency Division Multiplex
OFDM2A	Orthogonal Frequency Division Multi-hop Multi-Access
OFDMA	Orthogonal Frequency Division Multiple Access
OSS	Operator Shared Secret
OTA	Over-The-Air
P2MP	Point to Multi-Point
PAR	Previous Access Router
PCF	Point Coordination Function
PCM	Pulse Code Modulation
PCMCIA	Personal Computer Memory Card International Association
PCoA	Previous Care of Address
PDAs	Personal Digital Assistants
PDU	Protocol Data Unit
PEAP	Protected EAP
PEAQ	Perceptual Evaluation of Audio Quality
PER	Packet Error Rate
PESQ	Perceptual Evaluation of Speech Quality
PF	Proportionate Fair
PFMR	Proportional Fair with Minimum/Maximum Rate Constraints
PHS	Packet Header Suppression
PHY	Physical Layer
PKC	Public Key Certificates
PKM	Privacy Key Management
PM	Poll Me bit

PMIP	Proxy-MIP
PMK	Pairwise Master Key
PMM	Packet Mobility Management (protocol)
PMP	Point to Multipoint
PN	Packet Number
PoA	Point of Attachment
PPP	Point-to-Point
PPPoE	Point-to-Point Protocol over Ethernet
Pre-PAK	pre-Primary Authorization Key
PrRtAdv	Proxy Router Advertisement
PS	Privacy Sublayer
PSK	Pre-Shared Key
PSNR	Peak Signal to Noise Ratio
PSOR	PF Scheduling for OFDMA Relay Networks
PSTN	Public Switched Telephone Network
PTK	Pairwise Transient Key
PTP	Point To Point
PUSC	Partial Usage of Subchannels
QAM	Quadrature Amplitude Modulation
QoS	Quality of Service
QoS	Quality of Signal
QPSK	Quadrature Phase Shift Keying
RADIUS	Remote Authentication Dial-In User Service
RAND	Random Number
RC	Resource Controller
REG-REQ	Registration Request
REG-RSP	Registration Response
REQ	Request
RES	Result
RF	Radio Frequency
RLC	Radio Link Control
RNG-REQ	Ranging Request
RNG-RSP	Ranging Response
RNM	Reference Network Model
ROC	Rollover Counter
RP	Reference Point
RR	Round Robin
RRA	Radio Resource Agent
RRC	Radio Resource Control
RRM	Radio Resource Management
RRP	Registration RePly
RRQ	Registration ReQuest
RS	Relay Station
RSA	Rivest, Shamir, and Adelman
RSP	Response
RSS	Received Signal Strength

RSSI	Received Signal Strength Indication
RTG	Receive/Transmit Transition Gap
rtPS	Real Time Polling service
RtSolPr	Router Solicitation for Proxy Advertisement
SA	Security Association
SAID	SA Identifier
SAP	Service Access Point
SBC-RSP	SS Basic Capabilitiy response
SC	Single Carrier
SCN	Service Class Name
SCTP	Stream Control Transmission Protocol
SDU	Segment Data Units
SeS	Security Sublayer
SFID	Service Flow IDentifier
SGKEK	Sub-Group Key Encryption Key
SHA	Secure Hash Algorithm
SIM	Subscriber Identity Module
SINR	Signal to Interference-plus-Noise Ratio
SIP	Session Initiation Protocol
SIR	Signal to Interference Ratio
SMS	Short Message Service
SNIR	Signal to Noise + Interference Ratio
SNMP	Simple Network Management Protocol
SNR	Signal to Noise Ratio
SOFDMA	Scalable Orthogonal Frequency Division Multiple Access
SR	Superior Router
SS	Spectrum Sharing
SS	Subscriber Station
SSCS	Service Specific Convergence Sublayer
SSID	Service Set Identifier
STS	Sub-channels of a Time Slot
TCP	Transmission Control Protocol
TDD	Time Division Duplex
TDMA	Time Division Multiple Access
TEK	Traffic Encryption Key
TFTP	Trivial File Transfer Protocol
THBA	Two-level Hierarchical Bandwidth Allocation scheme
TLS	Transport Layer Security
TLV	Type-Length-Value
TPP	Two-Phase Proportionating
TR	Transmit Receive
TTG	Transmit/Receive Transition Gap
TTLS	Tunneled Transport Layer Security
TTP	Trusted Third Party
TXOP	Transmission Opportunities
UCD	Uplink Channel Descriptor

UDP	User Datagram Protocol
UGS	Unsolicited Grant Service
UGS-AD	Unsolicited Grant Service-Activity Detection
UL	Uplink
UL-MAP	Uplink MAP
UMTS	Universal Mobile Telecommunications System
UNA	Unsolicited Neighbor Advertisement
VBR	Variable Bit Rate
VCEG	Video Coding Experts Group
VHDA	Vertical Handoff Decision Algorithm
VHO	Vertical Handover
VNSP	Visited Network Service Provider
VoD	Video on Demand
VoIP	Voice over IP
W2-AP	WiMAX/WiFi Access Point
WBA	Wireless Broadband Access
WEIRD	WiMAX Extension to Isolated Research Data networks
WEP	Wired EquivalentPrivacy
WFPQ	Weighted Fair Priority Queuing
WFQ	Weighted Fair Queuing
Wibro	Wireless Broadband
WiFi	Wireless Fidelity
WiMAX	Worldwide Interoperability for Microwave Access
WiMESH	WiMAX Mesh
WLAN	Wireless Local Area Network
WMAN	Wireless Metropolitan Area Network
WRI	WiMAX Roaming Interface
WRR	Weighted Round Robin
WRX	WiMAX Roaming Exchange
WWAN	Wireless Wide Area Network
XDSL	X Digital Subscriber Line
XML	Extensible Markup Language
XRES	Expected Response

List of Figures

List of Tables

Part A

Introduction

1

Overview of End-to-End WiMAX Network Architecture

Dr Mohuya Chakraborty, Ph.D (Engg.)
Member IEEE, Head, Department of Information Technology, Institute of Engineering & Management, Salt Lake, Kolkata, India

Dr Debika Bhattacharyya, Ph.D (Engg.)
Member IEEE, Head, Department of Computer Science & Engineering, Institute of Engineering & Management, Salt Lake, Kolkata, India

1.1 Introduction

WiMAX, the *Worldwide Interoperability for Microwave Access*, is a telecommunications technology that provides for the wireless transmission of data in a variety of ways, ranging from point-to-point links to full mobile cellular-type access. The WiMAX forum describes WiMAX as a standards-based technology enabling the delivery of last mile wireless broadband access as an alternative to cable and Digital Subscriber Line (DSL).

WiMAX network operators face a big challenge to enable interoperability between vendors which brings lower costs, greater flexibility and freedom. So it is important for network operators to understand the methods of establishing interoperability and how different products, solutions and applications from different vendors can coexist in the same WiMAX network.

This chapter aims to assist readers in understanding the end-to-end WiMAX network architecture in detail including the different interface specifications and also the various interoperability issues of the WiMAX network. Section 1.1 gives an overview of different wireless communications technologies. WiMAX technology is introduced in section 1.2. Section 1.3 describes the concept of mobile WiMAX. An overview of the end-to-end WiMAX network architecture is discussed in section 1.4. Radio interface specifications

WiMAX Security and Quality of Service: An End-to-End Perspective   Edited by Seok-Yee Tang,
Peter Müller and Hamid Sharif
© 2010 John Wiley & Sons, Ltd

for WiMAX are discussed in section 1.5. Section 1.6 throws light upon the interoperability amongst the different WiMAX network vendors. The chapter concludes in section 1.7.

1.2 Wireless Primer

Wireless means transmitting signals using radio waves as the medium instead of wires. Wireless technologies are used for tasks as simple as switching off the television or as complex as supplying the sales force with information from an automated enterprise application while in the field. Wireless technologies can be classified in different ways depending on their range. Each wireless technology is designed to serve a specific usage segment. The requirements for each usage segment are based on a variety of variables, including bandwidth needs, distance needs and power. Some of the inherent characteristics of wireless communications systems which make it attractive for users are given below:

- *Mobility*: A wireless communications system allows users to access information beyond their desk and conduct business from anywhere without having wire connectivity.
- *Reachability*: Wireless communications systems enable people to be better connected and reachable without any limitation as to location.
- *Simplicity*: Wireless communication systems are easy and fast to deploy in comparison with cabled networks. Initial setup cost may be a bit high but other advantages overcome that high cost.
- *Maintainability*: Being a wireless system, you do no need to spend too much to maintain a wireless network setup.
- *Roaming Services*: Using a wireless network system you can provide a service any where any time including train, busses, aeroplanes, etc.
- *New Services*: Wireless communications systems provide new smart services such as the Short Message Service (SMS) and Multimedia Messaging Service (MMS).

1.2.1 Wireless Network Topologies

There are basically three ways to setup a wireless network:

- *Point-to-point bridge*: As you know a bridge is used to connect two networks. A point-to-point bridge interconnects two buildings having different networks. For example, a wireless LAN bridge can interface with an Ethernet network directly to a particular access point.
- *Point-to-multipoint bridge*: This topology is used to connect three or more LANs that may be located on different floors in a building or across buildings.
- *Mesh or ad hoc network*: This network is an independent local area network that is not connected to a wired infrastructure and in which all stations are connected directly to one another.

1.2.2 Wireless Technologies

Wireless technologies can be classified in different ways depending on their range. Each wireless technology is designed to serve a specific usage segment. The requirements for each usage segment are based on a variety of variables, including Bandwidth needs, Distance needs and Power.

- *Wireless Wide Area Network (WWAN)*: This network enables us to access the Internet via a wireless wide area network (WWAN) access card and a PDA or laptop. They provide a very fast data speed compared with the data rates of mobile telecommunications technology, and their range is also extensive. Cellular and mobile networks based on CDMA and GSM are good examples of WWAN.
- *Wireless Personal Area Network (WPAN)*: These networks are very similar to WWAN except their range is very limited.
- *Wireless Local Area Network (WLAN)*: This network enables us to access the Internet in localized hotspots via a wireless local area network (WLAN) access card and a PDA or laptop. It is a type of local area network that uses high-frequency radio waves rather than wires to communicate between nodes. They provide a very fast data speed compared with the data rates of mobile telecommunications technology, and their range is very limited. WiFi is the most widespread and popular example of WLAN technology.
- *Wireless Metropolitan Area Network (WMAN)*: This network enables us to access the Internet and multimedia streaming services via a Wireless Region Area Network (WRAN). These networks provide a very fast data speed compared with the data rates of mobile telecommunication technology as well as other wireless networks, and their range is also extensive.

1.2.3 Performance Parameters of Wireless Networks

These are the following four major performance parameters of wireless networks:

- *Quality of Service (QoS)*: One of the primary concerns about wireless data delivery is that, like the Internet over wired services, QoS is inadequate. Lost packets and atmospheric interference are recurring problems with wireless protocols.
- *Security Risk*: This has been another major issue with a data transfer over a wireless network. Basic network security mechanisms are such as the Service Set Identifier (SSID) and Wired Equivalent Privacy (WEP). These measures may be adequate for residences and small businesses but they are inadequate for entities that require stronger security.
- *Reachable Range*: Normally a wireless network offers a range of about 100 metres or less. Range is a function of antenna design and power. Nowadays the range of wireless is extended to tens of miles so this should no longer be an issue.
- *Wireless Broadband Access (WBA)*: Broadband wireless is a technology that promises high-speed connection over the air. It uses radio waves to transmit and receive data directly to and from the potential users whenever they want it. Technologies such as 3G, WiFi and WiMAX work together to meet unique customer needs. Broadband Wireless Access (BWA) is a point-to-multipoint system which is made up of base station and subscriber equipment. Instead of using the physical connection between the base station and the subscriber, the base station uses an outdoor antenna to send and receive high-speed data and voice-to-subscriber equipment. BWA offers an effective,

complementary solution to wireline broadband, which has become recognized globally by a high percentage of the population.

1.2.4   WiFi and WiMAX

Wireless Fidelity (WiFi) is based on the IEEE 802.11 family of standards and is primarily a local area networking WiMAX similar to WiFi, but on a much larger scale and at faster speeds. A nomadic version would keep WiMAX-enabled devices connected over large areas, much like today's cell phones. We can compare it with WiFi based on the following factors:

- *IEEE Standards*: WiFi is based on the IEEE 802.11 standard whereas WiMAX is based on IEEE 802.16. However both are IEEE standards.
- *Range*: WiFi typically provides local network access for around a few hundred feet with speeds of up to 54 Mbps, a single WiMAX antenna is expected to have a range of up to four miles. Ranges beyond 10 miles are certainly possible, but for scalability purposes may not be desirable for heavily loaded networks. As such, WiMAX can bring the underlying Internet connection needed to service local WiFi networks.
- *Scalability*: WiFi is intended for LAN applications, users range from one to tens with one subscriber for each Consumer Premises Equipment (CPE) device. It has fixed channel sizes (20 MHz). WiMAX is designed to support from one to hundreds of CPEs efficiently, with unlimited subscribers behind each CPE. Flexible channel sizes from 1.5 MHz to 20 MHz.
- *Bit rate*: WiFi works at 2.7 bps/Hz and can peak at up to 54 Mbps in a 20 MHz channel. WiMAX works at 5 bps/Hz and can peak up to 100 Mbps in a 20 MHz channel.
- *QoS*: WiFi does not guarantee any QoS but WiMAX will provide you with several level of QoS. As such, WiMAX can bring the underlying Internet connection needed to service local WiFi networks. WiFi does not provide ubiquitous broadband while WiMAX does. A comparative analysis of WiFi and WiMAX vis-à-vis different network parameters is given in Table 1.1.

1.3   Introduction to WiMAX Technology

WiMAX is a *metropolitan area network* service that typically uses one or more base stations that can each provide service to users within a 30-mile radius for distributing broadband wireless data over wide geographic areas. WiMAX offers a rich set of features with a great deal of flexibility in terms of deployment options and potential service offerings. It can provide two forms of wireless service:

- Non-Line-of-Sight (NLoS) service – This is a WiFi sort of service. Here a small antenna on the computer connects to the WiMAX tower. In this mode, WiMAX uses a lower frequency range (~2 GHz to 11 GHz) similar to WiFi.
- Line-of-Sight (LoS) service – Here a fixed dish antenna points straight at the WiMAX tower from a rooftop or pole. The LoS connection is stronger and more stable, so it's able to send a lot of data with fewer errors. LoS transmissions use higher frequencies, with ranges reaching a possible 66 GHz.

Table 1.1 WiFi vs WiMAX

Feature	WiMAX (802.16a)	WiFi (802.11b)
Primary Application	Broadband Wireless Access	Wireless LAN
Frequency Band	Licensed/Unlicensed 2 G to 11 GHz	2.4 GHz ISM
Channel Bandwidth	Adjustable 1.25 M to 20 MHz	25 MHz
Half/Full Duplex	Full	Half
Radio Technology	OFDM (256-channels)	Direct Sequence Spread Spectrum
Bandwidth Efficiency	<=5 bps/Hz	<=0.44 bps/Hz
Modulation	BPSK, QPSK, 16-, 64-, 256-QAM	QPSK
FEC	Convolutional Code Reed-Solomon	None
Encryption	Mandatory- 3DES Optional- AES	Optional- RC4 (AES in 802.11i)
Mobility	Mobile WiMAX (802.16e)	In development
Mesh	Yes	Vendor Proprietary
Access Protocol	Request/Grant	CSMA/CA

Source: Dr Mohuya Chakraborty and Dr Debika Bhattacharyya.

1.3.1 Operational Principles

A WiMAX system consists of two parts:

- A WiMAX Base Station (BS) – According to IEEE 802.16 the specification range of WiMAX I is a 30-mile (50-km) radius from base station.
- A WiMAX receiver – The receiver and antenna could be a small box or Personal Computer Memory Card International Association (PCMCIA) card, or they could be built into a laptop the way WiFi access is today. Figure 1.1 explains the basic block diagram of WiMAX technology.

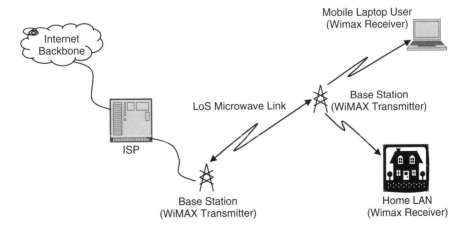

Figure 1.1 Operational principles of WiMAX technology. Source: Dr Mohuya Chakraborty and Dr Debika Bhattacharyya.

A WiMAX base station consists of indoor electronics and a WiMAX tower similar in concept to a cell-phone tower. A WiMAX base station can provide coverage to a very large area up to a radius of six miles. Any wireless device within the coverage area would be able to access the Internet. It uses the MAC layer defined in standard IEEE 802.16. This common interface that makes the networks interoperable would allocate uplink and downlink bandwidth to subscribers according to their needs, on an essentially real-time basis. Each base station provides wireless coverage over an area called a cell. Theoretically, the maximum radius of a cell is 50 km or 30 miles. However, practical considerations limit it to about 10 km or six miles. The WiMAX transmitter station can connect directly to the Internet using a high-bandwidth, wired connection (for example, a T3 line). It can also connect to another WiMAX transmitter using LoS microwave link. This connection to a second base station (often referred to as a backhaul), along with the ability of a single base station to cover up to 3000 square miles, is what allows WiMAX to provide coverage to remote rural areas. It is possible to connect several base stations to one another using high-speed backhaul microwave links. This would also allow for roaming by a WiMAX subscriber from one base station coverage area to another, similar to the roaming enabled by cell phones. A WiMAX receiver may have a separate antenna or could be a stand-alone box or a PCMCIA card sitting on user laptop or computer or any other device.

A typical WiMAX operation will comprise of WiMAX BSs to provide ubiquitous coverage over a metropolitan area. WiMAX BSs can be connected to the edge network by means of a wireless point-to-point link or, where available, a fibre link. Combining a wireless router with the WiMAX terminal will enable wireless distribution within the building premises by means of a WiFi LAN. Because of the relatively limited spectrum assignments in the lower-frequency bands, WiMAX deployments will usually have a limited capacity, requiring BS spacing on the order of two to three km. In lower density rural areas, deployments will often have a limited range, thus taking advantage of the full coverage capability of WiMAX, which can achieve NLoS coverage over an area of 75 sq km in the 3.5-GHz band.

WiMAX has been increasingly called the technology of the future. Belonging to the IEEE 802.16 series, WiMAX will support data transfer rates up to 70 Mbps over link distances up to 30 miles. Supporters of this standard promote it for a wide range of applications in fixed, portable, mobile and nomadic environments, including wireless backhaul for WiFi hot spots and cell sites, hot spots with wide area coverage, broadband data services at pedestrian and vehicular speeds, last-mile broadband access, etc. So WiMAX systems are expected to deliver broadband access services to residential and enterprise customers in an economical way.

WiMAX would operate in a similar manner to WiFi but at higher speeds, over greater distances and for a greater number of users. WiMAX has the ability to provide a service even in areas that are difficult for wired infrastructure to reach and with the ability to overcome the physical limitations of a traditional wired infrastructure.

1.3.2 WiMAX Speed and Range

WiMAX is expected to offer initially up to about 40 Mbps capacity per wireless channel for both fixed and portable applications, depending on the particular technical configuration chosen, enough to support hundreds of businesses with T-1 speed connectivity and thousands of residences with DSL speed connectivity. WiMAX can support voice and video as well as Internet data.

It is able to provide wireless broadband access to buildings, either in competition with existing wired networks or alone in currently unserved rural or thinly populated areas. It can also be used to connect WLAN hotspots to the Internet. It is also intended to provide broadband connectivity to mobile devices. Mobile devices are not as fast as fixed ones, but the expected characteristics are within the range of 15 Mbps capacity over a 3 km cell coverage area.

With WiMAX, users could really cut free from today's Internet access arrangements and be able to go online at broadband speeds, virtually wherever they like from within a Metro Zone. WiMAX could potentially be deployed in a variety of spectrum bands: 2.3 GHz, 2.5 GHz, 3.5 GHz and 5.8 GHz

1.3.3 Spectrum

There is no uniform global licensed spectrum for WiMAX, although the WiMAX Forum has published three licensed spectrum profiles: 2.3 GHz, 2.5 GHz and 3.5 GHz, in an effort to decrease cost. Economies of scale dictate that the more WiMAX embedded devices (such as mobile phones and WiMAX-embedded laptops) are produced, the lower the unit cost. (The two highest cost components of producing a mobile phone are the silicon and the extra radio needed for each band.) Similar economy of scale benefits apply to the production of Base Stations. In the unlicensed band, 5.x GHz is the approved profile. Telecom companies are unlikely to use this spectrum widely other than for backhaul, since they do not own and control the spectrum. In the USA, the biggest segment available is around 2.5 GHz, and is already assigned, primarily to Sprint Nextel and Clearwire. The most recent versions of both WiMAX standards in 802.16 cover spectrum ranges from at least the 2 GHz range through the 66 GHz range. The International standard of 3.5 GHz spectrum was the first to enjoy WiMAX products. The US license free spectrum at 5.8 GHz has a few WiMAX vendors building products. Licensed spectrum at 2.5 GHz used both domestically in the US and fairly widely abroad is the largest block in the US. Also, in the US and in Korea products are shipping for the 2.3 GHz spectrum range. Also in the US the 3.65 GHz band of frequencies now has WiMAX gear shipping to carriers. Elsewhere in the world, the most-likely bands used will be the Forum approved ones, with 2.3 GHz probably being most important in Asia. Some countries in Asia like India and Indonesia will use a mix of 2.5 GHz, 3.3 GHz and other frequencies. Pakistan's Wateen Telecom uses 3.5 GHz.

Analog TV bands (700 MHz) may become available for WiMAX usage, but await the complete roll out of digital TV, and there will be other uses suggested for that spectrum. In the US the FCC auction for this spectrum began in January 2008 and, as a result, the largest share of the spectrum went to Verizon Wireless and the next largest to AT&T. Both of these companies have stated their intention of supporting Long Term Evolution (LTE), a technology which competes directly with WiMAX. EU commissioner Viviane Reding has suggested re-allocation of 500–800 MHz spectrum for wireless communication, including WiMAX.

WiMAX profiles define channel size, Time Division Duplex (TDD)/ Frequency Division Duplex (FDD) and other necessary attributes in order to have inter-operating products. The current fixed profiles are defined for both TDD and FDD profiles. At this point, all of the mobile profiles are TDD only. The fixed profiles have channel sizes of 3.5 MHz, 5 MHz, 7 MHz and 10 MHz. The mobile profiles are 5 MHz, 8.75 MHz and 10 MHz. (Note: the 802.16 standard allows a far wider variety of channels, but only the above subsets are supported as WiMAX profiles.)

1.3.4 Limitations

A commonly-held misconception is that WiMAX will deliver 70 Mbit/s over 31 miles/50 kilometres. In reality, WiMAX can only do one or the other – operating over maximum range (31 miles/50 km) increases bit error rate and thus must use a lower bit rate. Lowering the range allows a device to operate at higher bit rates.

1.3.5 Need for WiMAX

WiMAX can satisfy a variety of access needs. Potential applications include:

- Extending broadband capabilities to bring them closer to subscribers, filling gaps in cable, DSL and T1 services, WiFi and cellular backhaul, providing last-100 meter access from fibre to the curb and giving service providers another cost-effective option for supporting broadband services.
- Supporting very high bandwidth solutions where large spectrum deployments (i.e. >10 MHz) are desired using existing infrastructure, keeping costs down while delivering the bandwidth needed to support a full range of high-value, multimedia services.
- Helping service providers meet many of the challenges they face owing to increasing consumer demand. WiMAX can help them in this regard without discarding their existing infrastructure investments because it has the ability to interoperate seamlessly across various network types.
- Providing wide area coverage and quality of service capabilities for applications ranging from real-time delay-sensitive Voice-over-Internet Protocol (VoIP) to real-time streaming video and non-real-time downloads, ensuring that subscribers obtain the performance they expect for all types of communications.
- Being an IP-based wireless broadband technology, WiMAX can be integrated into both wide-area third-generation (3G) mobile and wireless and wireline networks, allowing it to become part of a seamless anytime, anywhere broadband access solution.
- Ultimately, serving as the next step in the evolution of 3G mobile phones, via a potential combination of WiMAX and Code Division Multiple access (CDMA) standards called Fourth Generation (4G).

1.4 Mobile WiMAX

1.4.1 Overview of Mobile WiMAX

The Worldwide Interoperability for Microwave Access (WiMAX) standard, that is, the IEEE 802.16-2004 standard supports point-to-multipoint (PMP) as well as mesh mode.

In the PMP mode, multiple subscriber stations (SSs) are connected to one base station (BS) where the access channel from the BS to the SS is called the downlink (DL) channel, and the one from the SS to the BS is called the uplink (UL) channel. To support mobility, the IEEE has defined the IEEE 802.16e amendment, the mobile version of the 802.16 standard which is also known as mobile WiMAX. In mobile WiMAX battery life and handover are essential issues to support mobility between subnets in the same network domain (micromobility) and between two different network domains (macromobility).This new amendment aims at maintaining mobile clients connected to a MAN while moving around. It supports portable devices from mobile smart-phones and Personal digital assistants (PDAs) to notebook and laptop computers. IEEE 802.16e works in the 2.3 GHz and 2.5 GHz frequency bands.

During network entry a SS at first needs initial ranging to allocate CDMA codes in UL ranging opportunities. Then the SS is allowed to join the network to acquire correct transmitter parameters (timing offset and power level), a complete network entry process with a desired BS to join the network. After successful completion of initial ranging, the SS will request the BS to describe its available modulation capability, coding schemes, and duplexing methods. During this stage, the SS will acquire a downlink (DL) channel. Once the SS finds a DL channel and synchronizes with the BS at the PHY level, the MAC layer will look for downlink channel descriptor (DCD) and UCD (uplink channel descriptor) to get modulation and other parameters. The SS remains in synchronization with the BS as long as it continues to receive the DL-medium access protocol (MAP) and DCD messages. Finally, the SS will receive a set of transmission parameters from UCD as its UL channel. If no UL channel can be found after a suitable timeout period, the SS will continue scanning to find another DL channel. Once the UL parameters are obtained, the SS will perform the ranging process.

The second stage is authentication. At this stage, the BS authenticates and authorizes the SS. Then the BS performs a key exchange with the SS, such that the provided keys can enable the ciphering of transmission data. The third stage is registration. To register with the network, the SS and the BS will exchange registration request/response messages. The last stage is to establish IP connectivity. The SS gets its IP address and other parameters to establish IP connectivity. After this step, operational parameters can be transferred and connections can be set up.

1.4.2 Handover Process in Mobile WiMAX

Hard handover is definitely to be supported in mobile WiMAX networks. Hence, break-before-make operations may happen during the handover process. In other words, link disconnection may occur and throughput may degrade. Therefore, various levels of optimization are demanded to reduce association and connection establishment with the target BS. These optimization methods are not clearly defined in the IEEE 802.16e specification, so they should be supported in specific WiMAX systems and products.

On the contrary, soft handover is optional in mobile WiMAX networks. Two schemes, Macro-Diversity Handover (MDHO) and Fast Base Station Switching (FBSS) are supported. In case of MDHO, Mobile Station (MS) receives from multiple BSs simultaneously during handover, and chooses one as its target BS. As for FBSS, the MS receives from/transmits to one of several BSs (determined on a frame-by-frame basis) during

Table 1.2 LTE vs mobile WiMAX

Parameters	Mobile WiMAX	LTE
Access Technology	OFDMA(Downlink) OFDMA(Uplink)	OFDMA(Downlink) SC-FDMA(Uplink)
Frequency Band	2.3–2.4 GHz, 2.496–2.69 GHz, 3.3–3.8 GHz	Existing and New Frequency bands
Channel Bandwidth	5, 8.75, 10 MHz	1.25–20 MHz
Cell Radius	2–7 KM	5 Km
Cell Capacity	100–200 Users	More than 200 users at 5 MHz

Source: Dr Mohuya Chakraborty and Dr Debika Bhattacharyya.

handover, such that the MS can omit the decision process of selecting the target BS to shorten the latency of handover.

1.4.3 LTE vs. Mobile WiMAX

Mobile WiMAX is based on an open standard that was debated by a large community of engineers before getting ratified. The level of openness means that Mobile WiMAX equipment is standard and therefore cheaper to buy, sometimes at half the cost and sometimes even less. A parallel standardization effort is the Evolved Universal Mobile Telecommunications System (UMTS) terrestrial radio access network (E-UTRAN) also known as 3rd Generation Partnership Project (3GPP) Long Term Evolution (3GPP-LTE) launched by the 3GPP. There are certain similarities between the two technologies. First, both are 4G technologies designed to move data rather than voice. Both are IP networks based on Orthogonal Frequency Division Multiplexing (OFDM) technology – so rather than being rivals such as the Global System for Mobile Communications (GSM) and CDMA, they're more like siblings. However there are many differences between the two on various parameters such as frequency bands, access technology, channel bandwidth, cell radius and cell capacities. Table 1.2 provides a comparison of LTE and Mobile WiMAX.

1.5 Overview of End-to-End WiMAX Network Architecture

The IEEE 802.16e-2005 standard provides the air interface for WiMAX but does not define the full end-to-end WiMAX network. The WiMAX Forum's Network Working Group (NWG) is responsible for developing the end-to-end network requirements, architecture and protocols for WiMAX, using IEEE 802.16e-2005 as the air interface.

The WiMAX NWG has developed a network reference model to serve as an architecture framework for WiMAX deployments and to ensure interoperability among various WiMAX equipment and operators.

The end-to-end WiMAX Network Architecture has an extensive capability to support mobility and handovers. It will:

- Include vertical or inter-technology handovers – for example to WiFi, 3GPP, 3GPP2, DSL, or MSO – when such capability is enabled in multi-mode MS.

- Support IPv4 or IPv6 based mobility management. Within this framework, and as applicable, the architecture will accommodate MS with multiple IP addresses and simultaneous IPv4 and IPv6 connections.
- Support roaming between Network Service Provider (NSPs).
- Utilize mechanisms to support seamless handovers at up to vehicular speeds.

Some of the additional capabilities in support of mobility include the support of:

- Dynamic and static home address configurations.
- Dynamic assignment of the Home Agent in the service provider network as a form of route optimization, as well as in the home IP network as a form of load balancing.
- Dynamic assignment of the Home Agent based on policies, Scalability, Extensibility, Coverage and Operator Selection.

The network reference model envisions a unified network architecture for supporting fixed, nomadic and mobile deployments and is based on an IP service model. Figure 1.2 shows a simplified illustration of an IP-based end-to-end WiMAX network architecture [1].

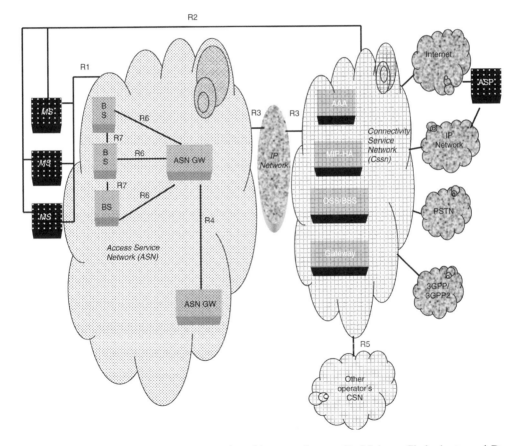

Figure 1.2 End-to-End WiMAX network architecture. Source: Dr Mohuya Chakraborty and Dr Debika Bhattacharyya.

The overall network may be divided logically into three parts:

1. MS used by the end user to access the network.
2. The Access Service Network (ASN), which comprises one or more base stations and one or more ASN gateways that form the radio access network at the edge.
3. Connectivity Service Network (CSN), which provides IP connectivity and all the IP core network functions.

The network reference model developed by the WiMAX Forum NWG defines a number of functional entities and interfaces between those entities. These entities are discussed briefly below.

- **Base Station (BS):** The BS is responsible for providing the air interface to the MS. Additional functions that may be part of the BS are micromobility management functions, such as handoff triggering and tunnel establishment, radio resource management, QoS policy enforcement, traffic classification, Dynamic Host Control Protocol (DHCP) proxy, key management, session management and multicast group management.
- **Access Service Network Gateway (ASN-GW):** The ASN gateway typically acts as a layer 2 traffic aggregation point within an ASN. Additional functions that may be part of the ASN gateway include intra-ASN location management and paging, radio resource management and admission control, caching of subscriber profiles and encryption keys, AAA client functionality, establishment and management of mobility tunnel with base stations, QoS and policy enforcement, foreign agent functionality for mobile IP and routing to the selected CSN.
- **Connectivity Service Network (CSN):** The CSN provides connectivity to the Internet, ASP, other public networks and corporate networks. The CSN is owned by the NSP and includes AAA servers that support authentication for the devices, users and specific services. The CSN also provides per user policy management of QoS and security. The CSN is also responsible for IP address management, support for roaming between different NSPs, location management between ASNs and mobility and roaming between ASNs.

The WiMAX architecture framework allows for the flexible decomposition and/or combination of functional entities when building the physical entities. The ASN interfaces the BS and the all-IP core network – the CSN. Typically the ASN includes numerous BSs with one or more ASN gateways. The ASN manages radio resources, MS access, mobility, security and QoS. It acts as a relay for the CSN for IP address allocation and AAA functions. The ASN gateway hosts the Mobile IP Home Agent.

The CSN performs core network functions, including policy and admission control, IP address allocation, billing and settlement. It hosts the Mobile IP Home Agent, the IP and Authorization, Authentication and Accounting (AAA) servers, and Public Switched Telephone Network (PSTN) and VoIP gateways. The CSN is also responsible for internetworking with non-WiMAX networks (e.g. 3G, DSL) and for roaming through links to other CSNs. Table 1.3 gives the reference network model interfaces.

In short, the ASN represents a boundary for functional interoperability with WiMAX clients, WiMAX connectivity service functions and aggregation of functions embodied

Table 1.3 Reference network model interfaces

Interface	Description
R1	Interface between the MS and the ASN. Functionality: air interface.
R2	Interface between the MS and the CSN. Functionality: AAP, IP host configuration, mobility management.
R3	Interface between the ASN and the CSN. Functionality: AAP, policy enforcement, mobility management.
R4	Interface between ASNs. Functionality: mobility management.
R5	Interface between CSNs. Functionality: internetworking, roaming.
R6	Interface between BS and ASN gateway. Functionality: IP tunnel management to establish and release MS connection.
R7	Interface between BSs. Functionality: handoffs

Source: Dr Mohuya Chakraborty and Dr Debika Bhattacharyya.

by different vendors. Mapping of functional entities to logical entities within ASNs as depicted in the NRM may be performed in different ways. The WiMAX Forum is in the process of network specifications in a manner that would allow a variety of vendor implementations that are interoperable and suited for a wide diversity of deployment requirements. A CSN providing IP connectivity services to the WiMAX subscriber(s) may comprise network elements such as routers, AAA proxy/servers, user databases and Interworking gateway devices. It may be deployed as part of a Greenfield WiMAX NSP or as part of an incumbent WiMAX NSP.

The ASN coordinates traffic across multiple BS, supports security, handoffs and QoS. The CSN manages core network operations through IP servers, AAA, VoIP and PSTN gateways, and provides an interface to legacy core networks and other operators' networks. The open IP architecture which is at the core of WiMAX marks a pivotal innovation among non-proprietary mobile technologies. It is set to decrease the complexity and cost to network operators, while increasing the flexibility in developing new services and applications and the freedom in selecting the best suited vendors. Furthermore, according to Rick Galatioto, Product Line Manager at Cisco, an ASN and IP solutions vendor, 'the adoption of an open IP architecture by network providers represents a crucial step towards empowering end users and giving them more control in choosing applications'. If network operators want to reap the full benefits that WiMAX and its all-IP architecture can deliver, they need to select carefully the ASN and CSN solutions that best suit their requirements and provide all the functionality required while avoiding unnecessary complexity in their network.

In short the end-to-end WiMAX Network Architecture has extensive support for scalable, extensible operation and flexibility in operator selection. In particular, it will:

- Enable a user to select manually or automatically from available Network Access Points (NAPs) and NSPs.
- Enable ASN and CSN system designs that easily scale upward and downward – in terms of coverage, range or capacity.
- Accommodate a variety of ASN topologies – including hub-and-spoke, hierarchical, and/or multi-hop interconnects.

- Accommodate a variety of backhaul links, both wireline and wireless with different latency and throughput characteristics.
- Support incremental infrastructure deployment.
- Support the phased introduction of IP services that in turn scale with an increasing number of active users and concurrent IP services per user.
- Support the integration of base stations of varying coverage and capacity – for example, pico, micro and macro base stations.
- Support the flexible decomposition and integration of ASN functions in ASN network deployments in order to enable use of load balancing schemes for efficient use of radio spectrum and network resources.

Additional features pertaining to manageability and performance of WiMAX Network Architecture include:

- Support a variety of online and offline client provisioning, enrollment and management schemes based on open, broadly deployable, IP-based, industry standards.
- Accommodation of Over-The-Air (OTA) services for MS terminal provisioning and software upgrades.
- Accommodation of use of header compression/suppression and/or payload compression for efficient use of the WiMAX radio resources.

It would be unfair not to mention a few WiMAX products here. The WiMAX portfolio includes:

- Aricent ASNLite – a standards-based, compact, and off-the-shelf, integrated ASN gateway software product that facilitates rapid development of WiMAX (802.16e) solutions.
- WiMAX Integrated Gateway (WING™) – A 'Network-in-a-Box' collapsed ASN and CSN solution comprising of an integrated Profile-C ASN-GW, AAA server and Home Agent. This solution serves the needs of rural, Tier 3, and enterprise deployments, and can run on any Commercial Off-The-Shelf (COTS) platform.
- eASN™ – A complete ASN gateway product supporting control plane, data plane and management plane functionalities. The product can be scaled to support up to 60 000 subscribers, and is ideal for medium to high density networks.
- sigASN™ – A control plane framework for Profile-C Macro/Micro/Enterprise ASN Gateway deployments. This framework can be used to develop ASN solutions for very high density networks.
- Base Station Framework – A Release-6 compliant control plane BS Framework for all types of Macro/Pico/Femto Profile-C Base Station deployments.

1.6 Radio Interface Specifications for WiMAX

1.6.1 Overview

The radio interface is the main building block of a WiMAX network and is responsible for most of the spectrum efficiency and cost savings that WiMAX promises. The IEEE 802.16 Working Group on Broadband Wireless Access Standards was established by the IEEE Standards Board in 1999, to develop standards for the global deployment of broadband Wireless Metropolitan Area Networks. The Workgroup is a unit of the IEEE 802 LAN/MAN Standards Committee.

Although the 802.16 family of standards is officially called WirelessMAN in IEEE, it has been commercialized under the name 'WiMAX' by an industry alliance called the WiMAX Forum. The mission of the Forum is to promote and certify compatibility and interoperability of broadband wireless products based on the IEEE 802.16 standards.

The first 802.16 standard was approved in December 2001. It delivered a standard for point to multipoint Broadband Wireless transmission in the 10–66 GHz band, with only an LOS capability. It uses a single carrier (SC) physical (PHY) standard. IEEE 802.16 standardizes the air interface and related functions associated with wireless local loop. 802.16a was an amendment to 802.16 and delivered a point to multipoint capability in the 2–11 GHz band. For this to be of use, it also required an NLOS capability, and the PHY standard was therefore extended to include OFDM and Orthogonal Frequency Division Multiple Access (OFDMA). 802.16a was ratified in January 2003 and was intended to provide 'last mile' fixed broadband access. 802.16c, a further amendment to 802.16, delivered a system profile for the 10–66 GHz 802.16 standard.

In September 2003, a revision project called 802.16d commenced aimed at aligning the standard with aspects of the European Telecommunications Standards Institute (ETSI) High Performance Radio Metropolitan Area Network (HIPERMAN) standard as well as laying down conformance and test specifications. This project concluded in 2004 with the release of 802.16-2004 which superseded the earlier 802.16 documents, including the a/b/c amendments.

An amendment to 802.16-2004, IEEE 802.16e-2005 (formerly known as IEEE 802.16e), addressing mobility, was concluded in 2005. This implements a number of improvements to 802.16-2004, including better support for QoS and the use of Scalable OFDMA. It is sometimes called 'Mobile WiMAX', after the WiMAX forum for interoperability, which is an industry-led organization formed to certify and promote broadband wireless products based upon the IEEE 802.16 standard.

1.6.2 802.16e-2005 Technology

The 802.16 standard essentially standardizes two aspects of the air interface – the physical (PHY) layer and the Media Access Control (MAC) layer. This section provides an overview of the technology employed in these two layers in the current version of the 802.16 specification (which is strictly 802.16-2004 as amended by 802.16e-2005, but which will be referred to as 802.16e for brevity).

1.6.2.1 PHY – The Physical Layer

802.16e uses Scalable OFDMA to carry data, supporting channel bandwidths of between 1.25 MHz and 20 MHz, with up to 2048 subcarriers. It supports adaptive modulation and coding, so that in conditions of good signal, a highly efficient 64 QAM coding scheme is used, whereas where the signal is poorer, a more robust BPSK coding mechanism is used. In intermediate conditions, 16 QAM and QPSK can also be employed. Other PHY features include integration of the latest technological innovations, such as 'beam forming' and

Multiple Input Multiple Output (MIMO) and Hybrid Automatic Repeat Request (HARQ) for good error correction performance.

OFDM belongs to a family of transmission schemes called multicarrier modulation, which is based on the idea of dividing a given high-bit-rate data stream into several parallel lower bit-rate streams and modulating each stream on separate carriers, often called subcarriers, or tones. Multicarrier modulation schemes eliminate or minimize InterSymbol Interference (ISI) by making the symbol time large enough so that the channel-induced delays (delay spread being a good measure of this in wireless channels) are an insignificant (typically, <10%) fraction of the symbol duration.

Therefore, in high-data-rate systems in which the symbol duration is small, being inversely proportional to the data rate, splitting the data stream into many parallel streams increases the symbol duration of each stream such that the delay spread is only a small fraction of the symbol duration. OFDM is a spectrally efficient version of multicarrier modulation, where the subcarriers are selected such that they are all orthogonal to one another over the symbol duration, thereby avoiding the need to have non-overlapping sub-carrier channels to eliminate intercarrier interference. In order to completely eliminate ISI, guard intervals are used between OFDM symbols. By making the guard interval larger than the expected multipath delay spread, ISI can be completely eliminated. Adding a guard interval, however, implies power wastage and a decrease in bandwidth efficiency.

'Beam forming' is an Advanced Antenna Technology (AAT) that ensures that radio power is concentrated where the WiMAX terminals are, adjusting the beam automatically as the terminals move around the coverage area. Beam forming enables dramatic reductions in the number of radio sites needed to provide coverage – in some instances by as much as 40% – while reducing interference and ensuring better indoor penetration of the radio signal.

'MIMO' is an AAT that combines the radio signals transmitted and received on separate antennas. The technique takes advantage of the multiple paths and reflections of a radio signal to strengthen radio communications, particularly in densely populated areas where signals can be degraded by buildings and other physical obstacles. MIMO antennae provide good NLOS characteristics (or higher bandwidth). MIMO also helps make radio links more robust, nearly doubling the capacity delivered in dense urban environments [2, 3].

1.6.2.2 MAC – The Media Access Control Layer

The 802.16 specifications describe three MAC sublayers: the Convergence Sublayer (CS), the Common Part Sublayer (CPS) and the Security Sublayer (SeS). There are a number of convergence sublayers, which describe the manner in which wireline technologies such as Ethernet, ATM and IP are encapsulated on the air interface, and the process of data classification, etc.

The CS aims to enable 802.16 to better accommodate the higher layer protocols placed above the MAC layer. The CS receives data frames from a higher layer and classifies the frame. On the basis of this classification, the CS can perform additional processing such as payload header compression, before passing the frame to the MAC CPS. The CS also accepts data frames from the MAC CPS. If the peer CS has performed any type of processing, the receiving CS will restore the data frame before passing it to a higher layer.

The CPS is the vital part of the 802.16 MAC that defines the medium access method. It provides functions related to network entry and initialization, duplexing, framing, channel access and QoS.

The SeS provides privacy to the subscribers across the wireless network. It also provides strong protection against theft of service to the operators. It describes how secure communications are delivered, by using secure key exchange during authentication, and encryption using AES or DES (as the encryption mechanism) during data transfer.

Further features of the MAC layer include power saving mechanisms (using Sleep Mode and Idle Mode) and handover mechanisms. A key feature of 802.16 is that it is a connection oriented technology. The SS cannot transmit data until it has been allocated a channel by the BS. This allows 802.16e to provide strong support for QoS.

1.6.3 Applications

Depending on the frequency band and implementation details, an access system built in accordance with this standardized radio interface specification can support a wide range of applications, from enterprise services to residential applications in urban, suburban and rural areas, as well as cellular backhauling. The specification could easily support both generic Internet-type data and real-time data, including two-way applications such as voice and videoconferencing. The technology is known as a Wireless MAN in IEEE 802.16. The word 'metropolitan' refers not to the application but to the scale. The design is oriented primarily toward outdoor applications. The architecture is primarily point-to-multipoint, with a base station serving subscribers in a cell that can range up to tens of km. Terminals are fixed or, in frequencies below 11 GHz, and are therefore ideal for providing access to buildings, such as businesses, homes, Internet cafes, telephone shops (telecentres), etc. The radio interface includes support for a variety of worldwide frequency allocations in either licensed or licence-exempt bands. At higher frequencies (above 10 GHz), supported data rates range over 100 Mbit/s per 25 MHz or 28 MHz channel, with many channels available under some administrations. At the lower frequencies (below 11 GHz), data rates range up to 70 Mbit/s per 20 MHz channel.

1.6.4 WiMAX Simulation Tools

The best features of WiMAX are the accurate calculation and optimization of the Radio Frequency and Capacity, Network planning. To achieve this you have to choose the best simulation tool which works on the WiMAX system. Some of the most well-known and widely used simulation and planning tools are: OPNET tool, Planet EV, EDX (Signal-Pro), Provision Communication, Radio Mobile (Freeware), Atoll, CelPlan, ICS Telecom, Asset 3G/WiMAX, Winprop, Volcano Siradel, NS-2 (Freeware), NS-3 (Freeware), Qualnet NCTuns Network Simulator and Emulator

1.7 Interoperability Issues in WiMAX

WiMAX, as with many new technologies, is based on an open standard. While standards increasingly play an essential role in driving implementation, success is not guaranteed.

The success of a standard-based technology depends on the strong interoperability amongst the operators, vendors and solution and content providers. As a standard-based technology, WiMAX enables inter-vendor interoperability resulting in lower costs, greater flexibility and freedom and faster innovation to operators.

A strong commitment to ensuring full interoperability, both through certification and ad-hoc testing between vendors takes place within the WiMAX network. The network operators must realize the process of establishing the interoperability and the underlying principles so that they understand how different products, solutions and applications from different vendors can coexist in the same WiMAX network.

The specifications developed by the NWG within the WiMAX Forum define the role of the ASN and CSN and ensure that WiMAX networks can internetwork with other networks, using WiMAX or other wireless or wired access technologies such as cellular, WiFi, DSL, cable or fibre. In addition, the NWG specifications are designed to enable network operators to enjoy the benefits of vendor interoperability at the infrastructure level, to rely on a consistent client interface and, if they so desire, to open their network to virtual operators, akin to existing cellular Mobile Virtual Network Operators (MVNOs).

To fulfill these goals, the NWG specifications define interfaces for the Reference Network Model (RNM) between key elements of the ASN and CSN, as shown in Table 1.3. To comply with the specifications, vendors are required to leave most of these open. As such when implementing ASN-GW one should also consider ASN profiles. Multiple ASN Profiles have been specified in WiMAX as a tool to manage diversity in ASN node usage and implementation. Release 1 of NWG Specifications on WiMAX supports three ASN Profiles: Profile A (Centralized ASN Model with BS and ASN GW in separate platforms through R6 interface), Profile B (Distributed ASN with the BS and ASN-GW functionalities implemented in a single platform) and C (It is like Profile A, except for RRM being non-split and located in BS) to accommodate varying network operator requirements and the vendors' preference for different network architectures as shown in Table 1.4.

Table 1.4   ASN profiles

Profile	Key Features
A	• Hierarchical model with more intelligence located at the ASN gateway. • The ASN gateway is involved in the Radio Resource Management (RRM) and hosts the Radio Resource Controller (RRC). It also handles handoffs between BSs. • Open interfaces: R1, R3, R4, R6.
B	• Flat, distributed model, with BSs playing a more substantial role in managing traffic and mobility. • The ASN network acts as a black box, with R6 being a closed interface. • Open interfaces: R1, R3, R4.
C	• Centralized model similar to A, but BSs are responsible for all the RRM, including the RRC and Radio Resource Agent (RRA), and the handoffs between BSs. • Open interfaces: R1, R3, R4, R6.

Source: Dr Mohuya Chakraborty and Dr Debika Bhattacharyya.

As Profile C operators are not tied into one vendor for BTS and ASN Gateway equipment, they can force prices down through playing off different suppliers against each other. They can also choose the suppliers that can best support the functionality and services they want to offer over their network rather than being tied to one vendor that might not be up to the job. Despite the apparent advantages of Profile C, some 'turnkey' vendors are still successfully tempting operators with the two other Profiles available between the BTS and ASN Gateway: Profile A and Profile B. As both Profiles can create vendor lock-ins, they stand in the way of progress with regard to WiMAX interoperability between multiple vendors and, potentially, lower equipment prices. Profile B does not define any interface between the BTS and the ASN Gateway, so it is possible for Profile B vendors to pursue proprietary solutions and lock in their customers. Due to increased customer demand, however, many of the big WiMAX suppliers that started out by supplying profile B equipment, including Cisco (through its acquisition of Navini Networks) are now shifting to Profile C.

Profiles A and C both use a hierarchical model with a topology similar to that used in cellular networks and that is well suited to support full mobility. In profile A, the RRM resides entirely at the ASN gateway and this increases its workload. Profile C instead relies on the BS for the RRM and effectively separates the radio functionality – residing in the BS – from the network management – residing in the ASN gateway. This contrasts with profile A where both functions coexist in the ASN gateway. The separation of the radio functionality and network management facilitates intervendor interoperability as it allows network operators to select a different vendor for each function and so avoid conflicts and duplications. In addition, fixed operators may decide not to deploy an ASN gateway and instead use their existing Broadband Access Server (BRAS) and AAA server with tunneling protocols such as Point-to-Point Protocol over Ethernet (PPPoE). Profile C facilitates this approach because it does not require a separate ASN gateway for the radio management functions. To better meet the operators' demand for flexibility, an increasing number of vendors have elected to support ASN Profile C or plans to do so and we expect it to become the dominant one.

In Profile B, more processing is required at the BS and this may increase their complexity and cost. This solution may be attractive to small network operators focusing on fixed or nomadic services. As Profile B essentially leaves the R6 interface (Table 1.3) closed, it can be implemented as a solution in which there is no ASN gateway (each BS performs the ASN gateway role) or with a proprietary ASN gateway that manages only BSs from the same vendor and acts as a black box. Network operators who want to deploy BSs from another vendor would only be able to do so by deploying another end-to-end ASN network for the new equipment. Interoperability among ASN elements (BSs and ASN gateways) is supported among all products that comply with the specifications for the same ASN profile.

1.8 Summary

WiMAX, the next-generation of wireless technology has been designed to enable pervasive, high-speed mobile Internet access to the widest array of devices including notebook PCs, handsets, smartphones and consumer electronics such as gaming devices, cameras, camcorders, music players and more. It has been observed that being 4G wireless tech-nology, WiMAX delivers low-cost, open networks and is the first all-IP mobile Internet solution enabling efficient and scalable networks for data, video and voice.

When using WiMAX devices with directional antennae, speeds of 10 Mbit/s at 10 km distance is possible, while for WiMAX devices with omni-directional antennae only 10 Mbit/s over 2 km is possible. There is no uniform global licensed spectrum for WiMAX, although three licensed spectrum profiles are being used generally −2.3 GHz, 2.5 GHz and 3.5 GHz.

With an end-to-end WiMAX network architecture, the WiMAX system simply becomes an extension of the IP network to the mobile user. Leveraging simple IP-based backhaul connections, service providers can very readily service a myriad of WiMAX base sites (e.g. large, medium, sectorized, omni, micro, pico) for varying coverage and capacity profiles addressing outside environments, inside buildings, fixed and fully mobile connections. Service Providers will very simply grow their networks based on system usage leveraging standard IP components

The high performance of WiMAX technology paired with the cost advantages offered by a distributed WiMAX network architecture brings WiMAX solutions within reach of oper-ators in all regions and segments. With WiMAX systems, markets, in the absence of basic voice connections, can leapfrog to VoIP, high-speed data, and video delivery – further bridging the digital divide – and markets seeking advanced, bandwidth-intensive, mobile communications can realize true personal broadband experiences.

References

[1] M. Paolini, 'Building End-to-End WiMAX Networks', Senza Fili Consulting, April 2007.
[2] A. Paulraj, R. Nabar and D. Gore, *Introduction to Space-Time Wireless Communications*, Cambridge University Press, Cambridge, UK, May 2003.
[3] G. G. Raleigh and J.M. Cioffi, 'Spatio-temporal Coding for Wireless Systems'. IEEE Trans. On Communication, **4**(3): 357–66, 1996.

Part B

Security

2

WiMAX Security Defined in 802.16 Standards

Slim Rekhis and Noureddine Boudriga
Communication Networks and Security Research Laboratory (CN&S),
University of the 7th November at Carthage, Tunis, Tunisia

2.1 Introduction

Recently, the use of the IEEE 802.16 standard to build Metropolitan Area Networks has gained a great deal of interest from ISPs as a possible solution for supporting broadband wireless communication with fixed and mobile access. The standard offers high throughput broadband connections and coverage with respect to WLANs, and provides a security sublayer which is responsible for secrecy, authentication and secure key exchange. IEEE 802.16 networks can be used to provide several applications including 'last mile' broadband connections, hotspot and cellular backhaul and high-speed connectivity. Several versions of 802.16 networks were released. While the first versions have shown some security weaknesses that were later corrected by the recently released versions, the security mechanisms of 802.16 still remain vulnerable and the limited deployment of such technology is insufficient to satisfy the demands of security.

This chapter analyses WiMAX (Worldwide Interoperability for Microwave Access) security as defined in the different released versions of the IEEE 802.16 standards. It gives an overview of the WMAN 802.16 networks and introduces the main security requirements to be met by a standard for broadband access. It then describes the security mechanisms to be ensured by the security sublayer as well as the vulnerabilities of the initial versions, namely those related to fixed WiMAX. In this chapter, the security amendments carried out in the recent versions of the mobile WiMAX, are described and analysed.

WiMAX Security and Quality of Service: An End-to-End Perspective   Edited by Seok-Yee Tang,
Peter Müller and Hamid Sharif
© 2010 John Wiley & Sons, Ltd

2.2   Overview of 802.16 WMAN Networks

This section reviews the basic WMAN network topology and its general features, the protocol architecture and the security sublayer content of the IEEE 802.16 standard. It shows, whenever they exists, the differences between the different stable revisions of 802.16, notably 802.16d and 802.16e.

2.2.1   IEEE 802.16 Standards and Connectivity Modes

A WiMAX[1] network topology is organized in a cellular-like architecture. The network is deployed to provide access to a large urban or rural area. Figure 2.1 shows the WiMAX network topology, where a cell is composed of one or several base stations, denoted by BSs, and a set of Mobile or Subscriber Stations, denoted by MSs or SSs respectively. Depending on the version of the IEEE 802.16 standard and the frequency employed, the Subscribers Stations may or not be in the Line-of-Sight of the Base Station antenna. To extend the network and connect to backhaul, the architecture supports the use of Repeater Stations (RSs).

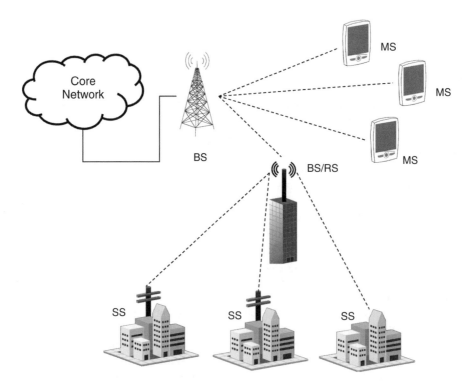

Figure 2.1   IEEE 802.16 standard's network topology.

---

[1] The term WiMAX refers to a marketing trend trademarked by the WiMAX forum to provide a description related to technology using the IEEE 802.16 standard.

The 802.16 specification has evolved during the last decade and undergone gradual expansion. The original specification [1] was approved in 2001, making the IEEE 802.16 a wireless MAN standard. It was developed to provide a high data rate and Point To Point (PTP) communication between fixed subscriber stations. The standard introduced the use of licensed frequencies ranging from 10 to 66 GHz so that interference is reduced. However, due to the short wavelength of the used signal, a Line of Sight (LoS) condition is required. Moreover, in this standard multipath propagation was not supported. A typical usage of this standard consists in connecting, via point to fixed point backhaul, a tower to a fixed location which is connected to a wired network. In this specification, directional antennae are used at both sides, and Frequency Division Duplexing (FDD) or Time Division Duplexing (TDD) is supported. Using highly directional antennae, the network could support a data rate of 32–134 Mbps with a channel of 28 MHz. The cell radius ranges between 2 and 5 kms. The security protection techniques are rudimentary and rely on antenna directivity to protect against intrusions.

Two amendments were later published. The first is the IEEE 802.16c [2] which was approved in December 2002. In this version, detailed system profiles or typical implementations regarding 10–66 GHz were added, and the errors of the previous version of the standard were corrected. The second is the 802.16a [3] standard. It broadened the use of WiMAX and introduced the use of licensed frequencies below 11 GHz for WiMAX operation. Typically, the used 2–11 GHz band contains licensed and unlicensed frequencies. The supported wavelength, which is longer compared to the initial WiMAX specification, allows the signal to traverse solid objects. Multi-path propagation is allowed and the network supports a Non-LOS (NLOS) propagation environment. In spite of the support of PTP backhaul two additional modes are introduced. The first is Point to Multi-Point (P2MP), where a set of subscribers may connect to a single BS. The second is the Mesh mode where a Subscriber Station (SS), which is not forced to connect only to a BS, may transmit to the neighbouring SS, thus extending network coverage and reducing system failure. Internet Service Providers (ISPs) may profit from 802.16a technology in order to connect rural regions, especially if the use of available low-rate wired infrastructure would limit connections capabilities for costumers. This specification requires the use of omni-directional antennae and Dynamic Frequency Selection (DFS). The latter technique allows one to avoid interference by switching dynamically to another Radio Frequency (RF) channel based on some measurements including the Signal to Interference Ratio (SIR). The used channel bandwidth ranges from 1.25 to 28 MHz. The IEEE 802.16a network, which uses Orthogonal Frequency Division Multiplexing (OFDM), supports up to 75 Mbps data-rate. The cell radius could reach 50 km but typically ranges from 5 to 10 km.

The IEEE 802.16d [4], also named 802.16-2004, replaced, improved and consolidated the original IEEE 802.16, 802.16a, and 802.16c standards. The IEEE 802.16d, which is considered as the first release of the WiMAX standard and a basis for WiMAX compatibility [5], introduces a real usage of WiMAX in fixed systems. In this version, new license free frequencies were introduced, notably frequencies below 11 GHz. Two frequency bands are supported, notably 2–11 GHz and 10–66 GHz. The Channel Bandwidth is scalable and ranges from 1.25 to 28 MHz. The bit rate is up to 75 Mbps using a channel of 20 MHz. The 802.16-2004 specification uses sectored omni-directional antennae in replacement of directional antennae, thus reducing the problems associated with precise antenna pointing. It also includes two-way authentication mechanisms between the subscriber stations and the base station. All of the above versions of IEEE 802.16 standard considered only fixed operations, and did not take into consideration the mechanisms and functionalities required in a mobile connectivity scheme. In fact, handover operations are not supported and modulation schemes are not designed to cope with a mobile environment and variation of channel conditions.

The most important amendment of WiMAX was done to support the mobility of user at vehicular speed (120 km/h). The basis of this version is described in 802.16e [6], which is also denoted by 802.16-2005 [7]. This standard operates on frequencies ranging from 2 to 6 GHz and includes protocols that allow mobility in the network. Proper handover mechanisms which support authentication are proposed. From the security perspective, a secure key exchange during authentication and data encryption using DES or AES is supported. Moreover, new security mechanisms are introduced to correct vulnerabilities in the 802.16-2004 version. The 802.16e specification introduced several enhancements at the physical and MAC layers, supports a channel bandwidth between 1.25 and 20 MHz, and provides a low/medium data rate (e.g., <15 Mbps using a channel of 5 MHz) for mobile and roaming users. By comparison with 802.16d, the number of supported users has increased and the Quality of Service (QoS) is better supported. The cell radius ranges between 2 and 5 km. The IEEE 802.16e standard facilitates and provides access to broadband Internet connections for laptops and PDAs integrating WiMAX adapters. In mobile WiMAX, several additional networking mechanisms are supported, including mutual authentication between mobile subscribers and the network, and transfer of security and quality of service during handover operations.

Several other standardization projects [8] have been undertaken by IEEE Working Group (WG), notably:

- IEEE 802.16f/i for mobile Management Information Base (MIB) support. These two versions have been merged in the IEEE 802.16 rev2 draft, which consolidates IEEE 802.16-2004, IEEE 802.16e, IEEE 802.16f, IEEE 802.16g and possibly IEEE 802.16i.
- IEEE 802.16g which defines the support of management of plane procedures and services.
- IEEE 802.16h which supports Wireless MAN for a license-exempt band.
- IEEE 802.16j and IEEE 802.16m for new air interface specification supporting mobile multi-hop relay features, functions and interoperable relay stations to enhance coverage (both cellular micro and macro cell coverage) and capacity of the network and provides high data rate for both fixed and mobile stations.

2.2.2   Network Architecture

WiMAX is based on IEEE 802.16 standards and the WiMAX forum Network Working Group (NWG) specification. The specification of the physical and MAC layer of the radio link, which is the focus of the IEEE 802.16 standards, is not sufficient to build an interoperable broadband wireless network. In fact, end-to-end services such as IP connectivity, QoS and security and mobility management are a requisite. In this context, the standardization and development of the end-to-end related aspects, including requirements, architecture and protocols, are beyond the 80.216 standards and are the responsibility of the WiMAX Forum's Network Working Group (NWG).

A Network Reference Model (NRM) [9], which provides a unified network, was developed to:

- promote interoperability between WiMAX operators and equipment by defining key functional entities and interfaces between them (referred as reference points over which a network interoperability framework is defined);
- support modularity and flexibility by allowing several types of decomposition of functions in topologies;
- support fixed, nomadic and mobile deployments and provide decomposition of access network and connectivity networks;
- share the network between several business models namely the Network Access Provider (NAP), the Network Service Provider (NSP) and Application Service Provider (ASP). The first owns the network and operations, the second provides IP connectivity and core network services to the WiMAX network and the last provides application services.

Figure 2.2 shows the important functional entities of WiMAX reference model, namely, the Subscriber Station (SS) or the Mobile Station (MS), the Access Service Network (ASN) and the Connectivity Service Network (CSN). The first two are owned by NAP while the last is owned by NSP. Since the architecture has changed as 802.16 versions have progressed, we stress on describing the model that goes with last operational version, namely the IEEE 802.16e (mobile WiMAX).

The mobile station is the equipment used by the end user to access the network. It could be a mobile station or any device that supports multiple hosts.

The ASN represents the WiMAX access network for subscribers, which provides the interface between the MS and the CSN. It comprises one or several Base Stations (BS) and one or several ASN gateways. The ASN is in charge of managing radio resources including handover control and execution, performing layer-two connectivity with the MSs, interoperating with other ASNs, relaying functionalities between the CSN and the MS to establish connectivity in IP layer and performing paging and location management.

The base station equipment is what actually provides the interface between the MS and the WiMAX network. It implements functionalities related to WiMAX PHY and MAC layers in compliance with 802.16 standard, and is characterized by a coverage radius and a frequency assignment. The number of base stations within an ASN is equal to the number of assigned frequencies and their deployment depends on the required bandwidth or the geographic coverage. The coverage radius of a BS ranges between 500 and 900 metres in an urban area, and is planned by operators to cover 4 km in a rural area [10]. Such a radius is highly comparable to the area covered by base stations in GSM and UMTS networks today. The base station is in charge of QoS policy enforcement, scheduling the uplink and downlink air link resources, managing the radio resources, classifying traffic, handling signaling messages exchanged with the ASN Gateway, establishing tunnels and managing keys. In order to enforce a load balancing or fault tolerance, a single BS may be connected to several ASN-GWs.

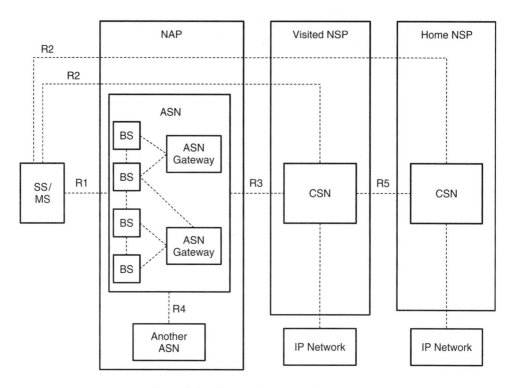

Figure 2.2   WiMAX network architecture.

The ASN-GW represents a layer-two traffic aggregation point within the ASN. It supports functions related to connection and mobility management, paging, DHCP Proxying/relaying, authentication of subscribers' identities, AAA (Authentication, Authorization, and Accounting) client functionality, service flow management, caching of subscribers' profile and encryption keys and routing to the CSN. The ASN-GW may also implement the Foreign Agent (FA) in mobile IP, which provides connectivity to mobile users who visit the network and store information about them. It advertises care-of addresses in order to route packets which are sent to the mobile node. The ASN-GW plays the role of authenticator and key distributor by transmitting signals to the AAA server and verifying the user credentials in the network re/entry using EAP (Extensible Authentication Protocol). A security context is created when the AAA session is established and keys are generated and shared between the MS and the BS. The AAA client in the ASN-GW collects accounting information related to flow including the number of bits transmitted or received, and duration. The ASN-GW is also responsible for managing profiles and policy functions which include but are not limited to the allowed QoS rate, and the type of flow. In addition a context per mobile subscriber and BS is maintained. This context, which includes subscriber profile, security context and characteristics of the mobile equipment, is exchanged between serving BSs during the handover.

The CSN represents the core network of WiMAX and is responsible for the transport, authentication and switching. It executes functions related to admission control and billing and enables IP connectivity to the WiMAX subscribers by allocating IP addresses to them. The CSN hosts also the DHCP and AAA servers, the VoIP (Voice over IP) gateways and the Mobile IP Home Agent (HA). The CSN is responsible for roaming through links to other and interconnecting with non-WiMAX networks including Public Switching Telephone Network (PSTN) and 3G cellular telephonic networks (it hosts gateways to these networks). The CSN enables Inter-CSN tunneling to support roaming between NSPs.

The set of reference points represents conceptual links that are used to connect two different functional entities. The first reference point, defined by R1, connects the MS to the BS. It represents the air interface defined in the physical layer and the Medium Access Control sublayer and implements the IEEE.16 standard. The second reference point, which is R2, is a logical interface that exists between the MS and ASN-GW or CSN. It is associated with authentication and authorization, and is used for IP host configuration management, and mobility and service management. The third reference point, say R3, is the interface between the ASN and CSN. It supports authentication, authorization, policy and mobility management between ASN and CSN. It also implements a tunnel between the ASN and CSN. The interface R4 exists between two ASNs and ensures the interworking of ASNs when a mobile station moves between them. The R5 reference point is an interface between two CSNs and is used for internetworking between home and visited NSP when a mobile station is in visited network. This interface carries activities such as roaming.

2.2.3 Protocol Architecture

The IEEE 802.16 protocol architecture is composed of two main layers as shown in Figure 2.3, namely the Physical (PHY) layer and the Medium Access Control (MAC) layer. The MAC layer is itself composed of three sublayers.

The first layer is the Service Specific Convergence Sub-layer (CS). It communicates with high layers, acquires external network data from CS Service Access Point (SAP), transforms them into MAC Segment Data Units (SDUs) and maps high level transmission parameters into 802.16 MAC layer flows and associations. Several high-level protocols can be implemented in different CSs. At present, there are two types of CS [11]: the Asynchronous Transfer Mode (ATM) Convergence Sub-layer which is used for ATM networks and services, and the Packet Convergence Sub-layer which is used for packet services including Ethernet, Point-to-Point (PPP) protocol, and TCP/IP.

The second layer is the Common Part Sub-layer (CPS). It represents the core of the standard and is responsible for system access, bandwidth allocation, connection management and packing or fragmenting multiple MAC SDU into MAC PDUs. Functions such as uplink scheduling, bandwidth request and grant and connection control are also defined. Using packing and fragmenting feature, in addition to the Packet Header Suppression (PHS), repetitive information, especially the datagram header, are deleted, thus saving the available bandwidth. In this context, it could be stated that the IEEE 802.16 standard does not fully respect the OSI model which requires appending a header to the forwarded datagram and guaranteeing transparency and independency between layers.

The third sublayer, which is the security sublayer, addresses the authentication, authorization, key establishment and exchange and encryption and decryption of data exchanged

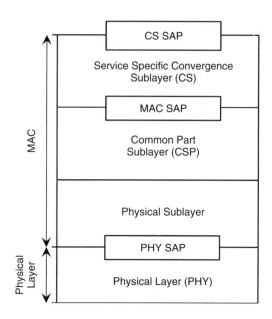

Figure 2.3 IEEE 802.16 protocol stack.

between the PHY and the MAC layers. The 802.16 standard offers several security measures to defeat a wide variety of security attacks. The details of security properties, mechanisms and threats will be discussed in greater detail later in this chapter. Roughly speaking, the secure key distribution is done through the use of Privacy Key Management (PKM) and the nodes identification is performed based on the use of X.509 technology. The security of connections is maintained based on the use of Security Associations (SA), which support several algorithms (e.g. AES or DES encryption) and can be of two types: data SA or authorization SA. Every SA gets a SA Identifier (SAID) by the BS. Authentication between the base station and mobile/subscribers station relies on the use of PKM.

2.2.4 Network Entry Procedure

To gain access to the network and perform initialization, the SS goes through a multi-step process [12].

First, an SS must scan for a suitable downlink signal from the BS and try to synchronize with it by detecting the periodic frame preambles. This downlink channel will later be used for establishing channels parameters. If a prior channel existed before, the SS tries to use the operational parameters already determined. Otherwise, it will scan the channel using all the bandwidth frequencies in the supported band.

Second, the SS looks for the Downlink Channel Descriptor (DCD) and the Uplink Channel Descriptor (UCD), which are broadcast by the BS and contain information regarding the characteristics of the uplink and downlink channels, the modulation type and the Forwarded Error Correction (FEC) scheme. The SS also listens for the uplink-map and downlink-map messages, which are denoted by UL-MAP and DL-MAP respectively, and detects their burst start times.

Third, the SS performs initial ranging which allows it to set the physical parameters properly (e.g. acquiring the timing offset of the network, requesting power adjustment). The SS performs the initial ranging by sending a Ranging Request packet (RNG-REQ) in the initial ranging contention slot. If the RNG-REQ message is received correctly by the BS, it responds with a ranging response packet (RNG-RSP) to adjust the SS transmission time, frequency and power and inform the SS about its primary management Connection ID (CID). The subsequent RNG-REQ and RNG-REP messages will be exchanged using this CID. Note that ranging is also made by the SS periodically and therefore the RNG-REQ will be sent in the data intervals granted by the BS in order to adjust power levels, time and frequency offsets. Once a ranging is completed, the SS informs the BS on its physical capabilities (e.g., modulation, coding scheme, half or full duplex support in FDD). At this stage, the BS can decide to accept or reject these capabilities.

Fourth, the SS requests an authorization to enter the network by executing the g protocol. The procedure of authenticating the SS and performing key exchange will be discussed in the description of the PKM protocol. Upon completion of this phase, the SS has a set of authentication and encryption keys.

Fifth, the SS registers to the BS by sending a registration request message. The BS responds with a registration response message which contains a secondary management CID and the IP version used for that connection. The reception of the registration response message means that the SS is now registered to the network and allowed to access it. Next, the SS invokes the Dynamic Host Configuration Protocol (DHCP) to get parameters related to IP connectivity (e.g., IP address), and obtains the network time of day. Optionally, the BS may proceed to set up connections for the service flows pre-provisioned during SS initialization.

2.3   Security Requirements for Broadband Access in WMAN Networks

The uses of broadband communication systems continue to increase as time goes by, and users may potentially use these networks for sending sensitive data or accessing business secrets or personal information. The security of broadband access in WMAN networks is an important concern for both the users and operators. Protecting the privacy of users' communications and guaranteeing the accurate billing and authentication of roaming users is an important issue that must be dealt with by these networks [13].

Since the IEEE 802.11 wireless networks were widely used and achieved a high level of success, a great deal of attention was given to the protection of these networks, which suffered from significant security weaknesses and proved to be extremely vulnerable, especially in their initial released standards and implementation. Several lessons regarding security protection were learned from these networks, which need to be exploited in the evolving broadband access technologies. According to [14], most of the IEEE 802.11 weaknesses can be organized into two classes: identity and Media-access control. The first class is related to the source address authentication of the MAC layer datagrams. An attacker could exploit, generate and forge different MAC control messages and exploit several weaknesses. The second class is related to the fair sharing of the transmission medium. The physical carrier sense mechanism could be attacked by sending several short and successive packets, forcing all other nodes to wait for their turn to broadcast. However, since the attackers transmit continuously, legitimate users may be deprived from accessing the medium. The attacker could also send few packets but use a forged high value of the datagram length field, thus gaining a long transmission period. The remaining nodes attached to the network will not use the carrier sense mechanisms during the estimated transmission time of that datagram, to see whether the medium is busy or not.

By way of example, consider the de-authentication attack. After authenticating himself to the network, a user goes through the association process prior to being able to exchange and broadcast data in the network. One of the messages used in the authentication and association steps is the de-authentication message. When a node receives this message from the base station that serves access, it detaches itself from the network. This message is typically used when there is, in the same covered area, several Wireless IEEE 802.16 networks and the user wants to switch between them. An attacker generates a datagram containing a de-authentication message by sending it to the broadcast address while forging the address of the base station. All the connected nodes that are able to receive such a message will be detached immediately. The determination of the address of the base station is not difficult, since most of them broadcast their presence in the network. Even if such a feature is hidden, it suffices for the attacker to listen to the traffic between some node and the base station. By repeating the de-authentication attack, the network can be brought to a denial of service. The reasons for the success of such an attack are twofold. First, the authentication technique of the de-authentication message is very weak and relies simply on the source address. Second, the message does not include a cryptographic protection (e.g., a Hash Message Authentication Code, HMAC) which may prevent malicious nodes from generating it.

Various security requirements should be involved in the design of a wireless broadband network.

The first and most important requirement is authentication, which can be of two types. The first type is the authentication of mobiles or stations by the network, the authentication of the base station by the users, or mutual authentication between the users and the networks.

The network operator needs to authenticate the mobile devices and users using adequate credentials and protect the network against several types of spoofing attack. Since users are mobile, they can roam from one cluster/cell to another. The base station, which is currently serving access, needs to communicate with the base station to which the user attempts to hand over or roam. In this context, authentication of the communicating base stations is an important property that should be guaranteed. On the opposite side, the users should be able to authenticate the base station to which they are attached and be sure that their privacy is guaranteed and the data they are sending is protected against tampering during transmission over the open medium. Preventing rogue base station attacks, which consist in spoofing a base station and attracting users (e.g., by sending stronger signal) to disconnect from the legitimate base station and connect to the malicious one, is an important issue. To protect against rogue base station attacks and prevent users from disclosing secret information when they connect to this fake station, two way authentication between users and the access point is a highly important feature to guarantee. The second type is the authentication to setup between nodes to help them authenticate each other independently of the used station. This is especially important if the Mesh or Adhoc access mode is used. The intermediate nodes used to route the traffic to the ultimate base station should not have complete access to the user data content and the security of the routing algorithms should be guaranteed.

Since users can roam from one network to another, the network should ensure that they are gaining access to an authorized service. Conversely, to protect users, an unauthorized user/device should be unable to gain access to any service in the network. In addition, users also need to be able to gain access to the correct credentials and services they are subscribed to. Guaranteeing the accuracy of the accounting mechanisms and preventing users from being overcharged is also an important security concern.

Owing to the openness of the wireless medium, information sent in plain view could easily be eavesdropped upon. Privacy should therefore be addressed in the security design of these networks. Privacy may not only address the content of the exchanged data, or records about the services used by users, but also the location of mobiles, or records of connections exchanged between operators. Note that a user may be able to release its location for emergency purposes or for signal reception troubleshooting. Another security requirement consists in protecting messages against replay attack, using a timestamp, or transient information.

2.4   Security Mechanisms in Initial 802.16 Networks

This section describes the security mechanisms introduced by the security sublayer of the IEEE 802.16 protocol. We focus especially on the description of those provided by the IEEE 802.16-2004 version, which supersedes all of the previous and earlier released versions (i.e., IEEE 802.16, and IEEE 802.16a/c). The components and mechanisms described below are provided by the privacy sublayer of the MAC layer. They are classified into security association handling, certificate using, PKM protocol-based authorization, key management and privacy guaranteeing.

2.4.1   Security Associations

A Security Association (SA) is a set of security information parameters which is used to maintain the state of security relevant to a communication. Typically, it is shared between the BS and one or many of its client SS (to support multicast). By means of an SA, an SS will be authorized for a WiMAX service. Three different SAs are defined by the standard: *primary*, *static* and *dynamic* [15], [16]. The primary security association is established by the SS in the initialization phase, while static SAs are configured on the BS. A Dynamic SA is generated by the BS and delivered to the SS and is used for dynamic transport connections. Since every SS can have several service flows, it may use several SAs. Every dynamic SA is created and destroyed dynamically in real time in response to the creation and termination of service flows. The BS has just to make sure that an assigned SA respects the characteristics of the type of the service to be accessed by the SS, and that an SS has access to only the SA which it is authorized to access. Each SS can have two or three SAs. The first is used on the secondary management channel. The second or the two last are used for either both the uplink and downlink channels or each one of them. Typically, all the downstream is protected using the primary SA.

Since a primary SA is unique per SS, the downstream cannot be protected using such type of SA in the case of multicast communication. Static or Dynamic SAs are used in this case.

Two types of SAs are supported by IEEE 802.16: authorization SA and data SA [17], but only the data SA is explicitly defined. Authorization SAs are used for authorizing the SS and establishing the data SA between BS and SS, while data SAs protect the transport connections. An Authorization SA consists of the following information [18]:

- A digital X.509 certificate to identify the SS.
- A 160-bit Authorization Key (AK) which is used for authorizing the use of IEEE 802.16 transport connections. This key is maintained secretly between the SS and BS. It is used to derive the Key Encryption Key (KEK).
- A 4 bits to identify the AK and differentiate between the successive AKs. It is denoted by AK sequence number.
- An AK life time representing the validity duration of the AK. Usually the AK lifetime is set to seven days, but may be configured between one and 70 days. The SS should request new key material within the duration of the AK's validity.
- A Key Encryption key (KEK) which is used by the BS to distribute Traffic Encryption Keys (TEKs). A TEK is located in the data SA and is encrypted using the TEK and sent to the SS. The SS will later use the KEK to decrypt the received encrypted TEK. A KEK is constructed as $KEK = Truncate_{128}(SHA1((AK|0^{44}) \oplus 53^{64}))$ where function $Truncate_{128}(seq)$ takes as a parameter the sequence seq and extracts the first 128 bits, | is the concatenation operator, $\oplus$ is the exclusive OR operator, SHA1(-) is the cryptographic hashing function using SHA-1 (Secure Hash Standard) algorithm, and x^y provides the sequence of bits obtained by repeating the byte x for y times.
- A Hash function based HMAC key which is used by for checking the authenticity and integrity of the key material exchanged between the BS and the SS. Two keys are described: an uplink HMAC key applied on the key distribution messages send by the SS to the BS, and a downlink HMAC key applied on the key distribution messages sent by the BS to the SS. They are constructed from the AK and are given by $Downlink\ HMAC\ Key = SHA1((AK|0^{44}) \oplus 3A^{64}))$ and $Uplink\ HMAC\ Key = SHA1((AK|0^{44}) \oplus 5C^{64}))$, respectively. The uplink HMAC key is applied on the key distribution messages send by the SS to the BS, and the downlink HMAC key is applied on the key distribution messages sent by the BS to the SS.
- A list of authorized data SA.

A Data SA consists of the following security information:

- A 16 bits SA Identifier (SAID).
- Two Traffic Encryption Keys (TEK), which are named TEKold (current used key) and TEKnew (used later when the TEKold expires) and are used to encrypt data.
- Two 2-bit identifiers, for identifying the TEKold and TEKnew respectively.
- A TEK life time describing the remaining validity period. Typically this value is equal to half a day, but it ranges from 30 minutes to seven days.
- Two 64-bit Initialization Vector (IV), one for each TEK, which are block of random numbers.

- An encryption algorithm which is used to protect the data exchanged over the connection. Typically, it is possible to use two algorithms in the IEEE 802.16-2004 version, including the Data Encryption Standard (DES) in Cipher Block Chaining (CBC) mode using a key of 56 bits and Advanced Encryption Standard (AES) in CCM (Counter with CBC-MAC) mode using a 128-bit key.
- An indication on the type of the SA, namely primary, static, or dynamic SA.

2.4.2   Use of Certificates

An X.509 certificate is used as a means of identification of the communicating parties, notably the BS and SS, and prevention of impersonation. Two types of certificates are considered by the standard: manufacturer's certificates, and SS certificates. However, the standard does not define the BS certificate. The X.509 certificate [19] profile defined in the IEEE standard requires the use of the following fields in the supported certificates:

- A version of X.509 certificates.
- The unique serial number identifying the certificate.
- The algorithm used by the issuer of the certificate to digitally sign the certificate. It stands for Rivest, Shamir and Adelman (RSA) encryption with SHA1 hashing.
- A certificate issuer name which identifies the authority that issued the certificate using the X.500 standard.
- The validity period of the certificate, which defines the period over which the public key is valid.
- The certificate subject, which represents the unique ID of the certificate holder. If the certificate is an SS certificate, the field contains the SS's MAC address.
- The Public Key of the subject named in the certificate. It mentions the value of the public key together with the algorithm identifier which specifies the cryptosystem related to the key and some related parameters. In the standard the cryptosystem described by the certificate is restricted to RSA encryption.
- The signature algorithm which is identical to the one used by the certificate issuer.
- The digital signature of the certificate. It represents the output of the digital signature algorithm, executed on the Abstract Syntax Notation. A Distinguished Encoding Rule (ASN1-DER) encodes the remaining content of the certificate.

The manufacturer's certificate can be a self-signed certificate or issued by a third party. The user certificate is typically created and generated by the manufacturer. This statement requires the SS to store securely and maintain in secret the private key associated with the public key of its certificate. Based on the standard specification, every SS should carry a unique X.509 digital certificate issued by the SS's manufacturer. To verify and validate the certificate provided in the SS, the BS has to validate the certificate path or chain and use the manufacturer's public key located in the manufacturer's certificate.

Some questions regarding the use of certificates remain open. First, there is no indication in the standard regarding the certification authority. While the use of a self-signed manufacturer's certificate could solve the problem, interoperability of SS and BS equipments may require a cross certification between different manufacturers' Certification

Authority (CA). The use of external authority certificates requires that the BS also verifies the revocation status of these certificates. A Certificate Revocation List (CRL) needs to be downloaded periodically into the BS and checked during the certificate chain validation; or an Online Certificate Status Protocol (OCSP) needs to be used for real time verification of the certificate's status. Second, since the SS obtains its certificate during manufacturing, and in order to prevent a device from being cloned, the private key is typically embedded within the device hardware. While an attack consisting in extracting the private key from the device has little chance of success, the standard should take into consideration the case of an SS's private key being compromised. The SS should be able to obtain a new certificate. Even if a dynamic certificate generation feature is integrated into the standard's specifications, there is no explicit indication regarding the protocol.

2.4.3   PKM Protocol

Security of connections access in WiMAX is done with respect to the Privacy Key Management (PKM) protocol. The protocol is responsible for the normal and periodical authorization of SSs and distribution of key material to them, as well as reauthorization and key refresh. It also manages the application of the supported encryption and authentication algorithms to the exchanged MAC Protocol Data Units (MPDUs).

The version of the PKM protocol, which will be described below, is that defined for use in the IEEE 802.16-2004 standard. This version was later extended to cope with mobility in the IEEE 802.16e standard.

The PKM protocol is comparable to a conventional a client/server model, where the SS proceeds as a client to request keying material and the BS responds to these requests, making sure that the client is authorized to get the key material associated with the services that he is authorized to access. PKM uses X.509 certificates and symmetric cryptography to secure key exchange between an SS and a BS. It is a three-phase based protocol, as shown in Figure 2.4. The remaining part of this section describes each of these phases.

2.4.4   PKM Authorization

The first phase of the PKM is the process of authorizing the SS by the BS. The details of this phase are shown by Figure 2.5. To connect with the BS, the SS sends an authentication message (denoted by AuthenticationInfMess) containing the certificate of SS vendor [20]. The design of the protocol assumes that any device issued by a recognized manufacturer can be trusted. If the security policy of the BS only accepts devices known in advance, the 802.16 standards allow one to ignore the first message.

Immediately after that, the SS sends an authorization Request Message (denoted by AuthorizationReqMess) to the attached BS, requesting an Authorization Key (AK). This information will be used as a shared secret. The message contains the following information:

- The SS certificate.
- A description of the cryptographic capabilities supported by the SS. Note that a cryptographic capability takes the form of a list of consecutive cryptographic suites, where

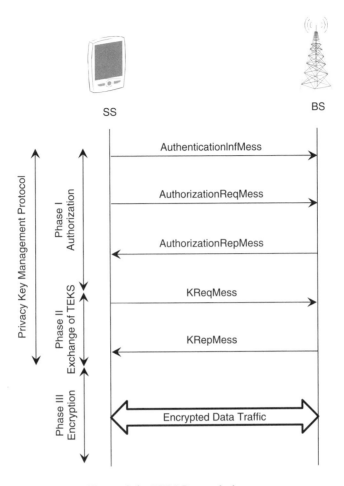

Figure 2.4   PKM Protocol phases.

each suite indicates a set of packet data encryption and packet data authentication algorithms.
- The security association identifier (SAID) of the SS's primary SA. This value is equal to the primary 16-bit Connection Identifier (CID) that the SS receives from the BS during the network entry and the initialization phase.

The SS will be authorized based on the verification of its certificate. The public key contained in the certificate will be used for constructing the third message. The BS verifies also whether it supports one or more of the cryptographic capabilities of the SS. The response of the BS to the SS is described by message 3, which is denoted by AuthorizationRepMess. The aim of this message is to instantiate an authorization SA between the two stations. It contains:

- The Authorization Key (AK) generated by the BS and encrypted using the SS public key contained in its certificate. A proper use of this AK shows an authorization regarding

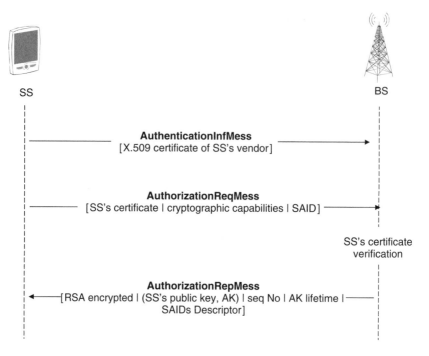

Figure 2.5   PKM authorization phase.

the access of the WiMAX channel. Note that no constraints on the generation of this key were defined by the standard.

- A 4-bit AK sequence number to differentiate between the consecutive Authorization Keys.
- The AK life time value.
- The SAIDs descriptor(s) as the identity and properties of the primary SA and zero or more existing static SAs for which the SS may be authorized to get the keying information. As stated previously, the SAID of the primary SA and the primacy CID are equal. The standard states that no dynamic SA could be identified in this message.

Upon reception of the AuthorizationRepMess message, the BS computes the KEK and the message authentication keys (*Downlink HMAC Key and Uplink HMAC Key*) using formulae described in Subsection **Security Associations**. These authentication keys will be used later in the exchange of key materials, especially the TEKs.

To maintain its authorization status, the SS has to refresh its AK periodically for the purposes of security. This is done by resending message 2 (i.e., AuthorizationReqMess message) prior to the expiration of the current operational AK. The reauthorization is thus similar to the authorization except that AuthenticationInfMess is not sent. In fact the BS knows the SS identity and the AK under use suffices to authenticate the SS. It could therefore be noted that both the SS and BS maintain at the same time two active AKs (i.e., the life time of the two AKs is overlapping) in order to avoid service interruption during reauthorization.

2.4.5 Privacy and Key Management

In the second phase of the PKM protocol, which is shown by Figure 2.6, the aim is to initiate the exchange of TEKs, and establish a data SA. The TEKs will be later used for encryption. As stated previously, the authorizationRepMess message contains, in addition to the SAID and properties of the SA, from zero to several static SAs for which the SS is authorized to obtain the key material. Therefore, the SS starts, in this phase, a separate state machine for each of the SAID identified in the authorizationRepMess message. Every state machine is responsible for managing (e.g. establishing, refreshing) the keying material associated with the related SAIDs.

Every SS sends periodically a Key Request Message (KReqMess) to the BS, asking it for the renewal of the TEK. This message is composed of:

- the AK sequence number which allows the BS to determine the Uplink HMAC Key used by the SS to generate the HMAC digest of this message;
- the SAID related to the SA whose keying material is requested. This SAID is related to the started TEK state machine;
- the HMAC digest produced by the application of the HMAC function on the message payload using the Uplink HMAC Key.

After making sure that the received SAID matches the SA at the SS and verifying the authenticity and the integrity of the KReqMess message by checking the HMAC digest, the BS responds to that message. It sends a key Reply Message (KRepMess) containing the new key material needed by the TEK state machine. At any time, the BS maintains two active key materials per SAID, which are denoted by TEK-Parameters in the KRepMess. A keying material includes:

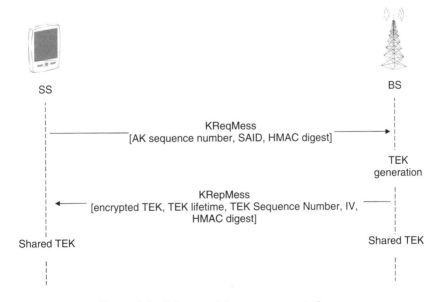

Figure 2.6 Privacy and key management phase.

- TEK encrypted with the KEKs using either the 3DES in EDE mode with 128 bits, RSA PKCS#1, or AES in ECB mode with 128 bits;
- the remaining lifetime of the TEK;
- the TEK sequence number;
- a 64-bit initialization vector.

The KRepMess message contains an AK sequence number, the SAID, the parameters related to the old TEK and the new TEK and an HMAC digest to ensure the SS that the message is sent by the BS without being tampered with. Note that the validity durations of the two TEKs overlap. In fact, the new TEK is being activated before the old TEK expires and the old TEK is destroyed after the activation of the new TEK. The lifetime of a TEK is also used by the SS to estimate when the BS will invalidate a previous TEK or request a new TEK.

If the SAID in the KReqMess message is invalid, the BS responds with a Key Reject Message containing the AK sequence number, the SAID and an error code with an indication regarding the reason of rejection and a HMAC digest. The SS could therefore resend another KReqMess message to get a new TEK.

Note that the current described phase of the PKM protocol could start by an optional message, denoted by RkeyMess, which precedes the KReqMess message. This message is sent by the BS to trigger rekeying before the SS requests it. It contains the AK sequence number, the SAID related to the SA whose keying materials are requested, and the digest of the message produced by the downlink HMAC key.

2.4.6   Data Encryption

After achieving the SA authorization and the TEK exchange, transmitted data between the SS and BS starts to be encrypted using the TEK. An encryption algorithm is used to encipher the MAC PDU. Note that, neither the CRC nor the MAC header is involved in encryption in order to guarantee the forwarding of the MAC PDU and support diverse services. In the MAC header, an Encryption Control (EC) field is set to 1 as an indication regarding the availability of an encrypted MPDS. In addition, the 2-bits Encryption Key Sequence (EKS) field indicates the used TEK. Encryption can be done by means of the Data Encryption Standard (DES) using Cipher Block Chaining (CBC) mode with 56 bits. Figure 2.7 summarizes the fundamental steps used to produce cipher text in the IEEE 802.16-2094 standard if DES-CBC is used. The initialization vector used to encrypt the MPDU is equal to the output of the SA Initialization Vector (IV) *xored* with the synchronization field in the PHY frame header. The DES-CBC encrypts the payload using the generated IV and the authenticated TEK.

2.5   Analysis of Security Weaknesses in Initial Versions of 802.16

Although the IEEE 802.16-2004 standard appears to be secure thanks to the integration of security functionalities in the security sublayer part of the MAC layer, several security weaknesses related to this version were discovered and are described by the literature. Most of them are related to authentication, privacy, key management and availability. This

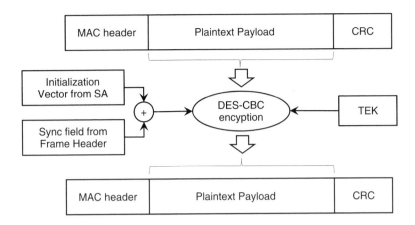

Figure 2.7   DES data encryption in IEEE 802.16-2004.

section describes the main security weakness related the IEEE 802.16 standard, showing potential attacks and the viable countermeasures to prevent them.

2.5.1   Physical-Level Based Attacks

In IEEE 802.16 standards, security is defined above the physical layer, leading the network to integrate the inherent vulnerabilities of wireless links. The physical layer is therefore vulnerable to jamming, scrambling and water torture.

A jamming attack [21], [22] consists of generating a strong noise (e.g., Gaussian noise) in order to interfere and reduce the capacity of the wireless channel. Such behaviour may compromise service availability, especially if the attacker is close to the base station. The Frequency Division Multiple Access (OFDM) and Scalable OFDM access (SOFDMA) used by WiMAX are not capable of handling such type of attack. A jamming attack requires specialized hardware. The risk associated with such an attack, which could be easily detected using radio spectrum analysers and monitoring equipments, is important. The location of the attacker can be detected using radio direction finding tools [23]. Law enforcement can also play an important role in stopping jammers. If mobile WiMAX is used, the attacker may change its position and make the anomaly monitoring and SS localization more difficult to achieve. Resistance to jamming attacks can be achieved by: (a) increasing the signal bandwidth. In this context, spreading techniques, including Frequency Hopping Spread Spectrum (FHSS) and Direct Sequence Spread Spectrum (DSSS), can be used; or (b) increasing the signal power. In this context, powerful transmitters and high gain transmission or receiving antennae can be used.

A scrambling attack aims to scramble control or management information selectively, thus disrupting the normal operation of the network, or taking control of it. A scrambling technique is a special type of jamming attack which occurs for a short length of time to attack specific frames or parts of frames. Scrambling is of great concern, especially when it targets messages which do not tolerate delay, including channel measurement report requests or responses. Scrambling attacks may also target slots reserved for data traffic sending, leading the target user to retransmit data. Consequently, their granted bandwidth is reduced. While conducting a scrambling attack is much more complex than jamming, jamming can be unintentional owing to natural noise interruption and the available period for the attack. A jamming attack can be detected based on the analysis of discrepancies regarding the system's performance.

Water torture is another kind of physical layer attack which consists of sending useless frames in order to drain the SS's battery or exhaust its resources. In this context, a suitable mechanism for detecting and discarding the bogus frames needs to be employed.

2.5.2 Attacks on Authentication

While, in the authorization phase of the PKM protocol, the mobile authenticates itself by sending its certificate, the BS does not. An attacker could pretend to be a legitimate BS (rogue BS attack) and thus attracts the MS. The attracted MS will try to attach itself to the rogue BS. If the Carrier Sense Multiple Access method is used, such as in IEEE 802.11 networks, the attacker would simply have to capture the identity of the BS, and wait until the medium becomes idles to send a message using the BS identity. However, in WiMAX networks, the potential use of Time Division Multiple Access (TDMA) complicates the attacks. The attacker has to capture the BS identity and wait until a time slot allocated to the legitimate BS starts. It transmits the message using a high signal strength in order to force the SS to discard the signal sent by the legitimate BS [24].

After trying to attach itself to the rogue BS, the SS executes the authorization phase of PKM protocol by sending the first two messages. Upon the receipt of the Authorization-ReqMess, the rogue BS sends back an AuthorizationRepMess to the SS. In this message, the attacker includes a self-generated AK, which is encrypted under the SS public key. Since the SS does not verify the authenticity of the received message, it accepts the received AK and proceeds with KEK and Uplink and Downlink HMAC keys generation. The attacker could therefore gain control over the SS communication. Using the credentials of the SS, the attacker could register himself at the BS, thus establishing a Man In the Middle (MIM) attack.

The main weaknesses related to the described attack are due to lack of mutual authentication [25]. To guarantee the ability of the SS to authenticate the BS, a possible solution may consist in appending to the AuthorizationRepMess message the signature and the certificate of the BS. To guarantee the freshness of the AuthorizationRepMess message and prevent replay attacks, it would be useful to add to that message the timestamp that the SS appended in the AuthorizationReqMess message. The aim is to protect the BS against replay attacks of the AuthorizationReqMess message and let the SS verify that the message, which it receives, corresponds to the sent request. Note that the appended timestamp should be taken into consideration during the signature of the Authorization-RepMess by the BS. The main drawback related to the use of timestamps consists of the need to synchronize the BS and SS. This should not be difficult since, in the IEEE 802.16 standard, the SS and BS synchronize with each other during the initial ranging before the start of the PKM protocol. The timestamps will be applied there.

To use timestamps as a solution against replay attacks, [26] proposed the introduction of the following modifications:

1. AuthenticationInfMess (SS → BS): Cert (Manufacturer(SS)).
2. AuthorizationReqMess (SS → BS): T_{SS} | Cert(SS) | Capabilities | SAID | SigSS.
3. AuthorizationRepMess (BS → SS): T_{SS} | T_{BS} | [Pre-AK]$_{pubSS}$ | LifeTime$_{AK}$ | SeqNo
   | SAID$_{List}$ | Cert(BS) | SigBS.

A timestamp, denoted by T_{SS} is generated in the SS and inserted in the AuthorizationRe-qMess message. The BS inserts in the AuthorizationRepMess message the SS's timestamp, received in the second message, along with a new generated timestamp, denoted by T_{BS}, in order to guarantee the message freshness and provide a countermeasure against replay attacks. The second and third messages are signed by the private key of the SS and the private key of the BS, respectively. The two produced signatures are denoted by SigSS and SigBS, respectively. By using both the BS timestamp and signature, the SS can verify that the received message is fresh and alive and corresponds to its request.

Using nonces in AuthorizationReqMess and AuthorizationRepMess messages instead of timestamps could be a viable solution. Note that the nonce in the AuthorizationRepMess may optionally be encrypted using the SS public key. Another solution could consist in replacing the AK by pre-shared AK and let the SS and BS derive the AK locally. A technique for guaranteeing the authenticity and integrity of the pre-AK sent between the SS and BS should be ensured. The authors in [17] proposed to modify the authorization phase of the PKM protocol as follows:

1. AuthenticationInfMess (SS → BS): Cert (Manufacturer(SS).
2. AuthorizationReqMess (SS → BS): NonceSS | Cert(SS) | Capabilities | SAID.
3. AuthorizationRepMess (BS → SS): NonceSS | NonceBS | [Pre-AK]$_{pubSS}$ | LifeTime$_{AK}$
   | SeqNo | SAID$_{List}$ | Cert(BS) | SigBS.

A random number, denoted by NonceSS, is generated by the SS and inserted in the AuthorizationReqMess message. The BS inserts in the AuthorizationRepMess message the received NonceSS along with a new generated nonce denoted by NonceBS. The proposed solution only allows the SS to make sure that the third message is fresh and corresponds to its request. To protect the BS from replay attacks on the AuthorizationReqMess message, a list of previously received nonces by the same SS should be used in order to detect replayed messages.

2.5.3   Attacks on Key Management

Similar to the authorization protocol, the key management phase of the PKM is vulnerable to the replay attack. In fact, in some situations, the SS cannot distinguish between a new and a reused data SA, especially as the KRepMess message does not include sufficient information that could be used to verify the replier's authenticity. However, since every SS maintains two keying materials, namely the oldTek and the newTek, the SS could easily detect whether the KRepMess message corresponds to its request. Therefore, the attacker has to conduct the replay attack when the SS requests the keying materials for the first time. Nevertheless, since the HMAC_KEY_D key, which is used to verify the integrity of KRepMess message, is computed based on the AK received in the authorization phase, the replayed KRepMess should be related to the same instance of the protocol. To conduct his attack, the adversary should save a copy of the KReqMess and KRepMess messages and then force, or wait for, the second phase of the PKM to be reset. After that, since the SS will request a new keying material, the attacker can replay the KRepMess message. This replay attack can be thwarted easily by forcing the PKM protocol to re-start from the authorization phase every time the second phase fails or is reset, so that a new AK is exchanged [26].

Another potential replay attack is possible against the KRepMess message which is used to send key material to the SS for a given SAID. The weakness exploited by the attack is related to the 4-bit sequence number that is used. Note that this sequence number provides a relationship between instances of the first phase and second phase of the protocol. If this sequence is used in a sequence buffer, its value ranges between 0 and 15. Owing to the tiny range of this value, the attacker could capture KRepMess messages, brute force (compute) the TEK, and replay the message so as to succeed later in decrypting the data traffic. However, to succeed in his attack, the adversary should also replay a correct HMAC value, meaning that a correct Downlink HMAC key should be used. Since this key is derived from AK in the first phase of the protocol, the replay attack could succeed if the sequence number is replayed within the same instance of the second phase of the PKM protocol. Otherwise, a coincidence between the AK values related to replayed message and to the AK value related to the current instance of the protocol under attack, should occur. However, owing to the randomness of the AK, this situation is highly infrequent.

The replay attack could also target the first optional message of the second phase of the PKM protocol. If an attacker replays the first message, the BS will assign and send new keying material using a KRepMess message. The legitimate SS, which is not aware of the attack, will think that it is the BS which requested the rekeying and sent the first optional message. As a consequence, this attack causes both the SS and BS to exchange keying material without intending to. The following solution was proposed by [26] to mitigate this attack:

1. RkeyMess (BS → SS): TBS | SeqNo | SAID | HMAC(RkeyMess)
2. KReqMess (SS → BS): TBS | TSS | SeqNo | SAID | HMAC(KReqMess)
3. KRepMess (BS → SS): TSS | TBS | SeqNo | SAID | OldTEK | NewTEK | HMAC (KRepMess)

Two timestamps, say TBS and TSS, are generated by the SS and BS respectively in the messages they send. If the BS sends the optional message RkeyMess, the TBS generated in that message will be included by the SS in the KReqMess message. The BS can omit it in KRepMess by setting it to 0. If the RkeyMess is not sent and the SS initiates the request, the timestamp TBS is set to 0 in KReqMess. The BS sets the TBS in KRepMess message corresponding to the TSS value received in the KReqMess message.

Intention to replay also affects the data SA definition. In fact, the standard uses 2-bit key identifiers in a circular buffer for identifying the TEKs. Typically, the key identifier space should support the use of a number of totally different key identifiers within the largest value of AK lifetime. Since an AK lifetime can reach 70 days, and the shortest TEK is equal to 30 minutes, a total number of 3360 different TEKs could be supported by the key identifier field. Unfortunately, this is not the case, since only four different values can be used by a 2-bit key identifier. The latter wraps from 3 to 0 on every fourth rekeying operation. An attacker could thus replay used TEKs, exposing both the TEK and subscribed data to a compromise.

2.5.4   Attacks on Privacy

To support encryption, the IEEE 802.16 standard includes DES in CBC mode as an encryption algorithm. This algorithm operates on blocks of data of 64 bits to perform each encryption or decryption operation and uses 56-bit DES keys (TEK) and a random Initialization Vector (IV). However, 56 bit key is not secure, based on today's computer capability, which easily allows one to try every possible key within a reasonable time (brute force attack). Moreover, the CBC-IV is predictable since the SA IV is public and constant for its TEK and the PHY synchronization field is highly repetitive. To mitigate the problem while maintaining the possibility of the use of DES, the only solution consists of generating each per-frame initialization vector randomly and inserting it in the payload. However, this solution increases the encryption overhead.

In [17], it is stated that the CBC modes using block cipher become unsecure after operating on $2^{n/2}$ blocks, where n represents the block size. Since DES uses blocks of 64 bits, it becomes unsecure if used on more than 2^{32} 64-bit blocks. As stated previously, the default TEK life time is half a day but could be used at maximum for seven days. If the used throughput allows one to produce the 2^{32} 64-bit blocks in a time period shorter than the TEK life time, the encryption scheme becomes highly vulnerable. Considering a throughput of 6 Mbps, the total of 2^{32} 64-bit blocks would be produced in 12 hours.

The data protection scheme using DES not only fails to provide strong confidentiality, it also fails to protect against replay attacks, nor does it allow for verifying the integrity. Based on the deficiencies regarding the use of DES, the standard allows for encrypting the message using the AES in CCM mode with 128 bit TEK key. Based on the use of packet number, protection against replay attacks is possible.

2.5.5   Attacks on Availability

Another technique consists of exploiting message replay so that the legitimate BS is declined [18]. Since the AuthorizationReqMess does not contain any field which guarantees its freshness, the PKM protocol becomes vulnerable to replay attack. After a legitimate SS sends the AuthorizationReqMess message, the attacker intercepts this message and stores it. It will then send the captured message repeatedly to the BS. While this operation does not allow the attacker to obtain the value of AK, it could burden the BS and force it to decline the authentic and legitimate SS [27].

To guard against such a type of attack, the AuthorizationReqMess should contain a time-stamp or nonce together with the digital signature of the SS (using its private key) to guarantee the authenticity and freshness of the message.

In the second phase of the PKM protocol, while the replay attack on the KRepMess message has little probability of success (as was explained in the section on 'Attacks on key management'), the BS remains vulnerable to an attack on the KReqMess message. In fact, in contrast with the KRepMess message, the KReqMess message does not contain keying materials (i.e., the old and new TEK) which allow the receiver to compare it with the previous received message. Since the BS cannot verify the freshness of the received message, it will generate and assign a new keying material (i.e., TEK) to the SS, even though the latter has not requested it. After that, it replies with a KRepMess message to the SS. If this situation occurs frequently, it could exhaust the resources of the BS.

Ranging Response messages (RNG-REP), exchanged during the node attachment to the network, are vulnerable to serious types of attack. In fact, these messages are not encrypted, cannot be authenticated and are stateless. Since an SS takes into consideration immediately the new parameters provided by the BS, RNG-REP messages can be misused by an attacker. Among the optional fields of this message, is the Ranging Status field, which is used to state whether uplink messages are received within acceptable limits by the BS. An attacker can detect the Channel ID (CID) that the victim SS is using and generate a spoofed RNG-REP message by setting the Ranging Status field to 2 (which corresponds to 'abort'). Note that the attacker could simply brute force the Channel ID by cycling through 65 536 possible values. Therefore the victim SS is prevented from conducting periodic ranging and is excluded from the network.

Based on the deficiencies in protecting the RNG-REP message, an attacker could use that message to override the uplink and downlink channel that the SS uses. If there is no BS operating on that channel, the SS will continue to scan the frequencies by listening for a minimum of 2 ms before moving to the next channel. Depending on the number of usable channels, the latter operation could take a considerable time. Note that after scanning all the channels, the SS will try to reuse the proper channel. If the attack is repeated, the SS will be unable to access the network and is induced into a denial of service.

2.6   Security Amendments in Recent Versions if IEEE 802.16

In IEEE 802.16e, the security sublayer was redefined in order to eliminate most of the security weaknesses found in the previous versions of this protocol and satisfy security requirements for mobile services. In this context, the security sublayer is enhanced, encryption methods are improved, mutual authentication mechanisms are introduced to protect against various types of replay and MiM attacks, pre-authentication between the MS and BS is enforced to reduce any potential interruption of services during handover operations, a key hierarchy is defined to allow an MS authenticating itself to an AAA server once independent of the number of BS it authenticates with handover and the PKM protocol is extended to version 2. Note that within the extended IEEE.16e security sublayer, the two versions of the PKM protocol are supported. Version 1 is simply extended to support new ciphering algorithms, including 3DES-EDE and AES-ECB for key material confidentiality and AES-CCM for MPDU confidentiality. On the other side, HMAC-SHA-1 is used for the protection of the integrity of key management messages.

In this section, we describe the security features in IEEE 802.16e by describing the enhancement introduced with regard to the IEEE 802.16-2004 version discussed in the third section. New properties and mechanisms provided by the use of PKMv2 are described in detail.

2.6.1   Authorization, Mutual Authentication and Access Control

The PKMv2 protocol supports mutual authentication and authorization, giving the opportunity to the SS and BS to authenticate each other's identity. Different modes of mutual authentication are supported: RSA based authentication, EAP based authentication [28], or RSA followed by EAP based authentication. RSA based authentication involves the use of X.509 Digital certificates together with RSA encryption. EAP involves symmetric cryptography and is based on the use of EAP (RFC 3748) which is an authentication credential carrier protocol for user authentication during remote or local network access. A Back-end authentication infrastructure, such as the AAA (e.g., RADIUS) architecture is used. The mobile can be authenticated using a credential issued by the operator (e.g. SIM card) or an x509 digital certificate. Several EAP authentication methods [29] are supported by PKM v2, including EAP-TLS (X509 based authentication), EAP-AKA (Authentication and Key Agreement), and EAP-CHAPv2 (Microsoft Challenge Handshake Protocol). The output of EAP exchange is 512-bit key, called the Master Session Key (MSK), which is the root of key hierarchy. Using this key, the MS and the BS derive a Pairwise Master Key (PMK), which is itself used to derive the AK.

We present in the following the RSA-based authentication which is a key transport protocol used by the authorization phase of the PKMv2 protocol [30]. In this version of the PKM protocol, a mutual authentication mechanism is used, and nonces are appended to protect against replay attacks. The authorization phase, which is described below, is composed of four steps, where the first is optional.

1. Authorization initiation (MS $\rightarrow$ BS): MS.manufacturer$_{Cert}$.
2. Authorization Request (MS $\rightarrow$ BS): N_{MS} | MS_{Cert} | Capabilities | BCID.
3. Authorization Reply (BS $\rightarrow$ MS): N_{MS} | N_{BS} | KU_{MS}(pre-AK, MSID) | SeqNo | Lifetime | SAIDs | BS_{Cert} | SIG_{BS} (Authorization reply).
4. Authorization Acknowledgement (MS $\rightarrow$ BS): N_{BS} | MS_{addr} | AK(N_{BS}, MS_{Addr}).

Similar to the PKMv1, it is the MS that initiates the authorization protocol. It can send an optional message containing the MS's manufacturer's certificate. The MS then sends an authorization request message containing its X509 certificate (containing a common name equal to the MS's MAC address), along with a nonce, denoted by N_{MS}, containing 64-bit random value that it generates. The message also includes the MS's capabilities (supported authentication and data encryption algorithms) and the Basic Connection Identity (BCID) which is equivalent to the CID and assigned to the MS when it entered the network and requested ranging. Note that this message is not protected and may be subject to forgery or modification. Further to the reception of the authorization request message, the BS sends back an Authorization Reply message containing the MS's nonce already received by the MS, a nonce generated by itself (denoted by N_{BS}), its certificate, a 256-bit pre-AK along with MS's identifier (MSID) encrypted with MS's public key and the Authorization key attributes including the key lifetime and sequence number and one or more SAIDs. The authorization reply message is signed by the BS. Note that the SAIDs in this message are optional in the case where an RSA authorization exchange will be followed by an EAP authentication exchange. The AK will be derived from the pre-AK with the BS and MS addresses. Since only an authorized SS is able to extract the Pre-AK, the MS authorization can be checked based on the possession of the pre-AK.

To let the BS confirm the authorization request message and be sure that the SS has made a genuine request for access to the network services, an authorization acknowledgement message is sent by the SS further to the reception of the authorization reply message by the BS. This message contains the nonce sent by the BS, its MAC address together with the BS's nonce and the MS's MAC address encrypted with the pre-AK. At the conclusion of this process, the SS and the BS are mutually authenticated.

In IEEE 802.16 networks, mobility is supported so that an MS can hand over from one visited BS to another. During the handover operation, an MS may use pre-authentication with the new BS instead of executing the entire authorization procedure from the outset. In fact, since the authentication protocol is based on the use of a public key infrastructure, it might be better to avoid such steps and accelerate the re-entry to the network by establishing a new authorization key in the MS and target BS based on a pre-authentication mechanism.

With regard to voice call migration, for example, the ITU recommends a period of time of less than 30 milliseconds, to spend between leaving the first BS and reestablishing the context at another BS. On the other side, a BS deployed in the network should have copy of the AKs or TEKs established between another BS and its attached MS. Otherwise, if a BS is compromised, the SS will also be compromised in all the BSs it visits during the same session. Since responding to these constraints may require extensive development, the mechanisms related to pre-authentication are beyond the scope of the 802.16e standard. Typically, if a Ranging Request message sent by an MS includes the new serving BSID, and if the BS to which the MS hands over or roams has already received a message from the backbone containing the MS's information, the re-authorization process of the PKM protocol should be used by the MS and the new BS to complete the network re-entry in handover scene.

2.6.2 TEK Three-Way Handshake

In mobile WiMAX, the weaknesses related to the exchange of key material were eliminated. The PKMv2 is therefore secured against several Man In the Middle attacks by means of the TEK-SA three-way handshake which includes the use of nonces, AK identifier and a message authentication code derived from the AK. The handshake protocol supports several functions including key activation, SA parameters negotiation, security negotiation confirmation and SA parameters refresh for network re-entry [31]. Therefore, the protocol is executed either further to the initial authorization or during handover as follows:

1. SA-TEK Challenge (BS $\rightarrow$ MS): N_{BS} | SeqNo | AKID | LifeTime |H-C/MAC(-).
2. SA-TEK Request (MS $\rightarrow$ BS): N_{MS} | N_{BS} | SeqNo | AKID | Capabilities | SecNego-Params | PKM Config | H-C/MAC(-).
3. SA-TEK Response (BS $\rightarrow$ MS): N_{MS} | N_{BS} | SeqNo | AKID | SA-TEKUpdate | Frame-No | SA-Descriptors | SecNegoParams | H-C/MAC(-).

The first message, which is denoted by the SA-TEK challenge, is sent by the BS to the MS. It includes a nonce randomly generated by the BS, a 64-bit AKID (AK Identifier) of the AK whose procession is being proved in the previous phase of the PKM protocol. The AKID is computed with the MS and BS identification. Such feature is highly important for the enforcement of mutual authenticated key confirmation [31]. Similar to the previous version, the message is authenticated using a HMAC or a Cipher Message Authentication Code (CMAC), which is computed using a key derived from the AK. The SeqNo is a

4-bit value indicating the key for the H-C/MAC. The use of a nonce allows proving the liveness of the sent message and provides protection against replay attacks.

The second message, denoted by SA-TEK Request, is sent by the MS. It includes the BS nonce received in the previous message, and a nonce generated by the MS. The SecNegoParams field includes the parameters negotiated in the two management control frames SBC-REQ/REP. The other fields are similar to those available in the previous version of the PKM protocol. Upon reception of this message, the BS checks whether the AKID and the BS nonce are valid, the MS capabilities are supported, and the HMAC/CMAC value is correct. If this is the case, it replies by sending the third message. Otherwise, it ignores the message.

The third message, which is titled SA-TEK Response, is sent by the BS to the MS. It includes the same BS and NS nonces received in the second message, and the SA-TEK Update parameters (this is especially importantly during handover if the BS needs to specify the security capabilities for the session established with the MS). Typically, these parameters contain encrypted TEKs, group of keys (GKEK) and GTEK. Upon successful reception and validation of this message by verifying whether the values of nonces are correct, the negotiation parameters are supported, and HMAC/CMAC value is as expected, the MS installs the new received TEK and its associated parameters.

2.6.3   Encryption and Key Hierarchy

With The IEEE 802.16-2004 standard, both the Data Encryption Standard in CBC mode with 56-bit keys, or the Advanced Encryption Standard (AES) in CCM mode with 128-bit keys, can be used to cipher the payload of MAC PDU. In the first method, the data to be ciphered is fragmented into different blocks where one of them is ciphered with the key. Using the second method, the data is fragmented into different 128-bit blocks. Every encrypted MAC PDU receives a prefix of 4 bytes, representing the packet number in accordance with the SA. An 8-bytes Integrity Checking Value (ICV) is later appended to the end of the payload. The packet number is not encrypted but included in the authentication of the ICV. The MAC PDU payload, together with the ICV, is encrypted with the Traffic Encryption Key (TEK) using AES in CCM mode. By including a packet number, AES provides a mechanism against replay attacks, so that any packet number received more than once within a predefined period of time will be discarded. Compared with DES or 3DES, AES is more secure but also more complex and slower.

In IEEE 802.16e data encryption is supported using AES in four additional modes, namely CBC mode with 128-bit keys, Counter mode (CTR) with 128-bit keys for Multicast Broadcast Services and Key-Wrap. The Encryption Control (EC) bit in the generic MAC header is used to state whether the MAC PDU is encrypted or not. Note that, neither the generic MAC header nor the generic and primary MAC management messages are ciphered in IEEE 802.16e. The authors in [32] describe the use of AES in CTR mode with CBC authentication code (CCM).

As for TEK encryption with KEK, the IEEE 802.16e standard supports four methods, namely 3DES in EDE mode using 128-bit keys, RSA encryption using 1024-bit keys, AES encryption in ECB mode using 128 bit keys, and AES Key Wrap using 128-bit keys. Only the last method is specific to 802.16e, the other are already existing in 802.16-2004. Contrarily to AES, which uses the full 128 bits of the KEK to encrypt the TEK, the 3DES EDE mode uses the first 64 bits of the KEK for encryption and the remaining

64 bits for decryption. The ciphering process is performed in three cycles. In the first cycle, the TEK is encrypted with the first 64 bits of the KEK, while in the second cycle the output of the first cycle is decrypted using the second 64 bits of the KEK. The third cycle consists of encrypting the output of the second cycle using the first 64 bits of the KEK. Note that, in Mesh WiMAX architecture, the TEK encryption is performed based on the use of the SS's RSA public key.

2.6.4   Multicast and Broadcast Service (MBS)

The IEEE 802.16e defines Multicast and Broadcast Service (MBS) as distributing data content (especially multicast traffic) to the MSs across multiple BSs, while ensuring a protection against theft of service. Such a mechanism can be used for multimedia applications. Therefore, the definition of a group of secret keys becomes a requisite. The traffic to be sent is encrypted using a group-wide session key called Group Traffic Encryption Key (GTEK). The GTEK is used to encrypt multicast data packets, and is shared by all the MSs in the multicast group. To distribute and update this key efficiently to all the MSs in the multicast group, a solution could consist of letting the BS distribute this key to every MS individually and securely whenever one of these three situations happens:

- a new MS joins the multicast group and wants to receive a GTEK;
- an MS leaves the multicast group so that it is no longer able to decrypt messages using the active GTEK; or
- the GTEK in use is about to expire and should be updated.

It is obvious that such a solution lacks scalability since unicast messages may be exchanged frequently for the purpose of key exchange. The MBS uses a Multicast and Broadcast Rekeying Algorithm (MBRA) to refresh traffic keying material. A Group Key Encryption Key (GKEK) is used for encrypting the new GTEK and multicasting it to all the MSs before the expiration of the current GTEK in use. The initial GKEK is generated randomly by the BS and encrypted with the KEK using the same algorithms applied for TEK encryption (i.e., Key-request and Key-reply messages are used and exchanged over the primary management connections). Four encryption methods are supported by the PKMv2, namely 3DES, RSA, AES in ECB mode and AES Key-Wrap. The GKEK encryption is performed during the SA-TEK 3-way handshake further to the initial authorization or re-authorization, or in the initial GTEK request exchange, or in the GKEK update of the PKMv2 protocol. The GKEK is also used to compute the HMAC/CMAC Key_GD which is used to authenticate and verify the integrity of broadcast messages, including the GTEK update. A GKEK is assigned to a Group Security Association (GSA) which contains keying material useful for securing multicast group communication. This keying material is used to secure a multicast group. Two Group-Key-Update-Command messages are used by the BS for distributing the traffic keying material, by following this protocol [33], [34]:

- Message 1: BS → SS: {GKEK}$_{KEK}$
- Message 2: BS → SS: {GTEK}$_{GKEK}$

The first message which is a key update command is sent periodically by the BS to the MS, through its primary management connection, to update the GTEK. The GTEK sent in such a message is encrypted using the Key Encryption Key, which is derived from the AK generated during the process of authorization. Between two GKEK updates, the BS sends message 2, which is a key-update-command message, to the MSs in the multicast group to update the GTEK. Such a message is sent through the broadcast connection and contains the GTEK encrypted using the GKEK. The reader might note that the protocol still lacks scalability since it requires sending message 1 in unicast mode.

While in the unicast communication, all the keying materials are derived from the Authorization Key (AK); in Multicast and Broadcast Services (MBS) communication, the key materials are derived from another key entitled MBS Authentication Key (MAK) which may potentially be provisioned by an external source including an MBS or an AAA server [35]. This MAK may be shared by all the members of the MBS group. A MBSGSA contains the MAK, which has functionalities equivalent to those provided by the AK but local to the MBGSA, the MGTEK (MBS Group Traffic Encryption Key) and the MTK (MBS Transport Key) which are used to protect indirectly and directly the traffic used by multicast and broadcast service, respectively.

The primary goal of the MBS is to protect the forwarded content while reducing, to the greatest extent possible, the amount of generated overhead. In this context, the AES-CCM, which is the encryption mode used mainly in IEEE 802.16e networks, generates a heavy overhead. Moreover, integrity checking, which is also a feature whose implementation introduces some delays in generating the multicast traffic, is considered unnecessary if MBS applications are used. To cope with the context of MBS application, IEEE 802.16 considers the use of AES-CCM mode if integrity and replay protection are required, and AES in counter mode if no integrity protection of data is deemed necessary. In the latter situation it may be necessary to append an increasing counter to each MPDU during encryption, since the WiMAX channels are lossy. Typically, a 32-bit counter, which is composed of 24-bit physical layer synchronization field and 8-bit Rollover Counter (ROC), is created and sent with every MPDU. To reduce the overhead, only the 8-bit rollover counter is appended to the MPDU [12].

2.6.5 Security of Handover Schemes

The IEEE 802.16e standard defines three handover schemes, namely Hard Handover (HHO), Macro Diversity Handover (MDHO) and Fast Base Station Switching (FBSS). Of these three schemes the first is mandatory, while the others are called soft handovers and are optional. The IEEE 802.16e standard supports three possible security settings for every handover scenario. These security settings are defined by the Handover optimization bit#1 and bit#2 in the RNG-RSP message, as follows:

- Bit#1=0 and bit#2=0: Re-authentication and TEK 3-way handshake execution.
- Bit#1=1 and bit#2=0: No re-authentication procedure is executed. The TEKs for all SAs are updated.
- Bit#1=1 and bit#2=1: No re-authentication or TEK 3-way handshake execution. The MS keeps using the TEKs established with the serving BS.

Using an HHO, the MS communicates only with one BS at a time, meaning that the MS cannot establish a connection with the second BS before it breaks its connection with the old BS. Every BS forwards periodically a Neighbour Advertisement Message (NBR-ADV) which includes information regarding the neighbour BSs (e.g. number of BSs, their BSIDs). After switching its link to the target BS, the MS re-executes procedures related to ranging, authentication and registration. While this handover scheme is simple, it introduces high latency which could be higher than 100 ms. In particular, the execution of a full EAP authentication may require about 1000 ms and makes the IEEE 802.16 network unsuitable for using applications such as transmission of video conference or data streaming.

In the MDHO scheme, a set of BSs could be involved in the handover, and form a list called a diversity set. The MS and the BS monitor the BSs continually in this list, choose one of them and register with it. In downlink communication, several BSs could transmit data to the MS which will perform diversity combining. In uplink communication, the data sent by the MS is received by several BSs, which will perform selection diversity. All BSs involved in such handover scheme are required to share or transfer MAC-context based information such as operational authentication and encryption keys used by the established connections.

Similarly to the MDHO, the FBSS handover scheme allows the MS and BS to maintain the diversity set and the MAC-context related information is also shared by all BSs involved in the handover. However, in downlink and uplink communication, the MS exchanges data with only the anchor BS.

2.7   Analysis of Security Weaknesses in 802.16e

This section describes the main security weaknesses described by the literature regarding the 802.16e version. The latter are related mainly to problems with the security and authenticity of management communication messages and key sharing in a multicast and broadcast service. In this section, we also describe the main suggested solutions by the different contributors to the literature.

2.7.1   Attacks on Authorization

An initial analysis of the authorization phase of the PKMv2 protocol shows that the first message (Authorization Request) is not protected against modification or forgery. This weakness had existed in the PKMv1. We have already shown that if an attacker captures such message, when it is sent by a legitimate MS, and repeatedly sends it, it could burden the BS and force it to deny access to a legitimate MS.

Even if the authorization request is signed, the protocol is still vulnerable. In fact, while nonces are sent back to each other in the subsequent replied messages, a straightforward reasoning could state that it is not necessary to check the timestamps of the exchanged three messages and the method could enforce mutual authentication and be used without the requirement for synchronized clocks. However, this is not the case and the protocol is vulnerable to interleaving attack [33] by which the attacker could replay the first message and answer the BS, by providing correct nonces, and using the attacked MS as an oracle. The attack is shown in Figure 2.8 and is described as follows, assuming that the first

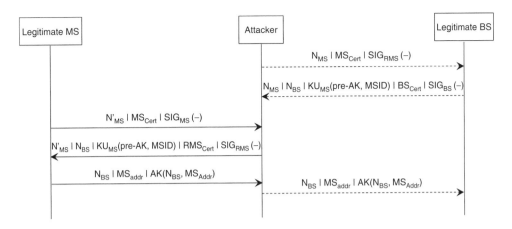

Figure 2.8 Impersonation attack on PKMv2's authorization phase.

message is signed by the MS. For the sake of clarity, we have omitted uncritical fields in the exchanged messages. Moreover the first message of the authorization phase, which is optional, is omitted.

The attacker participates in two sessions. In the first session (shown by dashed lines), it impersonates a legitimate SS, while in the second session (shown by continuous lines), it conducts a rogue base station attack. The attacker starts by sending to the BS a replayed message that it captured previously from the legitimate MS. Then, after receiving the BS response, the attacker finds itself unable to decrypt the pre-AK which was encrypted by the legitimate MS's public key. It will therefore be unable to send the authorization acknowledgement immediately since it cannot encrypt its address and the BS's nonce with the right AK. Consequently, the attacker uses the SS as an oracle to generate a correct acknowledgement message. It conducts a rogue base station attack and attracts the SS to connect to it and run a second instance of the PKM protocol. After sending the first message by the legitimate MS, the attacker replies to the MS by sending the BS nonce it received in the first session established with the legitimate BS. Similarly, it includes the pre-AK and the MSID received from the BS in the first session and encrypted with the legitimate MS's public key. However this message is signed with the attacker's certificate. To guarantee that the AK, which will be generated by the legitimate MS, and the AK generated by BS in the first instance of the protocol will be same, the attacker needs to impersonate the BS Address. This is easy to achieve and was shown in the description of attacks on PKMv1.

The MS replies to the attacker by sending its address and the legitimate BS's nonce, together with an encryption of these two values with the AK. The latter message received by the attacker from the legitimate SS will be replayed by the attacker and sent to the legitimate BS in order to finish the first session in which it impersonated a legitimate MS. While the PKMv2 uses AAA to allow a security session, the attacker could also forge and replay these messages to the MS.

To avoid this attack, a solution suggested in [33] would consist in adding the BS identity (BSID) to the last message and encrypt it, together with the SS address and BS nonce.

On the other side, the authors in [36] proposed to introduce timestamps on the exchanged messages, creating thus a hybrid solution which uses nonces together with timestamps.

2.7.2 Analysis of SA-TEK Three-Way Handshake

In [37] and [33] it was outlined that the SA-TEK three-way handshake protocol of PKMv2 is secure, even if the first message (i.e., the SA-TEK Challenge) can be subject to replay attack. In fact, this protocol has the similar form of the Needham Schroeder Secret Key protocol (NSSK) which was published in 1978 and has been studied in several security protocols validation frameworks. The latter protocol was proved to be secure after the implementation of some suggested revisions. To protect against replay of the SA-TEK Challenge message, the authors in [33] suggest adding timely information. Such a modification would be similar to those integrated in the Kerberos Protocol, which is also derived from NSSK. The SA-TEK three-way handshake protocol is not vulnerable to interleaving attacks since secret keys are used instead of public keys.

In [37] it was shown that the SA-TEK three-way handshake is not only secure but also over-secure due to the existence of security-based redundancy in this protocol. In fact it is not necessary to include the BS nonce, say N_{BS}, in the SA-TEK Key reply message. Since this nonce is generated by the BS, it does not guarantee anything to the SS. The nonce generated by the SS, say NBS, is sufficient to guarantee the confirmed freshness of the message sent by the BS.

2.7.3 Vulnerability to Denial of Service Attacks

In 802.16e networks, the network entry procedure, executed by a MS to attach itself to a BS, remains unprotected. Attackers can listen to the exchanged traffic and use the accessed information to forge ranging request (RNG-REQ) or ranging response (RNG-RSP) messages and manipulate, in consequence, different MS settings. Since such message is unauthenticated, the MS cannot determine its real source. An attacker may intercept and forge a RNG-REQ message by modifying the specified preferred downlink burst profile. It can also forge a RNG-RSP message to set the power-level of the MS to the minimum. The latter will trigger the initial ranging procedure repeatedly since it can barely transmit to the BS. In addition, the management communication between a MS and a Bs involves the sending of plaintext messages, and the origin of some management frames, sent in unicast or broadcast, is not authenticated. These message include some important unauthenticated messages [38]: authorization invalid (Auth-invalid) message, Mobile Traffic indicator (MOB_TRF-IND), Mobile neighbor advertisement (MOB_NBR-ADV), Fast Power Control (FPC), Multicast assignment request (MSCREQ), Power control mode Change Request (PMC-REQ), Mobile association Report (MOB_ASC-REP), Downlink burst file change request (DBPC-REQ), Ranging Request (RNG-REQ), and Ranging Response (RNG-RSP). These deficiencies make it possible for there to be several Denial of Service attacks [39].

The Mobile neighbour advertisement (MOB_NBR-ADV) message, which is sent by the currently serving BS to state the characteristics of the neighbour base stations, is not authenticated. An attacker could forge such a message to state the availability of a fake or a rogue base station, thus preventing the MS from performing an efficient handover or denying such an operation to it.

Fast Power Control (FPC) messages, which are sent by a BS to a MS asking it to adjust its transmission power, may be forged by an attacker to set the transmission power of an MS too low. The latter has to adjust its transmission power recursively to reach the BS again, leading to the sending of cumulated power adjustment messages. The attack could target several MSs at the same time period. Due to the use of CSMA, such cumulated sending may generate collisions in the uplink bandwidth request contention slots. The delay spent by the attacked MS in gaining correct transmission power gain may become excessively long. The attack may also drain the battery of the MSs and be considered to be a Water Torture attack.

The Auth-invalid message (Auth-Invalid) is sent from the BS to the MS if the AK shared between them expires or the HMAC/CMAC of some exchanged message in the Authorization phase indicates an unauthenticated message. Since the Auth-invalid message is unprotected (i.e. does not contain HMAC/CMAC digest), has a value that leads to a stateless rejection, and does not use the PKM serial number, it could be forged by an attacker to deny access to a legitimate user.

The Reset command (RES-CMD) message is typically sent by a BS to reset a non-responsive or malfunctioning MS. The MS will reset its MAC state machine. Contrary to the previous management messages, the RES-CMD can be authenticated by a MS. However, the attacker could force a BS to send this message to a target MS. To do so, it synchronizes with the network and receives the UL-MAP message to choose a victim CID and its burst file. Later the attacker transmits a signal at the time scheduled for the victim. The latter signal will be degraded or be completely unintelligible depending on the MS's signal strength. By forcing this operation to occur continually, the BS will send a RES-CMD command to the victim to reinitialize it.

Downlink Burst Profile Change Request (DBPC-REQ) is a message sent by the BS to the mobile subscriber to ask it to change the burst profile in order to cope with the variations of distance between the MS and BS and/or communication characteristics of the medium. An attacker could forge such a message to modify the burst profile deliberately (e.g. modification or modulation or encoding scheme) and disrupt the communication between the BS and the attacked MS.

During the handover operation, when a mobile station and a target BS are maintaining a network assisted association, the target BS does not directly send the RNG-RSP message to the MS, instead it forwards it over the backbone to the serving BS. The serving BS receives such message from all the neighboring target BSs, and aggregates all the copies into a Mobile Association report (MOB_ASC-REP) message. The message will be sent to the MS using the basic management connection. Such message, which includes the information (e.g. region for association at a 'predefined rendezvous time') useful to the MS for choosing the effective target BS, is not authenticated or protected against forgery. An attacker could forge a MOB_ASC-REP message so that it appears that no service is available from the target BSs. Such operation prevents the MS from being associated with the best candidate BS and forces it to continue benefiting from a downgraded service.

To secure the exchanged control messages, the authors in [40] proposed using the Diffie-Hellman key agreement protocol [41] in the initial ranging procedure. The random prime number 'p' and a primitive root 'q', which are used by the protocol to generate a shared secret key, are chosen from the ranging code that appears in the UL-MAP message.

2.7.4 Broadcasting and Multicasting Related Weaknesses

If the Multicast and Broadcast Service is used, data is distributed between MSs using the shared symmetric GTEK. Such a key is shared between all the members that belong to the same multicast group. Since the key is symmetric, each MS can not only decrypt the multicast traffic, but also encrypt it using the same key. An attacker could forge multicast traffic and send it to other MSs. The message has valid encryption and integrity code HMAC/CMAC. The users in the multicast group cannot determine the source of the traffic and assume always that it originates from the BS.

The MBRA protocol does not address forward and back secrecy. In fact, when a new MS joins the multicast group and receives from the BS the current GTEK it becomes able to decrypt all the previous messages that were multicasted during the current GTEK life-time. Additionally, the protocol does not prevent a SS, who leaves the multicast group, from continuing to decrypt the multicast traffic. In fact it remains able to receive the next GKEK ad/or decrypt the next GTEK. A trade-off between scalability and forward and backward secrecy should be considered during the choosing of the GTEK life-time. The standard recommends a value ranging from 0.5 hours to seven days, with a default value equal to 12 hours. A low value may reduce the BS overhead due to GTEK and GKEK update. However, it introduces lapses in forward and backward secrecy. In [34], the authors propose a solution with perfect secrecy. The MSs are organized into subgroups of comparable size. Instead of using a single GKEK, a hierarchy of Sub-Group KEKs (SGKEKs) is used. For a number of subgroups equal to N, every subgroup is asked to maintain a number of SGKEKs equal to k where k is given by $k = log_2(N)$.

To update GTEKs, the Multicast Broadcast Rekeying algorithm can be used. GTEKs are sent by the BS to all members of the multicast group, in an encrypted form using the shared GKEK. Every member who receives such message, decrypts it and updates the used GTEK. Since every member of the multicast group has the GKEK, it can use the Multicast Broadcasting Rekeying Algorithm (MBRA) to distribute a forged GTEK which has a valid encryption and authentication code. Consequently all the members of the multicast group will be forced to update their active GTEK. Further to such operation, none member will be able to decrypt the traffic that originates from the BS. This behavior is maintained until the next time the BS sends the Group Key Update message to update the current GTEK.

To mitigate the vulnerabilities related to key sharing in Multicast and Broadcast Service, the authors in [38] proposed two alternative solutions. The first consists in securely distributing the GTEK by the BS separately to each MS using the KEK shared between the BS and MS. The second consists in digitally signing the key update message used to redistribute the GTEK, instead of appending the HMAC. In [42], a Group-based Key Distribution Algorithm (GKDA) is proposed to provide a scalable and secure solution for key distribution in multicast groups.

2.7.5 Weaknesses in Handover Schemes

While the Handover optimization bits in the RNG-RSP message can be used to reduce latency, it also affects the security of the network once the handover is performed [43]. The more the latency is reduced, the more the security of the operation is reduced. For instance, setting bit#1 and bit#2 equal to 1 and 1, respectively, forces the network to keep using the same secret keys before and after the handover and prevents it from ensuring backward and forward secrecy. In fact if a malicious mobile station has compromised the security of the serving BS, it could also compromise the security of all the previous and following BSs. In the case where bit#1 and bit#2 are set to 1 and 0, respectively, the TEKs will be updated during the handover operation but the AK is preserved. Since the AK enables deriving the KEK and consequently obtaining the TEKs, a serving BS could use the unchanged AK to determine the updated TEK of the following target BSs. Forward secrecy cannot therefore be enforced. Given the weaknesses in the handover scheme, both bit#1 and bit#2 should be set to 0, so that no secret key can be reused in another BS after the handover operation.

2.8 Further Reading

In IEEE 802.16e, two modes of connections are defined: Point to Multi-Point (P2MP), and MESH. While in P2MP, a mobile can reach the base station in one hop, the MESH mode allows nodes, which are out of transmission range of the base station, to forward and receive traffic through their direct neighbours. Even if the MESH mode allows extending the network coverage, it suffers from many security vulnerabilities, most of them were studied in the Ad-hoc and Sensor networks [44]. When a node enters the network, it listens for a network descriptor message to generate a list of neighbours and available BS. From these neighbors, the new SS selects a sponsoring node. The latter will tunnel the PKM-REQ message to the base station or the authorization node, and the REG-REQ message to the registration node. It forwards back the received message to the new candidate node. After authorization, the new candidate node can establish links with its direct neighbours by following a challenge response process.

One of the main security vulnerabilities is related to the lack of encryption and authentication of the network descriptor message. In fact, since that message includes information such as neighbours, or the BS node ID and the corresponding hop count, a node can claim a shorter path to the BS and become a sponsoring node. Using this technique, the malicious node can create a sinkhole attack by luring the network entry traffic to it. Another attack on the network topology can be conducted by two colluding nodes which establish a secret channel between them. Later they tunnel the network descriptor messages from a node, say X, to a node, say Y, through the channel and replay it in another location of the network. Consequently, node X and Y can believe that they are neighbors while in reality they are far from each other. Additional security attacks on IEEE 802.16 MESH networks are discussed in [45].

While the PKMv2 protocol reduces the most important security vulnerabilities of PKMv1, it looks to have an exaggerated mixture of security features. In this context, a secure simplification of this protocol is shown in [37].

2.9 Summary

In this chapter we turned our attention to the description and analysis of WiMAX security protocols and mechanisms proposed by the different released versions on the IEEE 802.16 standard.

After discussing the IEEE 802.16 network architecture, protocol stack, connectivity modes and different amendments, we introduced a set of security requirements for designing wireless broadband networks and protecting both the users and operators of these networks. Two main versions of the IEE 802.16 standard were addressed in this chapter, notably the IEEE802.16-2004 and the IEEE 802.16-2005, which represent the standard used by the most popular implementation of fixed and mobile WiMAX networks, respectively. We started by describing the security features introduced by IEEE 802.16-2004. In this context, the PKM protocol was described and its different phases detailed, showing the use of security associations, X.509 certificates and cryptographic algorithms. After that, we analysed the security weaknesses of the standard, illustrating attacks on the physical layer, authentication, key management, privacy and availability. The analysis not only describes the attacks and the related vulnerabilities, but it also shows the different solutions that have been proposed to mitigate them. Further to that description, we presented the security amendments proposed by the mobile version of WiMAX network, namely the 802.16e. Enhancements in terms of authorization, authentication, access control, protection of encryption keys, together with security mechanisms brought up by the multicast and broadcast service and handover schemes, are detailed. The last part of this chapter illustrated the security weaknesses related to the IEEE 802.16e standard. Some of them existed in the 802.16d version but were not mitigated sufficiently by the newly released versions of the standard, while some were introduced mainly by the mobility and support of broadcasting and multicasting communication.

References

[1] 802.16-2001, I.S. IEEE Standard for Local and Metropolitan area networks Part 16: Air Interface for Fixed Broadband Wireless Access Systems. 2001; available from: http://ieeexplore.ieee.org/xpl/tocresult.jsp?isNumber=21532.

[2] 802.16c-2002, IEEE Standard for Local and metropolitan area networks - Part 16: Air Interface for Fixed Broadband Wireless Access Systems-Amendment 1: Detailed System Profiles for 10–66 GHz. 2002.

[3] 802.16a-2003. EEE Standard for Local and metropolitan area networks – Part 16: Air Interface for Fixed Broadband Wireless Access Systems – Amendment 2: Medium Access Control Modifications and Additional Physical Layer Specifications for 2–11 GHz. 2003; available from: http://ieeexplore.ieee.org/xpl/tocresult.jsp?isNumber=26891.

[4] 802.16-2004. IEEE Standard for Local and Metropolitan Area Networks Part 16: Air Interface for Fixed Broadband Wireless Access Systems. 2004; available from: http://ieeexplore.ieee.org/xpl/freeabs_all.jsp?isnumber=29691&arnumber=1350465&count=1&index=0.

[5] A. Knutas and J. Laakkonen, 'IEEE 802.16 – WiMAX. 2009', Lappeenranta University of Technology.

[6] 802.16-2004/Cor1-2005, e.-a.I.S., IEEE Standard for local and metropolitan area networks. Part 16: Air interface for fixed and mobile broadband wireless access systems. Amendment 2: Physical and medium access control layers for combined fixed and mobile operation in licensed bands and corrigendum. 2006.

[7] Standard, I – Part 16: Air Interface for Fixed and Mobile Broadband Wireless Access Systems Amendment 2: Physical and Medium Access Control Layers for Combined Fixed and Mobile Operation in Licensed Band. 2005; available from: http://standards.ieee.org/getieee802/download/802.16e-2005.pdf.

[8] N. Michiharu, C. Takafumi and S. Tamio, 'Standardization Activities for Mobile WiMAX' *Fujitsu Scientific and Technical Journal*, 2008. **44**(3): 285–91.

[9] M. Ergen, *Access Service Network in WiMAX: The role of ASN-GW*, in *Mobile Handset DesignLine*. 2007.

[10] A. Yarali and S. Rahman, *WiMAX Broadband Wireless Access Technology: services, architecture and deployment models*, in *Canadian Conference on Electrical and Computer Engineering (CCECE 2008)*. 2008: Ontario, Canada. 77–82.

[11] M.S. Kuran and T. Tugcu, 'Survey on Emerging Broadband Wireless Access Technologies' *Computer Networks*, 2007. **51**(11): 3013–46.

[12] T. Hardjono and L.R. Dondeti, *Security in Wireless LANs and MANs*. 2005: Artech House Publishers.

[13] N. Boudriga, *Security of Mobile Communications* 2009: Auerbach Publications.

[14] A.-V. Aikaterini, 'Security OF IEEE 802.16', in *Department of Computer and Systems Science*. 2006, Royal Institute of Technology.

[15] S. Adibi, G.B. Agnew and T. Tofigh, 'End-to-End (E2E) Security Approach in WiMAX: A Security Technical Overview for Corporate Multimedia Applications', in *Handbook of Research on Wireless Security*, Y. Zhang, J. Zheng and M. Ma, Editors. 2008, Information Science Reference.

[16] M. Bogdanoski, P. Latkoski, A. Risteski and B. Popovski, *IEEE 802.16 Security Issues: A Survey*, in *16th Telecommunication Forum TELFOR 2008*. 2008: Serbia, Belgrade.

[17] D. Johnston and J. Walker, 'Overview of IEEE 802.16 Security' IEE Security & Privacy, 2004. **2**(3): 40–8.

[18] E. Eren and K.-O. Detken, 'WiMAX Security – Assessment of the Security Mechanisms in IEEE 8.2.16d/e', in *The 12th World Multi-Conference on Systemics, Cybernetics and Informatics: WMSCI 2008*. 2008: Orlando, Florida, USA.

[19] R. Housley, W. Ford, W. Polk and D. Solo, 'Internet X.509 Public Key Infrastructure Certificate and CRL Profile', in *Request for Comments: 2459*, N.W. Group, Editor. 1999.

[20] S. Xu and C.-T. Huang. 'Attacks on PKM Protocols of IEEE 802.16 and Its Later Versions' in *3rd International Symposium on Wireless Communication Systems*. 2006. Valencia Spain.

[21] R.A. Poisel, *Modern Communications Jamming Principles and Techniques*. 2003: Artech House Publishers.

[22] M.J. Husso, 'Performance Analysis of a WiMAX System under Jamming', in *Department of Electrical and Communications Engineering*. 2006, HELSINKI UNIVERSITY OF TECHNOLOGY.

[23] U. Tariq, U.N. Jilani and T.A. Siddiqui, *Analysis on Fixed and Mobile WiMAX*. 2007, Blekinge Institute of Technology.

[24] M. Barbeau, 'WiMax/802.16 Threat Analysis', in *1st ACM international workshop on Quality of service & security in wireless and mobile networks* 2005: Montreal, Quebec, Canada

[25] T.F. Elrahman, *Security Technologies in Wireless Networks*. 2005; available from: www.aims.ac.za/resources/archive/2004/tayseer.ps.

[26] S. Xu, M. Matthews and C.-T. Huang, 'Security Issues in Privacy and Key Management Protocols of IEEE 802.16', in *44th annual Southeast regional conference*. 2006: Melbourne, Florida

[27] E. Eren, 'WiMAX Security Architecture – Analysis and Assessment', in *4th IEEE Workshop on Intelligent Data Acquisition and Advanced Computing Systems: Technology and Applications*. 2007. Dortmund, Germany.

[28] P. Urien and G. Pujolle, 'Security and Privacy for the Next Wireless Generation' *International Journal of Network Management*, 2008. **8**(2): 129–45.

[29] T. Otto, *Extensible Network Access Authentication*. 2006, University of Lübeck.

[30] A. Altaf, M.Y. Javed and A. Ahmed. 'Security Enhancements for Privacy and Key Management Protocol in IEEE 802.16e-2005', in *Ninth ACIS International Conference on Software Engineering, Artificial Intelligence, Networking, and Parallel/Distributed Computing*. 2008. Phuket, Thailand.

[31] T. Haibo, P. Liaojun and W. Yumin, 'Key management protocol of the IEEE 802.16e' *Wuhan University Journal of Natural Sciences*, 2007. **12**(1): 59–62.

[32] S.R. Bajgan, M. Pooyan, A. Khalilzadeh and R. Abdollahi, 'Security in link layer of WiMAX networks', in *World Congress on Science, Engineering and Technology*. 2008. Bangkok, Thailand.

[33] S. Xu, 'Security Protocols in WirelessMAN', in *College of Engineering and Computing*. 2008, University of South Carolina.

[34] C.-T. Huang and J.M. Chang, 'Responding to Security Issues in WiMAX Networks' *IT Professional*, 2008. **10**(5): 15–21.

[35] D. Pang, L. Tian, J.L. Hu, J.H. Zhou and J.L. Shi, *Overview and Analysis of IEEE 802.16e Security*, 2007; available from: http://citeseerx.ist.psu.edu/viewdoc/download?doi=10.1.1.89.2235&rep=rep1&type=pdf.

[36] A. Altaf, M.Y. Javed, S. Naseer and A. Latif, 'Performance Analysis of Secured Privacy and Key Man-
     agement Protocol in IEEE 802.16e-2005' *International Journal of Digital Content Technology and its
     Applications*, 2009. **3**(1): 103–9.
[37] E. Yuksel, H.R. Nielson, C.R. Nielsen and M.B. Orencik. 'A Secure Simplification of the PKMv2 Protocol
     in IEEE 802.16e-2005', in *Joint Workshop on Foundations of Computer Security and Automated Reasoning
     for Security Protocol Analysis*. 2007. Wroclaw, Poland.
[38] S. Naseer, M. Younus and A. Ahmed. 'Vulnerabilities Exposing IEEE 802.16e Networks to DoS Attacks:
     A Survey', in *2008 Ninth ACIS International Conference on Software Engineering, Artificial Intelligence,
     Networking, and Parallel/Distributed Computing*. 2008. Phuket, Thailand.
[39] A. Deininger, S. Kiyomoto, J. Kurihara and T. Tanaka, 'Security Vulnerabilities and Solutions in Mobile
     WiMAX' *International Journal of Computer Science and Network Security*, 2007. **7**(11): 88–97.
[40] T. Shon and W. Choi, 'An Analysis of Mobile WiMAX Security: Vulnerabilities and Solutions' *Lecture
     Notes in Computer Science*, 2007. **4658**: 88–97.
[41] W.D. Hellman, 'New Directions in Cryptography' *IEEE Transactions on Information Theory*, 1976. **22**(6):
     644–54.
[42] H. Li, G.B. Fan, J.G. Qiu and X.K. Lin, 'GKDA: A Group-Based Key Distribution Algorithm for WiMAX
     MBS Security' *Lecture Notes in Computer Science*, 2006. **4261**: 310–18.
[43] J. Hur, H. Shim, P. Kim, H. Yoon and N.-O. Song. 'Security Considerations for Handover Schemes in
     Mobile WiMAX Networks', in *IEEE Wireless Communications and Networking Conference*. 2008. Las
     Vegas, Nevada.
[44] D.G. Padmavathi and M.D. Shanmugapriya, 'A Survey of Attacks, Security Mechanisms and Challenges
     in Wireless Sensor Networks' *International Journal of Computer Science and Information Security*, 2009.
     4(1 & 2): 117–25.
[45] Y. Zhou and Y. Fang, 'Security of IEEE 802.16 in Mesh Mode', in *IEEE Military Communications
     Conference (MILCOM 2006)*. 2006: Washington, DC.

3

Key Management in 802.16e

Georgios Kambourakis and Stefanos Gritzalis
Laboratory of Information and Communication Systems Security
Department of Information and Communication Systems Engineering
University of the Aegean, Karlovassi, Samos, Greece

3.1 Introduction

Until now the IEEE 802.16 technologies, also known as WiMAX, may not have had the adoption rate of 802.11, but will likely be the predominant technology for Metropolitan Area Networks (MAN) deployments for the next decade. This is because WiMAX can support all-IP core network architecture, low latency, advanced Quality of Service (QoS) and sophisticated security [1]. The IEEE 802.16 working group on broadband wireless access standards, a unit of the IEEE 802 LAN/MAN standards committee (http://www/wirelessman.org/), is preparing and revising formal specifications for the global deployment of broadband Wireless MANs.

The 802.16 security model was designed initially to support authentication, confidentiality and integrity services. This is in contrast to the key management model used in the original 802.11 specification. The initial 802.16-2001 version of the standards [2] employs X.509 certificates for the authentication of devices on the network. Every subscriber device has its own certificate that identifies it uniquely to the 802.16 infrastructure. Of course, this allows providers to control which devices are authorized to use their networks. On the other hand, as in GSM networks, authentication is not mutual. Specifically, while the network can authenticate a subscriber device, the subscriber device has no way to authenticate the infrastructure. This leads to several attacks including man-in-the-middle, spoofing or replay attacks against the subscriber device. For link encryption, 802.16 uses

WiMAX Security and Quality of Service: An End-to-End Perspective Edited by Seok-Yee Tang,
Peter Müller and Hamid Sharif
© 2010 John Wiley & Sons, Ltd

DES [3]. Unfortunately, DES is a relatively weak algorithm by today's standards. Moreover, the way 802.16 uses DES is not as secure as it should be. The initialization vector used by 802.16 is predictable; this further weakens the confidentiality of the data.

The updated 802.16e standard [4], which is an amendment to 802.16-2004 [5], tries to rectify some of the aforementioned issues. So, 802.16e, similar to 802.11i [6], employs the Advanced Encryption Standard (AES) [7] as its core encryption algorithm. Furthermore, 802.16e supports the well-known Extensible Authentication Protocol (EAP) [8] for authenticating devices over the network. The EAP protocol, also used in 802.11i LANs, offers a plethora of authentication mechanisms in the context of the IEEE 802.1X framework [9].

Key management procedures in 802.16 are part of the Privacy Key Management (PKM) protocol [2, 4, 5] and define how the keys are created, which keys are available and for what purpose. Specifically, the keys used for the integrity protection of management frames and secure transmission of Traffic Encryption Keys (TEK) are produced from master keys. Master keys may be derived from two distinct sources (procedures), namely RSA and EAP or a combination of two. The key generation procedure based on Public Key Certificates (PKC) ends with a pre-Primary Authorization Key (pre-PAK), while the 802.1X/EAP procedure ends with a Master Session Key (MSK).

This chapter focuses specifically on the key management scheme of 802.16. Key derivation procedures and key hierarchy of PKM version 2 are examined and discussed thoroughly. Known vulnerabilities and countermeasures are identified and analysed. Some comparisons with IEEE 802.11i and Third Generation (3G) mobile networks standards are also provided.

3.2 Privacy Key Management Protocol

This section discusses the two versions of the PKM protocol that exist so far. More attention is paid to the PKMv2 because it is the strongest in terms of security and of course supersedes its predecessor, namely PKMv1. It is also worth noting that the PKM protocol is designed according to the Data Over Cable Service Interface Specification (DOCSIS) [10], which was originally used for a cable system. Also, this section provides a short description of the basic security components of 802.16 which is necessary in order to better comprehend the subsequent sections.

Security in IEEE 802.16 is twofold; the first goal is to provide confidentiality, that is, privacy and authenticity, across the wireless network, while the other is to provide access control to the network. Confidentiality is accomplished by encrypting connections between the Mobile Station (MS) and the Base Station (BS). Also, the BS protects against unauthorized access by enforcing encryption of service flows across the network. Apart from confidentiality, a WiMAX network should also support integrity. Data integrity assures the communicating parties that the received data is not altered or tampered with in transit by an adversary. Therefore, mechanisms are in place to ensure that both user data and signalling are protected from being tampered with while in transit. As already mentioned, the PKM protocol is used by the BS to control the distribution of keying data needed for the above-mentioned security services to MSs. Specifically, by using this key management protocol, the MS and the BS synchronize keying material. The BS also employs the same protocol to enforce conditional access to network services.

A crucial element of the security of 802.16 and especially of the PKM protocol is the so-called a Security Association (SA). Note that the same term is also used by the Internet Key Exchange (IKE) protocol, in the context of IPsec. An SA is defined as 'the set of security information a BS and one or more of its client MSs share in order to support secure communications across the network' [5]. Each SA has a unique identifier (SAID). It also contains a cryptographic suite identifier and possibly TEKs and initialization vectors. In any case, the exact content of an SA depends on the cryptographic suite it contains.

Within the IEEE 802.16 specification there exist three types of SAs: Primary, Static and Dynamic. A primary SA should be established by every MS during its initialization process. Its scope is the secondary management connection. It is to be mentioned here that within the 802.16 MAC layer, which is connection oriented, there exist two sorts of connection: management connections (i.e., basic, primary and secondary) and data transport connections. A basic connection used for short and urgent management messages is created for each MS when it joins the network. At MS initial network entry a primary connection is also created and used for delay tolerant management messages. The secondary management connection is used for IP encapsulated management messages, like those of the Dynamic Host Configuration Protocol (DHCP). On the other hand, transport connections are used for user traffic flows. They can be provisioned or established on demand.

Also, the primary SA is shared exclusively between an MS and the corresponding BS. The SAID of any MS's primary SA is equal to the basic Connection ID (CID) of that MS. The basic CID is the first static CID the BS assigns to an MS during initial ranging. Static SAs are administered within a BS. For instance, there is a static SA for the basic unicast service, but an MS may be subscribed for additional services. This results in a different static SA for each individual service. One the other hand, dynamic SAs are created on-the-fly when new service flows are started and they are destroyed when their flow is terminated. Note also that static SAs and dynamic SAs can be shared among several MSs when multicast is used.

An MS requests from the corresponding BS an SA's keying material by utilizing the PKM protocol. Every SA can be accessed by the authorized MS. Such authorization is provided by the BS in charge, as the case may be. All SA's keying data expires sometime. So, upon delivering SA keying data to an MS, the BS notifies the client about the expiration times. The MS tracks the keying material lifetimes constantly and before expiration updates them by querying the corresponding BS. Otherwise, the MS should repeat the network entry and initialization procedure. As already mentioned, key synchronization is also managed by the PKM protocol.

3.3 PKM Version 1

An MS utilizes the PKMv1 protocol [5] to acquire authorization and traffic keying material from the BS, and to support periodic reauthorization and key refresh. To do so, PKM uses X.509 digital certificates [11], the RSA public-key encryption algorithm [12] and strong encryption algorithms to perform key exchanges between the MS and BS.

There exist two Media Access Control (MAC) management frames for security policy negotiation. These are the PKM-REQ and PKM-RSP messages defined in [5]. At first, the PKM protocol uses public-key cryptography to establish a shared secret, namely the Authorization Key (AK), between the MS and the BS. After that, the AK is used to secure subsequent PKM exchanges of TEKs. This is an obvious advantage because the two parties are able to refresh their TEKs without repeating resource demanding public-key operations. The authentication mechanism using X.509 certificates guards against MS cloning and masquerading. Thus, the specification mandates that all MSs must have a build-in RSA private/public key pair and a factory-installed X.509 certificate or provide an internal algorithm to generate such key pairing on-the-fly. In the latter case, the MS must also provide a way to install a manufacturer-issued X.509 certificate after the generation of key materials. In the following we describe the PKMv1 protocol in more detail.

The authorization and key exchange procedure consists of two phases aiming to both authorize the MS and transfer the required keying material (i.e., AK and TEKs) from the BS to the MS securely. This procedure consists of five messages described in the following:

- *MS → BS (MS authentication information message)*: The MS begins authorization by sending this message to the corresponding BS. The message contains the MS manufacturer's X.509 certificate, issued by the manufacturer itself or by an external authority, for example, a Trusted Third Party (TTP). Note that this message is only informative, meaning that the BS may simply ignore it. Its only purpose is to provide a way for the BS to check initially the certificate of a client MS. That is why this message is omitted during the reauthorization process explained further down.
- *MS → BS (MS authorization request message)*: This message should follow an MS authentication information message directly. Actually, by sending this message to the BS, the MS applies for an AK; also, queries for any SAID corresponding to any static security SA the MS is authorized to access. The contents of this message are: (a) a manufacturer-issued X.509 certificate; (b) a description of the cryptographic algorithms the MS supports; this is a list of cryptographic suite identifiers that correspond to MS's cryptographic capabilities. Each of them specifies a pair of packet data encryption and authentication algorithms the MS supports; and (c) the MS's basic CID.
- *BS → MS (BS Authorization Reply message)*: Upon receipt of the previous second message, the BS performs the following actions: attests the identity of the client MS, decides in common with the MS the encryption algorithm and protocol support (if any, otherwise the procedure aborts), generates an AK, enciphers it using the MS's public key and finally sends it back to the MS in this third message. The contents of this message are: (a) The AK encapsulated with the MS's public key; (b) a 4-bit key sequence number (AKSeqNo), corresponding to different successive generations of AKs; (c) the key lifetime (AKLifetime); and (d) the SAIDs and properties of the single primary (basic unicast service) and zero or more static SAs the requesting MS is authorized to access and acquire keying data. Bear in mind that static SAs correspond to services the MS's user has subscribed for. Also note that dynamic SAs are not identified by this message. Every MS is bound to refresh its AK periodically by sending an authorization request message to the BS in charge. It is important to note that successive generations of AKs have overlapping lifetimes. Otherwise, service interruptions during reauthorization may be possible, that is, the MS may fail to renew the AK before it expires. Thus, both the MS and BS are capable of keeping up to two simultaneously active AKs during key refreshing periods.

After the authorization phase ends successfully, and in order for the MS to acquire TEKs, it must start a separate TEK state machine for each one of the SAIDs identified in the authorization reply message described above. TEK state machines are controlled by the MS per se. A TEK is also responsible for refreshing the keying material for any given SAID. The mechanism of acquiring/refreshing TEKs involves the two following messages:

- *MS → BS (MS Key Request message)*: The contents of this message are: (a) the AKSeqNo; (b) the SAID for which TEK parameters are requested; and (c) an keyed-Hash MAC (HMAC) message digest [13]; this is to protect the integrity of the message while in transit.
- *BS → MS (BS Key Reply message):* The BS responds to the previous message with a Key Reply message. This of course happens after the BS verifies the digest of the previous message. The contents of this message are: (a) the AKSeqNo; (b) the SAID for which TEK parameters will be sent; (c) TEK-Parameters 'older' generation of key parameters relevant to the SAID; (d) TEK-Parameters 'newer' generation of key parameters relevant to the SAID; and (e) an HMAC-Digest Keyed SHA message digest. The TEK-Parameters attribute is a multipart attribute containing: (i) the actual TEK; (ii) the TEK's remaining key lifetime; this information is important to the MS in order to estimate when the BS will invalidate a particular TEK and therefore when to schedule key updates; (iii) the key sequence number; and (iv) the Cipher Block Chaining (CBC) initialization vector used by the DES algorithm. Note that the TEK is encrypted (e.g. by 3-DES) using the appropriate Key Encryption Key (KEK) derived from the AK. The BS always maintains two active sets (generations) of keying material per SAID. The specification states that each generation becomes active halfway through the life of its predecessor and expires halfway through the life of its successor. So, a BS is supposed to include in this message both active generations of keying material corresponding to a given SAID.

3.4   PKM Version 2

The IEEE 802.16e standard [4], also referred to as mobile WiMAX, supports fixed and mobile services for both enterprise and consumer markets and remedies most of the security weaknesses of its predecessors. The security sublayer of 802.16e consists of two component protocols, namely the encapsulation protocol and key management. As already mentioned, here we focus on key management performed by the PKMv2 protocol. Actually, PKMv1 is a subset of PKMv2 in its function. The latter supports both mutual authentication and unilateral authentication and enables periodic re-authentication/reauthorization and key update. To do so, it employs either EAP in conjunction with an operator-selected EAP method such as EAP-TLS, or X.509 digital certificates (see the previous section) together with RSA public-key encryption or a mixed procedure starting with RSA authentication and followed by EAP authentication. With regard to EAP, several variations of authentication modes are realized: user-only authentication (user single EAP), device-only authentication (device single EAP), device and user authentication (single EAP) and device and user authentication (double EAP or back-to-back EAP; device authentication is executed first, user authentication follows). In contrast to its predecessor, PKMv2 offers strong encryption algorithms to perform key exchanges between an MS and the corresponding BS. After establishing a shared secret (the AK) between the MS and the BS, PKMv2 uses it to secure subsequent exchanges of TEKs between the two parties. The overview of PKMv2 provided in this section includes security negotiation, authorization/authentication, key derivation, handshake and key transportation.

3.4.1 Security Negotiation

Security policy negotiation is carried out by the SBC-REQ (Basic Capability Request) and SBC-RSP (Basic Capability Response) MAC management frames. These two compound messages of variable length transfer security negotiation parameters contained in the following fields (attributes): (a) PKM version support; (b) authorization policy support; (c) message authentication code mode; and (d) Packet Number (PN) window size. This is done upon the initial network entry or re-entry procedure of an MS. The first field indicates a PKM version, where a bit value of 0 indicates 'not supported' and 1 indicates 'supported'. Only one PKM version should be negotiated between the MS and BS. Regarding the second field there are several authorization choices, including no authorization, RSA-based authorization, EAP-based authorization, etc. The MS should support at least one authorization policy and inform BS of all supportable authorization policies in an SBC-REQ message. After that, the BS negotiates the authorization policy. If all bits of this second attribute included in the SBC-RSP message are zeroed, then no authorization is applied. The RSA-based authorization procedure of PKMv2 employs the Auth-Request/Reply/Reject/Acknowledgement messages. Also, the PKMv2 EAP-Transfer message is utilized during an EAP-based authorization procedure.

The third field carries the MAC mode, that is, the HMAC/Cipher-based MAC (CMAC) [14] that the MS supports. The MS should support at least one MAC mode and inform BS of all supportable MAC modes in an SBC-REQ message. After that, the BS negotiates the MAC mode and informs the MS using a SBC-RSP message. In a case where all bits of this attribute included in the SBC-RSP message are zeroed, no MAC mode is applied. As a result, MAC messages are not authenticated. The last field (PN) carries the size capability of the receiver PN window for SAs and management connections. This is essential as the receiver will track PNs within this window in order to thwart replay attacks.

3.4.2 Authentication/Authorization

In the following we focus on RSA-based authorization, an authenticated transport protocol, which is described within 802.16e. While a complete description of IEEE 802.1X framework is outside of the scope of this chapter, we do provide a short description of the EAP-based authentication mode. The MS-to-BS mutual authentication can occur in one of two modes of operation. In the first mode, only mutual authentication is applied, while in the second, mutual authentication is followed by EAP authentication. When the RSA-based authorization method is selected during security negotiation, the PKMv2 RSA-Request/Reply/Reject/Acknowledgement messages are used to generate and share a pre-PAK (Primary Authorization Key). The pre-PAK is mainly used to generate the PAK later on.

- *MS → BS (PKMv2 RSA-Request message)*: This message follows an Authentication Information message immediately (see section 3.3). By sending this message to the BS, the MS applies for an AK; also queries for any SAID corresponding to any static security SA the MS is authorized to access. The contents of this message are: (a) a 64-bit random number generated by the MS (MS_Random); (b) a manufacturer-issued X.509 certificate (MS_Cert), where the common name is the MS' MAC address; (c) the MS's basic CID refer to as the primary SAID; and (d) an RSA signature (SigMS), using the MS's private key, over all the other attributes in the message.
- *BS → MS (PKMv2 RSA-Reply message)*: Upon receipt of the previous message, the BS performs the following actions: attests the identity of the client MS, generates an AK (pre-PAK), encapsulates it using the MS's public key and finally sends it back to the MS using this message. The secrecy of the procedure is guaranteed by using random numbers. The contents of this message are: (a) an 64-bit random number generated by the MS (MS_Random); (b) a 64-bit random number generated by the BS (BS_Random): (c) the Encrypted pre-PAK, that is, RSA-OAEP-Encrypt(pre-PAK/MS MAC Address)$_{PubKey(MS)}$: (d) the key lifetime; (e) the key sequence number (AKSeqNo); (f) the BS's X.509 certificate (BS_Cert); and (g) an RSA signature (SigBS), using the BS's private key, over all the other attributes in the message.
- *MS → BS (PKMv2 RSA-Acknowledgement message)*: This message follows a PKMv2 RSA-Reply message or a PKMv2 RSA-Reject message. That is, via the current message, the MS demonstrates to the BS that it is alive. Its contents are: (a) a 64-bit random number generated by the BS (BS_Random); (b) an acknowledgement for the BS (Auth_Result_Code); success or failure; and (c) an RSA signature (SigMS), using the MS's private key, over all the other attributes in the message. If the Auth_Result_Code indicates failure then this message should also contain an Error-Code field describing the reason for rejecting the authorization request.

As with PKMv1 every MS is bound to refresh its AK periodically by sending an authorization request message to the BS in charge. So, successive generations of the MS's AKs have overlapping lifetimes (see section 3.3).

A successful execution of the EAP-based authentication mode – single EAP mode for simplicity – is described in the following. At the initial entry, the MS and the Authentication Server (AS) mutually authenticate each other using an EAP-based authentication method. The AS may be an Authorization Authentication Accounting (AAA) Diameter or RADIUS server. According to the 802.16e specifications the EAP authentication should follow the guidelines of [15] such as the mutual authentication support and protection against the man-in-the-middle attack. Examples of such strong methods are the EAP-TLS, EAP-TTLS, PEAP, EAP-AKA to mention just a few [16]. The product of the EAP exchange is the 512-bit Master Session Key (MSK), known to both the AS and the MS (also referred to as supplicant). After that, the MSK is securely transferred from the AS to the authenticator, that is, the BS. However, the message for the MSK distribution is not defined in the specifications.

3.4.3 Key Derivation and Hierarchy

The keys generated by the authentication and authorization processes, as described above, are used to protect the integrity of management frames and secure the transportation of TEKs. Thus, the PKMv2 key hierarchy defines specifically how the keys are generated and what keys are present in the system, as the case may be. As already mentioned, there are two primary sources of keying material corresponding to the authentication schemes supported; RSA-based and EAP-based. Bear in mind that the RSA-based authorization produces a pre-PAK, while the EAP-based yields in a MSK. All PKMv2 key derivations are based on the Dot16 Key Derivation Function (Dot16KDF) which is an AES counter (CTR) mode construction used to derive an arbitrary amount of keying material from source keying material [4]. The algorithm is defined differently depending on whether the MAC mode – which is negotiated during the security negotiation phase (see section 3.4.1) – is HMAC or CMAC. In case of CMAC the algorithm is as follows:

```
Dot16KDF(key, astring, keylength) { /* 'key' is a cryptographic key used by the
underlying digest algorithm (SHA-1 or CMAC-AES) */
      result = null;
      Kin = Truncate (key, 128);
      for (i = 0; i <= int((keylength-1)/128); i++) {
          result = result | CMAC(Kin, i | astring | keylength);
          /* 'astring' is an octet string used to alter the output of the algorithm
          'keylength' is used to determine the length of key material to generate */
      }
      return Truncate (result, keylength);
      // Truncate(x, y) is the rightmost y bits of a value x only if y ≤ x
}
```

In the case of HMAC the algorithm becomes:

```
Dot16KDF(key, astring, keylength) {
      result = null;
      Kin = Truncate (key, 160);
      for (i=0; i <= int( (keylength-1)/160 ); i++) {
          result = result | SHA-1( i| astring | keylength | Kin);
      }
      return Truncate (result, keylength);
}
```

A complete map of the 802.16e key hierarchy is depicted in Figure 3.1. For example, the EAP-based authentication process yields the MSK and then the other keys such as the Key Encryption Key (KEK) and HMAC/CMAC key are derived from the MSK. The MS and BS get a Pairwise Master Key (PMK) by truncating the MSK to 160 bits, and derive an authorization key (AK) from the PMK.

Multicast/Broadcast Service (MBS) is an integral part of 802.16e. MBS allows 802.16 providers to deliver multicast and/or broadcast services, for example, show video multicast service in a cell, to their subscribers. In fact, MBS is a mechanism for the distribution of data content across multiple BS from a centralized media server. However, for supporting such services group keys are required. It is implied that before receiving any MBS, an MS must register and authenticate with a BS via the PKM protocol. The Group TEK (GTEK) is used to encrypt multicast data packets and is shared among all MSs that belong to

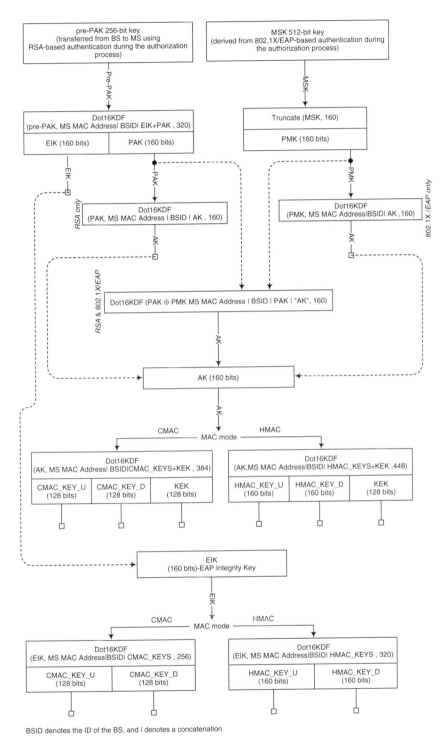

Figure 3.1   Message exchange and key derivation at MS initial network entry.

the same multicast group. This key is generated randomly by the BS or by a certain network node and is encrypted using same algorithms applied for TEK encryption. It is also transmitted to the MS employing multicast or unicast messages (i.e. through the primary management connection, except the PKMv2 Group Key Update Command which is transmitted over the broadcast connection). The Key Encryption Key (KEK) is used to encrypt a GTEK in a PKMv2 Key-Request and PKMv2 Key-Reply messages. Also, as explained in section 3.4.5, the Group KEK (GKEK) is used for GTEK encapsulation in a PKMv2 Group Key Update Command message. GKEK is generated randomly at the BS, encrypted with the KEK, and transmitted to the MS. There is one GKEK per Group SA (GSA).

Finally, there is a 128-bit MBS Traffic Key (MTK) used to encrypt (MBS) traffic data. It is defined as follows: MTK = Dot16KDF(MAK, MGTEK | 'MTK', 128). The generation and transport of the MAK (MBS AK) is provided by means defined at higher layers and remains outside the scope of the 802.16e standard. Also, the MGTEK is the Group TEK (GTEK) for the MBS. So, an MS can acquire the GTEK by exchanging the PKMv2 Key-Request message and the PKMv2 Key-Reply message with a BS or by receiving the PKMv2 Group Key Update Command message from a BS (see section 3.4.5).

3.4.4   Three-Way Handshake

After authorization, a three-way handshake takes place. That is, the three-way handshake is executed either at the initial network entry or at a re-entry procedure such as at a handover. The handshake protocol allows for many functions, such as SA keying para-meters negotiation and distribution, key activation, security negotiation confirmation, SA parameters refresh for network re-entry, etc. The SA-TEK-Challenge/Request/Response messages are used during the execution of the three-way handshake sequence. Before the handshake initiates, both the BS and MS derive a shared KEK and HMAC/CMAC keys from the AK. Therefore, any TEK distributed via the procedure (see SA_TEK_Update attribute in the third message below) are encrypted using the KEK. Note also that all message integrity checks are replay protected due to the nonces used.

- *BS → MS (PKMv2 SA-TEK-Challenge message):* Upon initial network entry or reau-thorization, the BS will send this message towards the MS. The MS should respond with a PKMv2 SA-TEK-Request (see next message) within a pre-defined time limit. If this is not the case, the BS retransmits the same challenge message for a pre-defined number of times before it initiates another full authentication, or simply drops the MS. The contents of this message are: (a) a 64-bits BS_Random; this is a fresh nonce for every new handshake; (b) the 4-bit AKSeqNo; (c) the 64-bit AKID of the AK (AKID) that was used for protecting this message: AKID = Dot16KDF(AK, AKSeqNo|MS MAC Address|BSID|'AK', 64); in the case of re-authentication this is the AKID of the new AK; (d) optionally, the PMK key lifetime; appears only when EAP-based method is used; and (e) a Message Integrity Code (MIC) over the contents of this message; this is an HMAC/CMAC digest against a forgery attack. The hash is calculated using a key derived from AK.
- *MS → BS (PKMv2 SA-TEK-Request message)*: Upon receipt of the previous message, the MS verifies its contents by checking the HMAC/CMAC digest. After that, the MS

sends the current message to the BS. Its contents are: (a) an 64-bits MS_Random; this is a fresh nonce for every new handshake; (b) the BS_Random contained in the previous message; (c) the AKSeqNo; (d) the AKID; identifies the AK used for protecting this message; (e) the cryptographic suites supported by the requesting MS (MS Security Capabilities); (f) the negotiated parameters (Security_Neg) contained in the insecurely negotiated SBC-REQ/RSP messages during the basic capabilities negotiation phase (see section 3.4.1); (g) the PKMv2 configuration settings which has a scope including Auth Reply, PKMv2 SA-TEK-response; it seems that this field is used to distinguish between the authorization and handshake procedures; and (h) a HMAC/CMAC MIC over the contents of this message. If the MS does not receive this message from the BS within a pre-defined time limit, it retransmits the request. This should be repeated until the MS reaches a maximum number of resends. After that, the MS may attempt to connect to another BS or start a full authentication.

- *BS → MS (PKMv2 SA-TEK-Response message)*: Upon receipt of the previous message, the BS confirms that the included AKID refers to an existing AK. Otherwise it drops the message. The BS also authenticates the message by checking its HMAC/CMAC. The BS verifies that the BS_Random contained in the SA TEK Request is equal to the value provided by itself in the first (challenge) message. If false, it ignores the message. Finally, the BS must verify that the MS's security capabilities – reported in the Security_Neg field – match those provided originally by the MS. If false, the BS informs this inconsistency to higher layers. If all checks are successful the BS responds to the corresponding MS using the current message. Its contents are: (a) the MS_Random; (b) the BS_Random; (c) the AKSeqNo; (d) a compound Type/Length/Value (TLV) list, namely SA_TEK_Update, each of which identifies the primary and static SAs, their SA identifiers (SAID) and additional properties of the SA that the MS is authorized to access and acquire keying data; this field is included only when an MS re-enters the network, for example, in case of a handover. For every unicast SAIDs this field contains all the corresponding keying material, that is, the TEK, the TEK's remaining key lifetime, its 2-bit key sequence number and the CBC Initialization Vector (IV). In the case of group or multicast/broadcast Group SAIDs (GSAIDs), this field contains all the keying material corresponding to a particular generation of a GSAID's GTEK, that is, the GTEK, the GTEK's remaining key lifetime, the GTEK's key sequence number and the CBC IV. All the above keys are encrypted with KEK. That is, for every active SA in previous serving BS, the corresponding TEK, GTEK, GKEK and associated parameters are included in this field. Thus, SA_TEK_Update provides a fast method for renewing active SAs used by the MS in its previous serving BS; (e) a Frame_Number field that contains a 24-bit absolute frame number in which the old PMK and all its associate AKs should be removed; (f) one or more compound SA-Descriptors; each of them specifies a SAID and additional properties of the SA. This attribute is only included when an MS initially enters the network; (g) the Security_Neg; this attribute confirms the authentication and message integrity parameters to be used (usually, the same as the ones insecurely negotiated in SBC-REQ/RSP); and (h) an HMAC/CMAC MIC over the contents of this message.

Upon receipt of the third message the MS authenticates it by checking its HMAC/ CMAC. The MS must also verify the BS's security negotiation parameters encoded in the Security_Neg attribute. Thus, this attribute is checked against the security negotiation parameters provided by the BS through the SBC-RSP message. If false, the MS should report the discontinuity to the upper layers. However, the MS may continue the communication with the BS by adopting the security negotiation parameters encoded in the SA-TEK Response message. If all checks return true, the MS installs the received TEKs and associated parameters accordingly. Later on, in the case of MS re-association with the same BS, the two parties are not bound to repeat an RSA-based or EAP authorization being that the AK has not expired. Instead, they can initiate a three-way handshake in order to renew the TEK. Moreover, as discussed in the next section, a given MS exchanges PKMv2 Key-Request/Reply messages with the BS in charge in order to acquire or refresh the TEK. During this phase the KEK and integrity keys used remain unchanged.

3.4.5 Key Delivery

After the handshake, the MS should run a TEK delivery protocol instance, that is, a TEK state machine, for each authorized SAID having a data flow that requires traffic encryption. This implies new TEK and related parameters and/or GTEK/GKEK and related parameters for MBS. The protocol uses the PKMv2 Key-Request/Key-Reply/Key-Reject messages. Also, periodically, TEK state machines send Key-Request messages to the BS, in order to refresh keying material for the corresponding SAIDs. More specifically:

- *MS → BS (PKMv2 Key-Request message)* that contains: (a) the AKSeqNo; (b) the SAID or GSAID for which keying material is requested; (c) an MS_Random; and (d) an HMAC/CMAC MIC over the contents of this message. This first message is optional and is sent only if the BS considers it necessary to refresh the key before the MS requests it. The BS responds to the current message with a Key-Reply message that carries the BS's active keying material for a specific SAID.
- *BS → MS (PKMv2 Key-Reply message)* that contains: (a) the AKSeqNo; (b) the SAID or GSAID for which keying material is delivered; (c) 'older' generation of TEK-Parameters relevant to SAID or GTEK-Parameters for MBS; (d) 'newer' generation of TEK-Parameters relevant to SAID or GTEK-Parameters for MBS; (e) GKEK-Parameters 'older' generation of GKEK-related parameters for MBS; (f) 'newer' generation of GKEK-related parameters for MBS; (g) the MS_Random same as in the previous message; and (h) an HMAC/CMAC MIC over the contents of this message.

Also, if the BS rejects the MS's traffic keying material request it responds using a PKMv2 Key-Reject message. For refreshing a GTEK key, a BS may also transmit a PKMv2 Group-Key-Update Command message including (pushing) an encrypted – using the GKEK – fresh GTEK to all the MS group members. Actually, the 802.16 standard defines two types of the PKMv2 Group-Key-Update Command message: the GKEK Update Mode and GTEK Update Mode. The first one is used for refreshing the GKEK, while the second one is used for refreshing the GTEK for MBS. These two messages contain a counter, namely the Key Push Counter, for protection against replay attacks. More specifically, the Multicast and Broadcast Rekeying Algorithm (MBRA) is executed as follows: the BS through its primary management connection sends a Key Update Command sporadically for the GKEK update mode to each MS. This message contains the

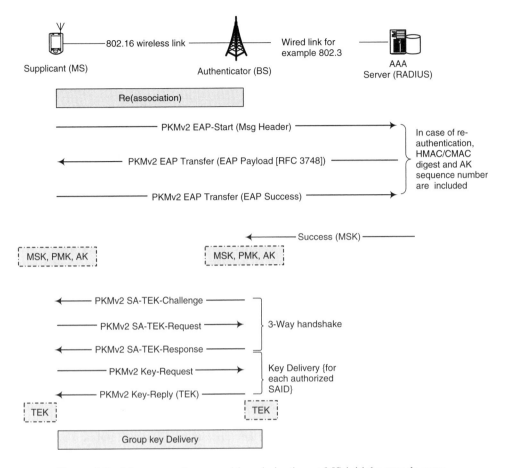

Figure 3.2   Message exchange and key derivation at MS initial network entry.

new GKEK encrypted with the KEK, which is derived from the AK established during authentication. Then, the BS transmits a Key Update Command for the GTEK update mode through the broadcast connection. The latter contains the new GTEK encrypted with the corresponding GKEK. A successful execution of a single EAP mode authentication at MS initial network entry, followed by three-way handshake and keying material delivery is depicted in Figure 3.2.

3.5   Vulnerabilities and Countermeasures

In this chapter we will not address physical layer attacks such as jamming or more general attacks such as cloning; instead we focus on vulnerabilities or weaknesses that directly or indirectly relate to or stem from the PKM protocol. Once more, our analysis concentrates on PKMv2. For PKMv1 security issues the reader can refer to [17]–[23]. It is stressed that with the publication of the Mobile WiMAX amendment, most of the PKMv1 vulnerabilities have been solved. Also, we suggest that the reader to refer to [24] for a secure simplification of the PKMv2 protocol. According to the authors, the simplified version of the protocol is guaranteed to be secure, but without various redundant fields, thus making it more effective.

3.5.1   Authorization

The PKMv2 RSA-Acknowledgement message (see section 3.4.2) seems to remedy a well-known weakness of PKMv1. Without this message, an attacker could replay the first message from an earlier session and associate with a BS. The BS would then proceed to a three-way handshake with the MS [25]. Moreover, during the authorization/authentication phase we note that the encrypted MS MAC Address field identifies the legitimate receiver of the PKMv2 RSA-Reply message. On the other hand, the third message does not identify the intended receiver explicitly. This issue may however be vulnerable to an interleaving attack, as described in [22, 23, 26]. Such an attack unfolds as follows: (a) the aggressor masquerades as a legitimate MS and exchanges the first two messages of the PKMv2 protocol with the legitimate BS; (b) after that, he masquerades as a legitimate BS and starts another PKMv2 protocol instance with the impersonated MS; (c) the aggressor employs the last message of the second PKMv2 protocol instance sent by the impersonated MS to reply to the BS. This message is actually the third message of the first PKMv2 protocol occurrence in step (a). The aforementioned attack can be prevented by simply adding a BSID in the PKMv2 RSA-Acknowledgement message before the SigMS attribute.

As presented in section 3.4.2, the PKMv2 RSA-Request message is well protected with a nonce and SigMS making it immune to a Denial of Service (DoS) attack. However, the same message can be utilized to trigger a replay attack because the BS has no way to perceive from nonce if this message is fresh or not. Many researchers have suggested replacing nonces with timestamps to remove this vulnerability [27]. During PKMv2 RSA authentication, the BS has to perform public key operations. However, public key encryption and signature requires considerable resources, so the BS may be paralyzed if flooded with false requests during a well-managed DoS attack [19].

Last but not least, the PKMv2 protocol takes it for granted that certificates are issued in the approved manner, that is, no parties with different public or private key pairs are certified to carry the same MAC address. If this assumption does not hold, each party can pretend to be the other. Thus, the specification must highlight the assumption explicitly that every certified MAC address is unique.

3.5.2   Key Derivation

According to the PKMv2 specification there is no session identifier to be used in the key derivation procedure. Consequently, the key derivation procedure cannot guarantee that the session key is fresh and unique among different sessions. The issue here is that the BS is responsible exclusively for guaranteeing that every key is fresh and unique. Research in the topic suggests that session key derivation should include as input a unique session identifier, that is, an (MS_Random/BS_Random) pair [26].

Another issue stems from the fact that all the bits in an AK are contributed by the BS. So, the MS must assume that the BS always produces a fresh, universally unique AK. This also implies that the random number generator installed in every BS is cryptographically secure. If not, it could leak the AK and inevitably the TEK. Researchers recommend that both the MS and the BS should contribute for producing an AK. For example, [21] propose that AK may be derived as: HMAC-SHA1(BS's AK/some MS random value).

3.5.3 Three-Way Handshake

The three-way handshake relies on the shared AK. As discussed in section 3.4.4, an AK may be the product of either the PKMv2 RSA Authentication protocol or EAP or both. The authorization/authentication phase mutually authenticates parties and establishes a common secret. A single principal does not play the roles of both MS and BS simultaneously, or there is an easy reflection attack [25]. It is also assumed that an attacker cannot drop packets, but he could record, replay and inject messages. All the messages involved in the three-way handshake afford a replay protection and are authenticated at the destination. So, without knowing the shared key, it is impossible to fool either the MS or the BS. After retrying to send a message the maximum allowed number of times the MS or the BS drops the connection if it gets no response from the peer. Thus, a connection release implies that the peer is down or the service is not running.

3.5.4 Key Delivery

A normal 802.16 MAC header does not include MAC addresses; by contrast with other networks here the 48-bit MAC address serves as an equipment identifier. Instead, in order to identify connections to equivalent peers in the MAC of the BS and the MS, the 16-bit CID is used. The key-delivery protocol as described in section 3.4.5 does not consider the identities of the two peers. However, this is performed via the CID. A BS uses the CID to differentiate between MSs and check the authorization status of a given MS. Therefore, actually the protocol only authenticates the BS to MS, not the opposite. So, as discussed in [26] the problem with this approach lies in TEK derivation. A TEK is generated by a BS and is used directly for encryption. However, this mechanism has a drawback which has been pointed out in [21]. The problem is that security is solely under the control of one participant, that is, the BS. The authors in [26] suggest that the AKE2 protocol [28] be used for key transportation for unicast connection. This protocol achieves mutual authentication and by adding another nonce of 'Alice' in its last message key mutual control is also realized.

After the authorization/authentication phase, the MS initiates a separate TEK machine for each of the authorized SAIDs. So, for each authorized service, the MS and BS will maintain locally a TEK state machine. This is done, however, without taking into account the usage status of the authorized services. For example, think of a user who participates in a teleconference, then switches off the service to surf the Internet, and after that initiates other services. According to the 802.16e standard such a client has to retain many TEK state machines even if they are not used. According to [26] this turns into an advantage when a user changes services among authorized ones, that is, no need to re-initiate a TEK state machine. On the opposite side, an MS has to maintain all TEK state machines for each authorized service. So, in a 802.16 mobile environment, where processing power and memory resources on the client side matters, this is an important issue. The authors in

[26] propose a solution to this problem by changing the trigger condition. That is, when a service is demanded by a user, the TEK state machine starts to run.

Another potential replay attack on the key delivery mechanism of 802.16e is identified by [21]. According to the specifications the key sequence number of a TEK has a length of only two bits. This sequence number is part of the TEK parameter within the PKMv2 Key-Reply message. An attacker is able to capture TEK messages and replay them to gain information needed in order to decrypt data traffic [27, 29]. This vulnerability can be mitigated by increasing the sequence number length. This will ensure that an adequate amount of TEK sequence numbers can be generated and transmitted within the longest validity duration of an AK. Considering 70 days as maximum lifetime duration of an AK and 30 minutes for the minimum lifetime duration of a TEK, a data SA could theoretically consume 3.360 TEKs over a complete AK-Lifetime [21]. Note, that AK lifetime is calculated as: AK lifetime = MIN(PAK lifetime, PMK lifetime). Also, the specification [4] suggests that PMK lifetime should not exceed the 24 hours time duration.

The MS in the key delivery protocol is assumed to be secure from replay attacks because the Old-TEK in a current PKMv2 Key-Reply message should be the New-TEK in the previous message. Also, the BS is able to verify whether a replayed Key-Request message coming from an MS is fresh or not by checking the MS_Random attribute contained in the corresponding Key_Request message.

3.5.5 Attacks on Confidentiality

Upon MS initial network entry communication parameters and settings with the corresponding BS are exchanged. Such information includes configuration settings, power settings, security negotiation parameters, mobility parameters, vendor information, MSs capabilities, etc. Contrary to the IEEE 802.16-2004 standard the network entry procedure between a BS and an MS in 802.16e affords integrity protection. This is because the 802.16e uses Short-HMAC tuple for protecting ranging (RNG-REQ/RSP) messages. Of course, this stands if the MS shares a valid security context with the target BS. If true, then the MS will conduct initial ranging with the target BS by sending a RNG-REQ including a Short-HMAC tuple. The designers chose Short-HMAC because most management messages are very short and a lengthy hash would severely increase the overall size of the message.

On the other hand, the management message exchange during network entry remains unencrypted and the information can be accessed simply by eavesdropping passively on the radio-link. Also, after initial network entry the management communication over the basic and primary management connections does not afford confidentiality. Note, however, that most of the management messages are transmitted using these connections. Encryption is only provided for key transfer messages. This also stands for the attribute that holds the transferred key; not for other attributes which remain in cleartext. By gathering, and later on, analysing management information an aggressor can build detailed profiles about MSs, for example, capabilities of devices, security settings, associations with BSs and so on. Moreover, by analysing such data a listening adversary may be able to determine the user movement and get a rough estimation of the position of the MS. The mapping of CID and MAC address may also be revealed by monitoring the MAC address sent in ranging or registration messages. After that, it is possible to associate the intercepted information to user equipment [30, 31].

Initial network can be protected by introduction of Diffie-Hellman Key exchange during initial ranging as suggested by [20]. By doing so, the two peers will create private/public key pairs and exchange public keys. Upon that all RNG-REQ/RSP messages can be protected using the exchanged public keys.

3.5.6 MBS Attacks

The MBS of 802.16e enables the distribution of data to multiple MS with one single message, thus saving cost and bandwidth. Broadcast messages are encrypted symmetrically with a shared key known to each member of the same group. Also, every member can decrypt the traffic using the same key. Message authentication is also based on the same shared key. However, by doing so, every group member is able to encrypt and authenticate messages as if they originate from the legitimate BS. The distribution of the GTEK when the MBRA is used is another important issue. Specifically, as discussed in section 3.4.5, GTEK is encrypted with the GKEK and broadcast to all group members. Note that the GKEK is also a shared key known to every group member. Therefore, using the GKEK, a malevolent insider is able to create fake encrypted and authenticated GTEK Key Update Command messages and distribute another GTEK. As a result, it is no longer possible for group members to decrypt MBS traffic coming from the legitimate BS. The insider can force MSs to accept the forged key in a number of ways, as described in [31]. If the system is not implemented properly, the key contained in the last subsequently transmitted GTEK update command messages may replace the original. Consequently, all the adversary has to do is send its GTEK update command message after the BS broadcasted a Key Update message. If the implementation is by the standard, the keys of both messages are accepted. So, the insider could falsify certain parts of the BSs GTEK update command message, making the receiving MS discard it. After that, the attacker can transmit its own GTEK update command message to the MS. Even worse, considering the fact that MBS is unidirectional, the BS is not able to detect that the MS has different GTEKs.

In a nutshell, the MBRA has two major problems. The first has to do with scalability, because it needs to unicast to each MS. Second, it is not able to cope with the issue of backward and forward secrecy. When a newcomer receives the current GTEK, it can decrypt all previous multicast messages transmitted during the same GTEK's lifetime. Also, an MS which leaves the group is able to receive the next GKEK or decrypt the next GTEK. MBRA is similar to the Group Key Management Protocol (GKMP) [32], which does not provide a solution for maintaining forward secrecy except by creating an entirely new group without the leaving member. Therefore, this scheme is not scalable for large dynamic groups.

One solution to this problem is to forbid broadcast key updates. The GTEK update command message could be transmitted unicastively to every MS, in a similar way as the GKEK update command message (the PKMv2 Group Key Update Command message for the GKEK update mode is carried over the Primary management connection). The key should then be encapsulated using the (unique) KEK of the MS. Public key cryptography is another option. Following this approach the GTEK update command message can be encrypted with the GKEK and broadcast, but is additionally signed using the private key of the BS. Any MS that receives a GTEK update command message can verify the signature using the public key of the BS and acquire the GTEK. However, the public key approach has the obvious disadvantage of performance; a symmetric solution can be processed very fast and protects from outsiders. A third option is to generate GTEKs as part of a hash chain, as described in [31]. Moreover, [33] propose another scheme which can be used for secure distribution of keys in groups.

A rekeying algorithm is introduced in [22], called Elapse, supporting perfect secrecy so as to address the problems of MBRA. Elapse is based on subgrouping the MSs so that the GKEK is not maintained by unicasting to individual MS but by broadcasting to subgroups.

Furthermore, GTEK lifetime affects scalability as well as forward/backward secrecy. The standard does not contain any directions on GTEK lifetime, although we can assume that GTEK lifetime is the same as that of TEKs, given the fact that GTEK is a special kind of TEK. So, based on the specifications we can infer that the lifetime of GTEK is 12 hours by default, 30 minutes minimum and seven days maximum. With regard to join and leave events, increased GTEK lifetime leads to much greater gaps in backward and forward secrecy, because a greater number of messages are encrypted via the given GTEK [22]. Extensive analysis is provided in [34] on MBS in 802.16 and several schemes are proposed that address this issue.

3.5.7 Mesh Mode Considerations

The two primary access modes operated in WiMAX are Point-to-MultiPoint (PMP) and Mesh. The PMP mode is a centralized polling mode that consists of a BS and several MSs. The BS polls MSs with polling modes of unicast, multicast and broadcast polling. On the other hand, the Mesh mode operates among MSs and does not require the presence of a BS., Mesh is similar to ad hoc networks, in which each MS acts as a wireless router for forwarding packets.

Authentication in mesh topology is a multi-hop version of the procedure defined for the Point-to-MultiPoint (PMP) mode. Even so, in mesh mode there must be a unique node that plays the role of BS, so that each MS has to perform authentication. Any node entering the network will use a multi-hop connection to communicate with the BS if the latter is not directly reachable. The 802.16e standard does not make any changes corresponding to the mesh authentication legacy specifications, but brings in new frames and new EAP-based authentication procedures. Whilst EAP methods are potentially secure, 802.16e relies on the fact that the path between the authenticator, that is, the sponsor node which is a proxy to the BS, and the authentication server that is, the BS, must somehow be secured. But as pointed out in [19] there is no such requirement for the UDP tunnel used in 802.16e. Thus, the way certain actions mandated by EAP methods will be performed remains ambiguous. This is especially the case with insecure EAP methods, like MS-CHAP, when deployed over an unprotected channel. Also, the mechanism requires that the sponsor node is authenticated to the network. If not, it would not be able to tunnel messages. It is to be noted that in mesh mode every node receives the same AK, namely the Operator Shared Secret (OSS). However, in case of EAP, the specification does not specifically define how the OSS is generated [19].

3.5.8 Handovers

The 802.1X/EAP framework does not provide a solution for low latency security estab-
lishments during a handover and re-authentication. To reduce the delay associated with the
SA re-establishment, the 802.16e specification proposes transferring the link security keys
from one BS to the next. Specifically, as described in [4], the specification provides different scenarios based on three distinct security settings of handover optimization bit#1 and
#2 contained in an RNG-RSP message. Each scenario maintains different level of hand-
over effectiveness and security level, but there is a trade-off between them. In the case of
bit#1=1 and bit#2=0, the TEKs are refreshed, but not the AK because re-authentication
is not performed. Thus, forward secrecy is at stake since the serving BS can decipher
the fresh TEKs for any subsequent target BSs. This is possible using the KEK derived
from the unchanged AKs. In a case where both bits are equal to 1, all existing secret
keys before the handover will be reused after it finishes. Unfortunately, this results in a
domino effect; if a single BS is compromised, all the previous and following BSs can be
compromised as well. Clearly, for the domino effect to be avoided a secret key must not
be reused in other BSs. Only if both bits are set to 0 the re-authentication and three-way
handshake are triggered upon a handover.

Pre-authentication is provided by 802.16e by means of facilitating a fast re-entry by
establishing of an AK in the MS and the target BS, yet the specification does not define a
pre-authentication scheme. Towards this, [35] proposes a pre-authentication scheme which
results in establishment of an AK in the MS and the target BS proactively, that is, before
the handover takes place. In the proposed scheme, it is assumed that both bits are set
to 0 so that the re-authentication and three-way handshake are performed to avoid any
security rollback attacks.

3.6 Comparisons with 802.11/UMTS

Whilst second generation (2G) mobile networks, wireless networks (e.g., GSM, 802.11)
and fixed WiMAX support only unilateral MS-to-network authentication, 3G networks
(e.g., UMTS, 802.11i, 802.16e) require, or at least recommend, network-to-MS authentication as well. This is also the case with 802.16e. By doing so, man-in-the-middle attacks,
exploiting false BS, are made hard to achieve.

The security architecture of 802.16e to a large extent originates from IEEE 802.11i
specifications [6]. First of all, security in both specifications is built around the SA concept. Second, the 802.11i standard introduces the 802.1X/EAP framework for mutual
authentication, especially for large networks (enterprise mode), where security matters.
Similarly, EAP is introduced in 802.16e to be used complementarily or jointly with RSA-
based authentication (see also section 3.4). Note also that the ad-hoc mode of 802.11i,
which is comparable to the mesh mode of 802.16e, also employs the 802.1X/EAP framework for access control. 802.1X/EAP provides a generic framework for network access
authentication. This framework allows an authenticator to mutually authenticate a peer
(supplicant) and establish common master secret keys, which can be used to secure communications. For example, in 802.11i the PMK is used to derive the Pairwise Transient
Key (PTK) and after that, several other keys to provide access link security. On the other
hand, 802.16e uses MSK to derive PMK as discussed in section 3.4.3. Also, the key

derivation functions are similar in logic in both IEEE standards. That is, 802.11i uses a Pseudo Random Function for key derivation, namely PRF-n, where n is the number of (key) bits produced at the output, while 802.16e uses a Dot16KDF as described in section 3.4.3. The inputs of the aforementioned functions and key length may vary as well, depending on the security mechanism employed. For instance, 802.11i PMK size is 256 bits, while the length of MSK in 802.16e is 512 bits. Additionally, 802.11i provides an option for Pre-Shared Keys (PSK), but this method is suggested for personal use (personal mode) only. Moreover, some incompatibilities with standard EAP practices do exist in 802.16e. For example, as suggested by [36], the back-to-back EAP method is specified incompletely within 802.16e and thus is incompatible with the more generic approaches defined by the IETF, such as EAP-TTLS, PEAP, EAP-after-EAP.

The 802.16e three-way handshake is also comparable to the 802.11i four-way handshake [6]. However, the four-way handshake is basically a key-refresh protocol, while the three-way handshake is a key distribution protocol [25]. More specifically, the EAPOL-key used in the four-way handshake message integrity check is derived using the nonces exchanged in the protocol. Because of that, the first message of the four-way handshake does not carry a MIC, making it vulnerable to DoS. This, however, does not apply to the three-way handshake as the MIC is based on the AK. Thus, the nonces do not participate in key derivation. It is suggested in [25] that there is a method for refreshing the AK in the context of three-way handshake by using the nonces exchanged; a MIC could be calculated over the contents of the first message using the old version of AK. The four-way handshake limits the use of the PMK by generating a new key (PTK). This is not the case with the three-way handshake.

The 802.11i also supports multicast/broadcast communication. The Group Transient Key (GTK) (established during a four-way handshake) used within a specific group may need to be refreshed. When an MS leaves the network, the GTK also needs to be updated. This is to preserve forward secrecy. To manage the updating, 802.11i defines a group key handshake that consists of a two-way handshake: (a) the AP sends the new GTK to each MS in the network; the GTK is encrypted using the KEK assigned to that MS and is protected using a MIC; and (b) the MS acknowledges the new GTK by replying to the AP. While this procedure is similar to the MBRA described in sections 3.4.5 and 3.5.6 some differences exist. For instance, a 802.11i GTK is unicast to each MS, while a GTEK is multicast. Last but not least, 802.11i provides a mechanism for generating a secret transient key, called STAkey, by the AP for direct MS-to-MS communication. Both stations must be associated with the same AP. Such a mechanism is not defined within 802.16e. A brief comparison between the 802.11i and 802.16e security mechanisms is given in Table 3.1.

2G mobile networks and 3G mobile networks use different Authentication and Key Agreement (AKA) mechanisms than those of 802.11i/16e. Of course, compared to the 2G mechanism, the 3G AKA provides substantially longer key lengths and mutual authentication. AKA typically runs in a UMTS Subscriber Identity Module or a CDMA2000 (removable) User Identity Module. In 3G networks, AKA is used for both radio network authentication and IP multimedia service authentication purposes. For integration with 802.11i, and presumably with 802.16, GSM networks use the EAP-SIM method [37] which is based on the standard GSM AKA. For the same purpose, UMTS and CDMA2000

Table 3.1 A brief comparison of 802.11i and 802.16e security mechanisms

	Authentication	Key Management	Encryption	Secure Multi-cast/Broadcast
802.16e	RSA-based or EAP-based or a combination of the two. Establishes an AK/MSK and one or more SAs.	For every SA authorized, the MS runs a TEK state machine to handle the secure exchange/update of TEK based on keys derived from AK/MSK.	Data Encryption Standard-Cipher Block Chaining (DES-CBC), AES in Counter with CBC-MAC (AES-CCM).	MBRA.
802.11i	Uses PSK or 802.1X/EAP. In case of EAP establishes a PTK from PMK using the four-way handshake.	Derives PTK from the PMK and after that two EAPOL-keys and a temporal key to secure both EAPOL handshakes and user data.	Wired Equivalent Privacy (WEP) based on RC4, Temporal Key Integrity Protocol (TKIP), Counter mode Cipher block chaining Message authentication code Protocol (CCMP) based on AES.	Four-way handshake for establishing a GTK (derived from Group Master Key (GMK)). Two-way handshake for refreshing the GTK. A STAkey exchange for station-to-station peer communication is also provided.

systems use another EAP method, namely EAP-AKA [40], which is based on the original UMTS/CDMA2000 AKA. Though UMTS AKA and EAP-AKA are almost identical, they differ in the transport method of the AKA protocol; the Packet Mobility Management (PMM) protocol in case of UMTS and EAP in case of 802.11. The former does not afford a fast re-authentication function, while EAP-AKA does offer such an option. Generally, however, the introduction of AKA inside EAP allows for several new applications: (a) the use of the 3G mobile network authentication infrastructure in the context of wireless LANs/MANs; and (b) relying on AKA and the existing infrastructure in a seamless way with any other technology that can use EAP.

Contrary to IEEE 802.11/16 standards, authentication and key derivation in current mobile networks relies on the fact that the identity module and the user's home network have agreed on a secret key beforehand. The authentication process initiates by the home network which generates an Authentication Vector (AV), based on the secret key and a sequence number. The AV contains a random number RAND, an authentication token AUTN used for authenticating the network to the identity module, an expected result part XRES, a 128-bit session key for integrity check IK and a 128-bit session key for encryption CK. The RAND and the AUTN are then transmitted to the identity module. The latter verifies the AUTN, based on the (same) secret key and the sequence number. If this process is successful that is the AUTN is valid and the sequence number used to generate AUTN is within a certain margin, the identity module produces an authentication result RES and sends it to the home network. The home network verifies the RES and, if the result is correct, IK and CK can be used to protect link layer communications. In the case of EAP-AKA, the EAP server obtains the authentication vectors (from the home network), compares RES and XRES, and uses CK and IK in key derivation.

With wireless technologies mushrooming, authentication and authorization of mobile users in heterogeneous access technology environments will be a major issue to be addressed by EAP methods. Besides, as already pointed out, EAP is utilized by both IEEE and 3G mobile networks standardization groups. Of course, as discussed, different network technologies use different authentication mechanisms. In this context, the qualities of security, lightweight demands on processor and memory, as well as interoperability across networks and network types and topologies will all be essential in EAP methods moving forward [16]. Several researchers address this topic by proposing secure EAP methods for heterogeneous access network integration. For example, the authors in [38] modify EAP methods to include the adoption of trusted hardware and introduce two authentication schemes capable of advancing authentication developments on wireless city networks. A new key management method is described in [39], called Handover Keying (HOKEY) for access to 802.16 and 802.16 handovers for 3G subscribers. The deployment of EAP-AKA for allowing access of UMTS subscribers to the 802.16 network is also addressed in [39].

3.7 Summary

Owing to the natural characteristics of wireless communication, anyone within range can intercept or inject frames, making wireless communication much more vulnerable to attacks than its wired equivalents. In this chapter we focused on the PKM protocol which is directly associated with the key management procedures of IEEE 802.16. Concentrating on PKMv2, several aspects of PKM functionality were examined including initial authorization /authentication between a BS and MS, key derivation and hierarchy, key distribution, etc. Also, most of the current vulnerabilities were revealed and some countermeasures, as proposed in the literature, were discussed. Analysis showed that while the 802.16e standard remedies the security issues of fixed WiMAX to a great degree, there is still room for additional refinement. A comparison of Mobile WiMAX with WiFi and 3G security mechanisms was also provided. In this context, the 802.1X/EAP framework seems to be the most promising solution for the integration of heterogeneous access technology networks.

References

[1] K. Lu, Y. Quian, H.-H. Chen and S. Fu, 'WiMAX Networks: From Access to Service Platform', *IEEE Network*, pp. 38–45, May/June 2008, IEEE Press.

[2] IEEE Std 802.16-2001 (2001) IEEE Standard for Local and Metropolitan Area Networks, Part 16: Air Interface for Fixed Broadband Wireless Access Systems, Approved 6 December 2001, IEEE Press.

[3] FIPS PUB 46-3 (1999) 'Data Encryption Standard (DES)', Federal Information Processing Standards Publication, Reaffirmed 1999 October 25, US Department of Commerce/National Institute of Standards and Technology.

[4] IEEE Std 802.16e (2006) IEEE Standard for Local and Metropolitan Area Networks, *Part 16:* Air Interface for Fixed and Mobile Broadband Wireless Access Systems, Amendment 2: Physical and Medium Access Control Layers for Combined Fixed and Mobile Operation in Licensed Bands and Amendment and Corrigendum to IEEE Std 802.16-2004, published Feb. 2006, IEEE Press.

[5] IEEE Std 802.16-2004 (2004) IEEE Standard for Local and Metropolitan Area Networks, Part 16: Air Interface for Fixed Broadband Wireless Access Systems, Revision of the IEEE 802.16-2001 Std, IEEE Press.

[6] IEEE Std 802.11i (2004b) IEEE Standard 802.11i-2004. Amendment to IEEE Std. 802.11, 1999 Edition, Amendment 6: MAC Security Enhancements, Part 11: Wireless LAN MAC and PHY Layer specifications, IEEE Press.

[7] FIPS 197 (2001) 'Announcing the Advanced Encryption Standard (AES)', Federal Information Processing Standards Publication 197, Nov. 2001.

[8] B. Aboba, L. Blunk, V. Vollbrecht and J. Carlson, 'Extensible Authentication Protocol (EAP)', IETF RFC 3748, June 2004.

[9] IEEE Std 802.1X-2004 (2004c) '802.1X IEEE Standard for Local and Metropolitan Area Networks, Port-Based Network Access Control', Revision of IEEE Std 802.1X-2001, IEEE.

[10] CableLabs (2005) 'DOCSIS 1.1 Baseline Privacy Plus Interface Specification', Data-Over-Cable-Service Interface Specifications CM-SP-BPI+-I12-050812, Cable Television Laboratories, Aug. 2005.

[11] R. Housley, W. Polk, W. Ford and D. Solo, 'Internet X.509 Public Key Infrastructure Certificate and Certificate Revocation List (CRL) Profile', IETF RFC 3280, April 2002.

[12] RSA Laboratories (2002) 'PKCS #1 v2.1: RSA Cryptography Standard', June 2002.

[13] H. Krawczyk, M. Bellare and R. Canetti, 'HMAC: Keyed-Hashing for Message Authentication', IETF RFC 2104, Feb. 1997.

[14] NIST SP 800-38B (2005) 'Recommendation for Block Cipher Modes of Operation: The CMAC Mode for Authentication', NIST, May 2005.

[15] D. Stanley, J. Walker and B. Aboba, 'Extensible Authentication Protocol (EAP) Method Requirements for Wireless LANs', IETF RFC 4017, March 2005.

[16] R. Dantu, G. Clothier and A. Atri, 'EAP methods for wireless networks', *Computer Standards & Interfaces* **29** (2007): 289–301, Elsevier Science.

[17] E. Eren and K. Detken, 'WiMAX-Security – Assessment of the Security Mechanisms in IEEE 802.16d/e' in Proceedings of 12th World Multi-Conference on Systemics, Cybernetics, and Informatics, USA, July 2008.

[18] M. Barbeau, 'WiMAX/802.16 Threat Analysis', in Proceedings of Q2SWinet, pp. 8–15, Oct. 2005, Canada, ACM Press.

[19] L. Maccari, M. Paoli and R. Fantacci, 'Security analysis of IEEE 802.16', in Proceedings of IEEE ICC, pp. 1160–5, June 2007, IEEE Press.

[20] T. Shon and W. Choi, 'An Analysis of Mobile WiMAX Security: Vulnerabilities and Solutions', in Proceedings of NBiS 2007, LNCS 4658, pp. 88–97, Springer.

[21] D. Johnston and J. Walker, 'Overview of IEEE 802.16 Security', IEEE Security and Privacy, **2**(3), pp. 40–8, IEEE Press.

[22] C.-T. Huang and J.M. Chang, 'Responding to Security Issues in WiMAX Networks', *IEEE IT Professional* **10**(5): 15–21, Sept. 2008, IEEE Press.

[23] S. Xu and C.-T. Huang, 'Attacks on PKM Protocols of IEEE 802.16 and its Later Versions', in Proceedings of 3rd International Symposium on Wireless Communication Systems, pp. 185–9, Sept. 2006, Spain, IEEE Press.

[24] E. Yuksel, H.R. Nielsen, C. Rosenkilde Nielson and M.B. Orencikin, 'A Secure Simplification of the PKMv2 Protocol in IEEE 802.16e-2005', in Proceedings of Joint Workshop on Foundations of Computer Security and Automated Reasoning for Security Protocol Analysis, pp. 149–64, Poland, IMM publications, 2007.

[25] A. Datta, C. He, J.C. Mitchell, A. Roy and M. Sundararajan, '802.16e Notes – Mitchell Group', June 2005.

[26] T. Haibo, P. Liaojun and W. Yumin, 'Key Management Protocol of the IEEE 802.16e', *WUJNS*, **12**(1): 59–62, 2007, Springer.

[27] S. Xu, M. Matthews and C. Huang, 'Security Issues in Privacy and Key Management Protocols of IEEE 802.16', in Proceedings of the 44th ACM Annual Southeast Regional Conference, pp. 113–18, March 2006, ACM Press.

[28] M. Bellare and P. Rogaway, 'Entity Authentication and Key Distribution', *Advances in Cryptology*, pp. 232–49, LNCS 773, Springer, 1993.

[29] E. Eren, 'WiMAX Security Architecture – Analysis and Assessment', in Proceedings of IEEE International Workshop on Intelligent Data Acquisition and Advanced Computing Systems: Technology and Applications, pp. 673–7, Sept. 2007, Germany, IEEE Press.

[30] S. Naseer, M. Younus and A. Ahmed, 'Vulnerabilities Exposing IEEE 802.16e Networks To DoS Attacks: A Survey', in Proceedings of 9th ACIS International Conference on Software Engineering, Artificial Intelligence, Networking, and Parallel/Distributed Computing, pp. 344–9, IEEE CS Press.

[31] A. Deininger, S. Kiyomoto, J. Kurihara and T. Tanaka, 'Security Vulnerabilities and Solutions in Mobile WiMAX', *IJCSNS International Journal of Computer Science and Network Security* **7**(11): 7–15, Nov. 2007.

[32] H. Harney and C. Muckenhirn, 'Group Key Management Protocol (GKMP) Specification', IETF RFC 2093, July 1997.

[33] H. Li, G. Fan, J. Qiu, J. and X. Lin, 'GKDA: A Group-Based Key Distribution Algorithm for WiMAX MBS Security', in Proceedings of PCM 2006, LNCS 4261, pp. 310–18, 2006, Springer.

[34] S. Xu, C.-T. Huang and M.M. Matthews, 'Secure Multicast in WiMAX', *Journal of Networks*, **3**(2): 48–57, Feb. 2008, Academy Publisher.

[35] J. Hur, , H. Shim, P. Kim, H. Yoon and N.-O. Song, 'Security Considerations for Handover Schemes in Mobile WiMAX Networks', in Proceedings of IEEE WCNC 2008, pp. 2531–6, March/April 2008, IEEE Press.

[36] D. Johnston, 'Deletion of Double EAP mode', IEEE 802.16 Broadband Wireless Access Working Group, IEEE C802.16maint-07/44, April 2009.

[37] H. Haverinen and J. Salowey, 'Extensible Authentication Protocol Method for Global System for Mobile Communications (GSM) Subscriber Identity Modules (EAP-SIM)', IETF RFC 4186, Jan. 2006.

[38] Y.-T. Chen, A. Studer and A. Perrig, 'Combining TLS and TPMs to Achieve Device and User Authentication for Wi-Fi and WiMAX Citywide Networks', in Proceedings of IEEE WCNC, pp. 2804–9, March–April 2008, IEEE Press.

[39] M. Nakhjiri, 'Use of EAP-AKA, IETF HOKEY and AAA Mechanisms to provide access and handover security and 3G-802.16m Internetworking', in Proceedings of IEEE PIMRC 2007, Sept. 2007, pp. 1–5, Greece, IEEE Press.

[40] J. Arkko and H. Haverinen, 'Extensible Authentication Protocol Method for 3rd Generation Authentication and Key Agreement (EAP-AKA)', IETF RFC 4187, Jan. 2006.

4

WiMAX Network Security[1]

Luca Adamo, Romano Fantacci and Leonardo Maccari
Department of Electronics and Telecommunications – University of Florence

4.1 Introduction

The possible usage scenarios of a WiMAX service network are extremely various. The network can support static and mobile users roaming in a metropolitan area, the kind of traffic can be the typical Internet browsing or real-time traffic with stringent QoS constraints. The components of a WiMAX network must assure that all these needs are fulfilled. To reach this goal, all the layers of the OSI stack are required to co-operate in order to guarantee security, session establishment, fast handover and correct quality levels.

The IEEE 802.16 family of standards is focused on the radio and MAC layers so it doesn't give any indication on how to manage the networking back-end. Nevertheless a WiMAX network can be composed of many distinct and heterogeneous elements: mobile stations, base stations, gateways, authentication servers that need to cooperate in a multi-user, multi-terminal and even multi-operator scenario. In order to define an interoperable network organization the WiMAX Forum [1] has produced a set of documents that specify the network elements, organization and configuration that must be deployed in a WiMAX network ([2], [3]).

[1] The topic presented in this chapter has been subject of investigations partially supported by the Italian National Project *Wireless multiplatfOrm mimo active access netwoRks for QoS-demanding muLtimedia Delivery* (WORLD) and FIRB *Integrated System for Emergency* (InSyEme) under Grant 2007R989S and RBIP063BPH.

In this chapter an overview of this organization will be given, based on the WiMAX Forum specification 1.2. The focus of our description will be the standards, the technical challenges and the solutions for mainly three issues:

- integration of authentication techniques and management of AAA (Authorization, Authentication, Accounting);
- IP addressing and networking issues;
- distribution of the QoS parameters.

These topics will be analyzed not from a MAC layer perspective but from the point of view of the network manager and of the interaction between the access network and the back-end.

4.2   WiMAX Network Reference Model

In order to understand completely the relationships between all the network components a logical representation of a WiMAX network must be introduced. Such a scheme is provided by the WiMAX Forum in [2] under the name of NRM (*Network Reference Model*) and distinguishes the *logical domains*, the *functional entities* and the *Reference Points* (RPs) as reported in Figure 4.1.

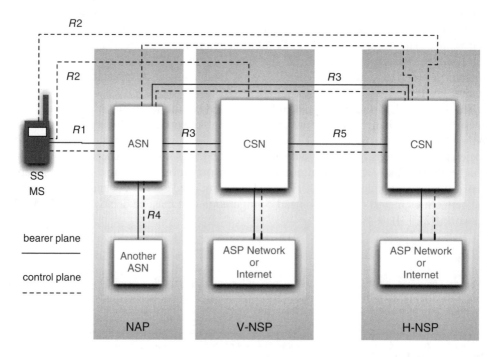

Figure 4.1   WiMAX network reference model.

The NRM is a logical model rather than a precise definition of a network architecture; the goal is to allow a variety of implementation solutions while maintaining an overall interoperability among different realizations of functional entities. For this reason no assumptions are made on the implementation of the functional entities. For some of them, however, guidance is provided through the so called *profiles* that we will describe later on. The following sections illustrate the most relevant components of the NRM.

4.2.1 Functional Entities

The main functional entities depicted in Figure 4.1 are:

- MS, *Mobile Station*. The MS is the generic device used by the subscriber to access the WiMAX network. The same device can be used by more than one user and the same user can access the network with more than one MS. Some configuration parameters can depend on the couple MS, user.
- ASN, *Access Service Network*. The ASN represents a boundary for functional interoperability with WiMAX clients and WiMAX connectivity services. It is mainly responsible for handling the layer 2 connectivity plane, forwarding all the AAA messages towards the H-NSP (Home Network Service Provider), relaying layer 3 service messages (e.g DHCP and Mobile IP). The logical decomposition of an ASN is shown in Figure 4.2. The two most relevant components of the ASN are the radio BS and the ASN-GW. The ASN-GW is the gateway to the IP network and the end-terminal of RP3 as described later on.

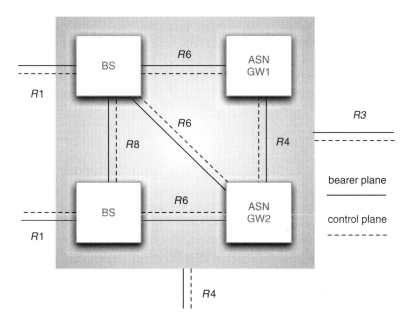

Figure 4.2   WiMAX network reference model, ASN decomposition.

- CSN *Connectivity Service Network*. The CSN is the entity entitled to the management of the IP layer 3 connectivity of the subscribers terminals. More specifically it covers the following tasks:
  - IP address provisioning;
  - gateway towards other networks;
  - performing AAA functions;
  - handling the *Inter-ASN Mobility* through the use of *Mobile IP*.

A MS, ASN or CSN are made up of logical functional entities that may be realized in a single physical host or may be distributed over multiple physical hosts.

4.2.2   Logical Domains

A logical domain can be seen as a group of functions that can be associated in a single domain. In Figure 4.1 three logical domains are presented.

- NAP, *Network Access Point*. The NAP is the physical point used by the subscriber terminal to access the network; from a logical point of view the ASN that is currently serving the MS is part of the NAP.
- H-NSP *Home Network Service Provider*. The H-NSP is the WiMAX service provider with which the WiMAX subscriber has a Service Level Agreement. This business entity authenticates and authorizes subscriber sessions and is responsible of the billing and charging procedures even in a roaming scenario where the subscriber is moving through various NSPs.
- V-NSP *Visited Network Service Provider*. A visited NSP is a WiMAX service provider that a subscriber uses to access the network in a roaming scenario even if there is no Service Level Agreements among the two parts. If the V-NSP has a roaming relationship with the H-NSP the V-NSP can be used to forward AAA messages from the subscriber to the H-NSP thus gaining access to the network on a foreign domain. The range of services provided to the subscriber by the V-NSP depends on the roaming relationship between that V-NSP and the subscriber's H-NSP.

4.2.3   Reference Points

Referring to Figure 4.1 a RP is the end-point of the communication between two functional entities and it constitutes the standard interface that must be used to achieve over-all interoperability among components of different manufacturers. The WiMAX Forum specifications [2] list several RPs in the NRM and also provides three examples of implementation profiles. Each profile includes only a subset of these RP. We describe here the most relevant RPs defined in the NRM, to help the reader to understand the rest of the chapter:

- *Reference Point R1*. R1 consists of the protocols and procedures used on the air interface between MS and ASN as defined in [4], [5] and [6]. It is the radio and MAC WiMAX interface.

- *Reference Point R3*. R3 is an interface between an ASN and a CSN (operated either by a H-NSP or a V-NSP). These two functional entities use this interface to vehiculate AAA messages, policy enforcement messages and Mobile IP mobility management capabilities. This interface is in practice an IP link supporting the RADIUS protocol and the gateway for user data.
- *Reference Point R4*. R4 is responsible for the communications between two ASNs that are managed by the same CSN. For instance, when a MS performs an handover between two ASNs into the same NSP the AAA phase could be avoided using communication over this interface.
- *Reference Point R5*. R5 defines and assures inter-networking functions between CSNs operated by H-NSP and V-NSP. This interface in practice will be an IP path that could be routed across dedicated or even public networks.
- *Reference Point R8*. Within an ASN with multiple BSs, R8 is the reference point used to support fast and seamless handover of the MSs.

4.2.4 ASN Profiles

Referring to the logical architecture shown in Figure 4.2 three possible implementation options are proposed and analyzed by WiMAX Forum specifications. These options are called *Profiles* and differ for the number of RPs exposed and for the grouping of some of the logical functions. In Figures 4.3 and 4.4 we report two ASN profiles proposed by the WiMAX Forum with the intent to guide the implementation of the ASN in two of the most common configurations. In the figures we omitted some of the modules that WiMAX Forum describes but that are out of the context of this review.

Profile A, shown in Figure 4.3 will be used in a configuration where a single ASN-GW is responsible to manage multiple BSs. This configuration is suitable in scenarios where it is necessary to cover a large area or to serve a high number of users. Mobility is completely handled by the ASN-GW that controls subscribers handovers over R4 reference point (ASN-anchored mobility).

Profile B shown in Figure 4.4 describes a configuration with a physical co-location of ASN-GW and BS where all the functions of the ASN are included in a single entity with an advantage in terms of complexity reduction. The drawback is that this solution implies a 1:1 ratio between ASN-GW and BSs and for this reason is appropriate only for scenarios with a small number of users dislocated in a localized area.

The function decomposition depicted in both Figure 4.3 and Figure 4.4 evidences the presence into the ASN of the entities used to support AAA procedures, MIP (Mobile IP) compliant mobility and DHCP (Dynamic Host Confguration Protocol). The different location of these entities is the most important element to be noticed when comparing the two profiles.

The choice of a profile must be guided by the extension of the coverage area and impacts on the costs and on the software and hardware solutions. For instance, using profile A the network manager may adopt hardware and software from different vendors, and could face configuration inconsistencies or incompatibilities. Using profile B these issues are avoided (R1 and R4 are standard and widely used interfaces) but scalability is limited.

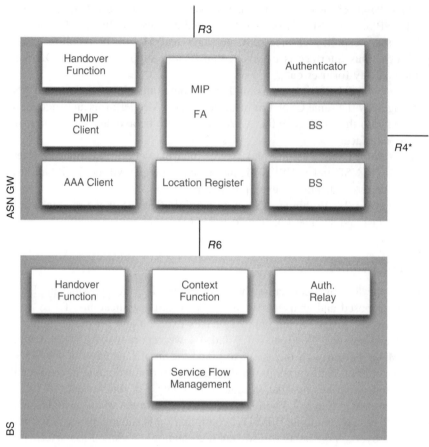

* ASN Anchored Mobility shall be possible over R6 and R4

* Support for this function is optional

Figure 4.3   WiMAX network profile A for ASN implementation.

4.3   The RADIUS Server

In any network configuration model it is included an *AAA server* where AAA stands for *Authorization, Authentication, Accounting*. The de-facto standard protocol for AAA is RADIUS (Remote Authentication Dial-In User Service) [7]. A successor of RADIUS is the DIAMETER protocol [8] that is currently under definition and will be supported in the future by the WiMAX Forum. The AAA server is used at any user login and plays three functions:

- Authorization: in this phase the received request is parsed and its validity is checked. Possible actions that can be taken are to accept the request (that is passed to the following modules), to reject the requests or to forward it to another server. A request could be rejected because the user that generated it was not allowed to, or because

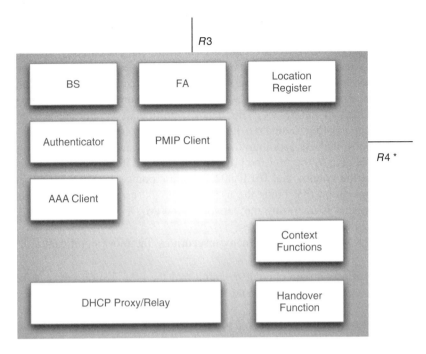

Figure 4.4 WiMAX network profile B for ASN implementation.

the ASN from which it comes has not correctly signed the packet. A request could be forwarded because the AAA server observes that it is not able to satisfy it. A request carries a username in the canonical form *user@domain*, if the server is not responsible for the specified domain it will forward the request to another authoritative server. This is particularly important in large metropolitan networks where users might be roaming from one operator to another. Since only the user's original operator owns the database containing the authentication credentials (i.e. password or certificates) the request must be forwarded to the H-NSP. The same applies to QoS parameters; the visited operator is not aware of the QoS properties that the user has negotiated with his own operator so they must be received from the H-NSP.

- Authentication: the user has to show that he is in possession of valid credentials to enter the network. At the same time the network must present valid credentials to the user in order to assure him that he is entering the correct network. This procedure is particularly delicate and will be explained more in detail.
- Accounting: this function is necessary to the manager of the network to record the activities of the users. The server receives from the ASN accounting packets that include the number of users, the start and end of user sessions or the resources occupied. All this information will end in a database that will be used mainly for billing purposes.

4.3.1 Authentication in WiMAX Infrastructure

The IEEE 802.16 standard supports two distinct authentication procedures, the first one was introduced with the *d* amendment of the standard and it is called PKMv1 (Private Key

Management version 1). Starting from IEEE 802.16e a new one has been proposed called PKMv2. PKMv1 doesn't rely on the use of an AAA server, the authentication algorithms are designed ad-hoc for WiMAX and it has been shown to be insecure [9] [10]. In PKMv1 each subscriber station is equipped with a couple of digital certificates that can be used to perform the authentication: one is inserted from the manufacturer and another that is user-dependent. The first one is signed by a CA owned by the manufacturer and is used to verify the hardware address of the device. This detail is particularly important and innovative compared to other existing standards. It is in fact very common for Ethernet or WiFi networks to perform address spoofing also at MAC layer, since most of the commercial network interfaces give to the user the possibility to change its own MAC address. As an example, most of the commercial IEEE 802.11 access points support MAC filtering and web-based authentication that binds the used credentials (username and password) with the MAC address that is accessing the network. In WiMAX this weak security measure is strengthened using digital certificates.

Nevertheless PKMv1 is subject to many insecurities, the most significant are:

- The authentication is unidirectional, meaning that the base station never authenticates itself with the client. The client could be led to enter a rogue network placed by an attacker.
- There is no form of authentication for some of the packets, this leads to the possibility for an attacker of replying authentication packets or personifying the base station.
- The lack of support for a centralized server is a great limitation to the deployment of large area networks where a set of base stations are managed by the same network administrator.

To address these issues PKMv2 has been introduced that adopts the IEEE 802.1X [11] port-based access control standard. Briefly, IEEE 802.1x determines the three roles that perform an authentication, a *supplicant* that is the client, an *authenticator* that is embedded in the access infrastructure (an ASN in WiMAX) and the *authentication server* that is the AAA server and resides in the CSN. In IEEE 802.1x the *authenticator* doesn't have any specific role into the authentication and acts as a proxy to the AAA server that owns the credentials of the clients and can verify them. No specific authentication algorithms are described in IEEE 802.1x, the EAP protocol [12] is used to transport a set of existent authentication protocols that can be password or certificate based.

The WiMAX Forum gives precise directives for the application of IEEE 802.1x standard into WiMAX networks: mandatory authentication methods are TLS [13] (Transport Layer Security), TTLS (Tunneled Transport Layer Security) [14] and AKA (Authentication and Key Agreement) [15], authentication must be bidirectional, support for OCSP (Online Certificate Status Protocol) [16] protocol is requested and keying material must be generated during the authentication to be reused in other security procedures, such as DHCP [17] or MIP [18]. Given the presence of digital certificates on the devices, user and device authentication is encouraged.

One of the turning points introduced by the WiMAX Forum architecture is the presence of the AAA server not only for authentication but for the distribution of user-based attributes to other network elements. The authentication is the only standard procedure that can be used by the client, the ASN and the CSN to negotiate parameters for MAC and IP layers. When the authentication has success the AAA server will attach to the Access-Accept RADIUS frame more RADIUS *attributes* (that's the terminology used by RADIUS protocol) that will be used by other networking elements. These attributes mostly deal with:

• QoS parameter that will be used by the BS to assure sufficient layer II resources.
• IP parameters, such as a static IP address, or the address of a DHCP server or MIP home agent address.

The possible configurations of a large area WiMAX network, and consequently the variables needed to configure an MS are many and theoretically an ASN could support them all. When a terminal enters the network the BS must be informed on which is the kind of authentication requested for the client, if it owns a static IP address or it uses DHCP, if it supports mobile IP and which in case are the addresses of its home or foreign agent. These parameters can be statically configured on each ASN with great scalability limitations. In a large network these parameters must be provided to the ASN elements when the authentication takes place, since the authentication is the only standard procedure that involves the CSN of the network. It is crucial to understand the role of an AAA server in a WiMAX network because it is responsible to distribute these pieces of information.

One more detail to focus on is that the presence of embedded digital certificates into the terminals can be used to perform a two step authentication, one for the device and another for the user. The TTLS protocol for instance is made up of two phases, the so called *outer authentication* that is certificate based and can be unidirectional or bidirectional and the *inner authentication* that is generally used to transport user/password challenges. Using TTLS the AAA server is not only authenticating an user but it is authenticating the user on a specific terminal. Some of the network parameters could be depending on the user, some other could be linked to the terminal or to the couple (user, terminal). This way the same account could be used for a home connection or for a mobile connection, with distinct quality and billing features.

4.4   WiMAX Networking Procedures and Security

This section provides an overview of the most important networking procedures used in a WiMAX architecture. First of all we discuss the handover phase, then the IP address provisioning problem is addressed presenting both DHCP and Mobile IP and the requirements that these two protocols introduce on the AAA implementation. The last part of the section is devoted to QoS considerations and to an analysis of the authentication mechanism.

4.4.1   Handover Procedure

A mobile station in WiMAX is free to change its point of attachment to the network to perform an handover. Two distinct kinds of handover can be performed by an MS: ASN-anchored or CSN-anchored. Referring to Figure 4.1, in the first case the mobile station is moving from a BS to another that resides in the same ASN. The ASN-GW to which each BS is connected is the same, so that there is no need to update routing tables or network address. In this case the handover is managed by the entities into the ASN and does not have impact on the IP layer. In the second case the MS is moving from an ASN to another. The CSN connected to both the ASNs could be the same, if the new ASN belongs to the H-NSP or another if the new ASN belongs to a V-NSP. In this second case the new CSN will behave as a proxy towards the previous one and the result will be logically the same.

Recall that an ASN is not a monolithic component but it is composed of many sub-modules. Those sub-modules can be physically placed in the same host as depicted in Figure 4.4 or separated as in Figure 4.3. Changing the point of attachment for a MS means changing the BS but depending on the chosen configuration might imply the change of other functional entities such as the authenticator or the MIP FA (Mobile IP Foreign Agent). In the following part of the chapter we will always refer to CSN-Anchored mobility since it has direct consequences on the networking layer.

A WiMAX network should be able to deal with roaming users that need to keep their sessions alive when they perform a handover. Two prerequisites to maintain the sessions active are: to keep the same IP address across handovers and to update the reverse routes from the remote destination to the MS. This tasks can be performed using the Mobile IP protocol for MSs that have support for this standard. If the MS supports only DHCP, then the same result can be obtained combining the DHCP protocol for the MS and an instance of MIP protocol that runs only on the back-end. These two options and their security implications will be described in the rest of this chapter.

WiMAX aims to be the first wireless mass technology that is targeted to mobile, IP-based, real-time application for large area networks. The WiMAX Forum addressed some networking issues using a composition of standard protocols adapted to resolve specific WiMAX problems. One of these issues is that there is no standard and commonly used way to move configuration parameters from a centralized server to a MS. In general, whenever a MS enters a network (for its first time or during an handover) the authentication procedure is the only phase in which the user, the mobile terminal, the ASN and the CSN are forced to communicate. This is the only moment in which authentication, networking and IP parameters can be moved from the management back-end to the rest of the network. In practice this happens with the following scheme:

- The parameters for a specific terminal or end-user are stored in a DB that the RADIUS server has access to.
- The parameters are moved to the authenticator using RADIUS attributes over R3 and possibly R5.
- From the authenticator, using custom protocols they are moved to other network elements, for instance, QoS parameters are moved to the BS together with the MSK key that has been negotiated during the authentication.
- If some of them must be transmitted to the MS they are inserted into DHCP extensions that will be used for the MS over R1.

Note that the RADIUS protocol is the only standard mean to move parameters from the CSN to the ASN. Into the ASN the endpoint that manages them is the IEEE 802.1x authenticator, that is encharged to dispatch them to the other entities, such as the DHCP server or the BS. This composition of protocols is completely backward compatible but has two limitations: it is quite complicate and it is fully concentrated into the authentication phase. This means that if later on in the communication some parameters need to be changed (such as QoS parameters) there is now no standard way of renegotiating them if not triggering a new authentication. Depending on the authentication protocol used, on the network configuration and on the network load the authentication can last up to a few seconds in which the MS is disconnected, since it has no valid cryptographic keys active. Summing up, the renegotiation of a parameter is a costly operation that has not been addressed in the present version of the WiMAX Forum specification. In the future version (version 2.0) higher level protocols for this task will be designed.

4.4.2 DHCP

An MS that doesn't support the MIP protocol may have a static preconfigured IP address or may use DHCP protocol to obtain a new one at each handover. Even in the first case the handover can not be transparent to the IP layer since other network parameters need be updated. In particular on the client side the MS will have to update its route to a new default gateway or the IP address of a Domain Name Server (DNS) present in the new network. There is no other standard protocol to move these parameters to the MS if not using standard DHCP protocol, so that even a MS configured to have a static IP address will need to use DHCP protocol.

At the end of a successful authentication the ASN is still not aware if the MS is MIP-compliant or not. The distinction is based on the MS actions, if the first packet sent by the MS is a DHCP DISCOVER frame, the ASN will conclude that it is a DHCP compliant station, if the frame is a MIP registration request it will conclude that it is a MIP-compliant station. In the first case it will have to activate DHCP and possibly PMIP procedures as described in the next section.

DHCP protocol can be configured in a WiMAX network in two ways, using a DHCP Relay or a DHCP proxy. In the first case the correspondent entity into the ASN doesn't act as a DHCP server but forwards the request to a remote server. In the second case it's the ASN itself that answers to the DHCP REQUEST frame and assigns the IP address to the MS. Recall that if the session must be kept active the IP of the MS must be the same before and after the handover. If the DHCP server is acting as a Relay its role will be to forward the DHCP REQUEST to another server that must be the same one that the MS used in its first authentication. The DHCP Relay may not know the IP address of this server, so this information must be included in some way into the information that the ASN receives at the end of the authentication, that is, in a RAIDUS attribute. This configuration parameter could also be pre-configured in the ASN but such a policy is not usable in multi-operator networks, in which the client may be coming from a foreign network.

Similarly, if the ASN is configured to behave as a DHCP proxy it will need to answer to the DHCP DISCOVER with the same IP address that the MS was using before the handover. A DHCP DISCOVER could be carrying a specific option that is used by the client to ask to be assigned a specific address that is already using. For security reasons this decision can not be based on the IP that the MS requests for itself since the DHCP frames are unauthenticated and an MS could be trying to achieve someone else's address.

Again, the IP address to be assigned to the MS will be included into a specific RADIUS attribute included in the Access-Accept packet coming from the CSN.

4.4.3 Security Issues

In the depicted scenario an attacker that is able to intercept and modify the DHCP packets that are forwarded from a DHCP relay to a remote DHCP server could produce severe security problems. For instance, he could change the default gateway to perform man in the middle attacks, or modify the DNS server in order to redirect the MS to fake websites. This possibility is concrete in WiMAX networks since there is no a-priori trust between the DHCP relay and server. It can not be assumed that other security measures such as virtual private networks are deployed to secure those paths.

To address this issue the DHCP protocol has a security extension that allows to authenticate the frames from a relay to a server. This security extension is based on the knowledge of a symmetric shared key called DHCP-RK. From this pre-configured key specific ones are derived to secure each single session. To be able to support dynamic redirection of the DHCP requests the DHCP-RK key must be dynamically moved into the DHCP Relay and into the DHCP server when needed. Again this is achieved using RADIUS attributes, but in a more complicated manner. The first complication resides in the fact that RADIUS is used for MS authentication but the DHCP-RK key is not MS-related. The key is related to the couple (DHCP Relay, DHCP Server) and can be used to configure more than a MS. Moreover, the remote DHCP server is not in the logical path of the authentication, so it will not receive the DHCP-RK with an AAA message during the authentication. The WiMAX Forum specifies the following behaviour, as depicted in Figure 4.5:

1. During the authentication the CSN generates a random key and internally assigns that key to the DHCP server involved into this session.
2. Together with the Access-Accept packet the ASN receives a RADIUS attribute containing the DHCP server IP, the DHCP-RK, the lifetime of the key and a unique ID.
3. The DHCP relay in the ASN will use this key to secure the DHCP DISCOVER frame that is sent to the server. The DHCP server will receive a DHCP DISCOVER including the Auth. suboption based on a key with the specified ID.
4. If it owns the key it will use it to verify the signature, otherwise it will issue a RADIUS request to the RADIUS server asking for the key. The RADIUS server will in turn respond with the DHCP-RK.
5. The DHCP server can answer to the request and the DHCP procedure be completed.

A few issues deserve to be detailed. Step 1 is performed when the RADIUS server is using a DHCP server for the first time, or when the lifetime of a previous key has expired. Similarly, step 4 is performed by the server if it has not interacted with the Relay before, or if the lifetime of the key has expired. The DHCP server has to embed a RADIUS client to be able to satisfy the request and the RADIUS protocol is used in an unconventional way. Generally an authenticator has an embedded RADIUS client to transport authentication packets that come from a supplicant. The supplicant uses an username to authenticate against the RADIUS server. In this case there is no real

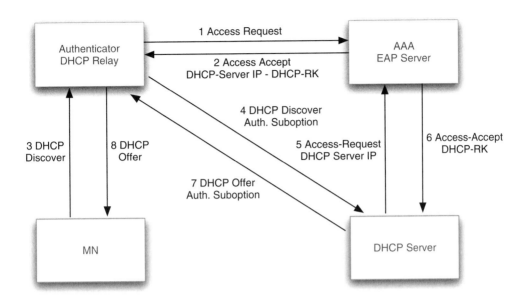

Figure 4.5   Key exchange for DHCP protocol key management.

user that has to perform any authentication, but the RADIUS frames carry requests that come directly from the a unique entity comprised of client and authenticator. This issue, together with the need for the server to store a key and redistribute it later on (recall that RADIUS is in principle a stateless protocol) has an impact on current RADIUS implementations.

4.4.4 Mobile IP Protocol

Mobile IP [18] is an Internet Engineering Task Force (IETF) standard communications protocol described in IETF RFC 3344 and IETF RFC 4721 whose main intent is to allow a user to have a permanent IP address while moving and changing his point of access to the network. This feature is highly valuable for all those connection oriented services that need a permanent connection during the user roaming through the network.

In the previous section we have explained how it is possible using DHCP to keep the same IP address during an handover. This is necessary to keep sessions active but it is not sufficient. The other end of the communication will receive IP packets but its replies will be routed to the home network of the node. The mobile IP protocol takes care of forwarding the packets from the home network to the visited network.

The Mobile IP protocol resolves the session persistence problem using two addresses for every terminal: an *home address* and a COA *Care-of-Address*. The home address is the IP address the node has got from his HA (Home Agent) on his home network; this IP address is the one that will remain fixed while roaming. The COA is the address given to the node by a FA (*Foreign Agent*), a mobility agent that has in charge the IP provisioning of the terminals attached to a V-NSP.

The way Mobile IP works can be briefly resumed as follows:

1. At the time of its first authentication an MS gets an IP address from the HA in his home network. Until it remains under the coverage of this network the datagrams to and from that node are routed simply using this address and the Mobile IP protocol is not used.
2. When a mobile terminal moves away from his home network to a different one it searches for a FA and receives a COA from this agent.
3. After the COA assignment, the MS starts a Mobile IP registration towards his HA using the Mobile IP RRQ *Registration Request* standard procedure. This registration has the intent to inform the HA about the current COA that should be used to reach the terminal on the visited network. To confirm that the registration has been received the HA sends back a RRP *Registration Reply*.
4. The path to and from the MS must be changed after the registration in the foreign domain. When the mobile terminal wants to communicate to a remote terminal it uses a direct route from its new network. When a correspondent node wants to communicate with the mobile node it has to use the *triangular routing* technique sending the IP datagram to the home address of this node. The HA will receive this packet, determine the current COA of the node from the last registration procedure, and send the packet to the right FA that finally forwards the datagram to the node.

From the security perspective Mobile IP defines an AEE *Authentication Enabling Extension* to both RRQ and RRP packets. The goal of this extension is to create secure channels between the MN and either a HA or a FA and between the FA and the HA. These three links are secured using three shared keys (MN-FA, MN-HA and FA-HA) that are generally preconfigured into the mobility agents. In WiMAX a dynamic way of generating MIP keys is introduced, that will be discussed in section 4.4.6 and 4.4.8.

The WiMAX Forum specifies that mobile user terminals IP provisioning problem should be addressed using the Mobile IP protocol to guarantee session persistence during subscriber mobility across multiple domains. Terminals are classified according to their mobile IP compliancy being divided into two groups, Proxy-MIP and Client-MIP terminals, PMIP and CMIP from now on.

4.4.5 PMIP

A Proxy-MIP terminal is a mobile terminal that completely lacks Mobile IP support but that still needs persistent connection during roaming and seamless handovers between multiple BSs like it happens for CMIP terminals. When a PMIP terminal accesses a foreign network it sends a DHCP DISCOVER in order to get an IP address of that network. To allow reverse routing the WiMAX Forum identifies a special entity called *PMIP Mobility Manager* that is in charge of handling the Mobile IP registration procedure for the user terminal. This entity intercepts the DHCP DISCOVER coming from the terminal and performs a MIP registration with the terminal HA. When this registration has been completed the PMIP mobility manager uses the DHCP protocol to assign the IP address to the mobile terminal. This procedure is transparent both for the user terminal (that simply use DHCP) and also for the HA that receives and handles normal MIP RRQ and RRP.

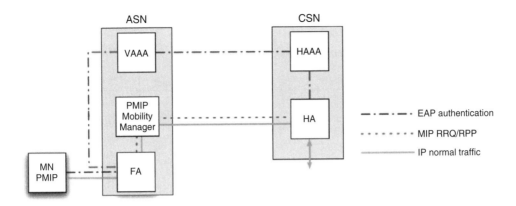

Figure 4.6   NRM and packet flows end-points, PMIP case.

The PMIP Mobility Manager receives also information about the authenticating client including its HoA *Home of Address* (address on the home network), the address of its HA and other additional details. Again, these parameters are moved to the authenticator using appropriate RADIUS attributes.

Figure 4.6 shows the different logic data-path used by MIP signalling traffic, AAA traffic and normal data traffic in a PMIP client connection scenario.

4.4.6   PMIP Security Considerations

The differences between PMIP and CMIP terminals have an impact also on the AAA authentication procedures and the security mechanism described by the WiMAX Forum. Figure 4.7 describes how the keying material needed to authenticate the MIP AEE is derived by the various entities involved.

When a terminal receives beaconing traffic that notifies the existence of a foreign network it wants to access, it will perform IEEE 802.1x authentication with its H-NSP. The RADIUS Access-Accept message carries also attributes used for MIP. Among those attributes we cited the HA address, but also an MSK key that has been produced by the authentication. The authenticator will store the MSK and share it with the co-located PMIP Mobility Manager. From that key will be generated the MIP-RK and a HA-RK key (see Figure 4.10). The PMIP Mobility Manager is now capable of generating MIP RRQ on behalf of the terminal using a MN-HA key (see Figure 4.10) to protect the request message with an AEE extension. Figure 4.7 shows what happens in the home network when a RRQ generated by the PMIP Mobility Manager is received by the HA. The HA lacks the MN-HA and HA-RK keys needed to authenticate and processes the MIP RRQ. These keys can be retrieved using a RADIUS Access-Request to the AAA. The AAA verifies if the request is correct and sends back the requested cryptographic material through a RADIUS Access Response. This procedure is basically the same with the one used by the DHCP server to retrieve the DHCP-RK. Note that with PMIP the MIP client and FA are co-located in the ASN, so that there is no need for an MN-FA key.

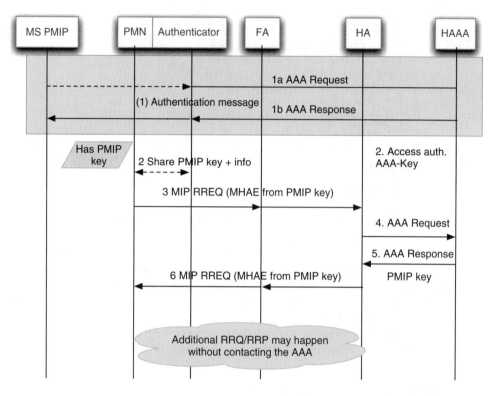

Figure 4.7   PMIP key generation and transfer – message sequence.

4.4.7   CMIP

A CMIP terminal is a mobile terminal RFC 3344 compliant with support for all the Mobile IP standard procedures as briefly described in 4.4.5. Figure 4.8 shows the different logic data-path used by MIP signalling traffic, AAA traffic and normal data traffic.

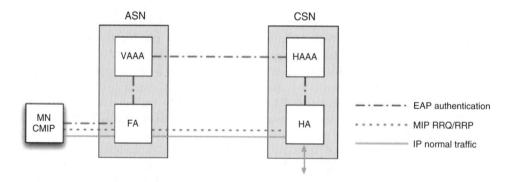

Figure 4.8   NRM and packet flows end-points, CMIP case.

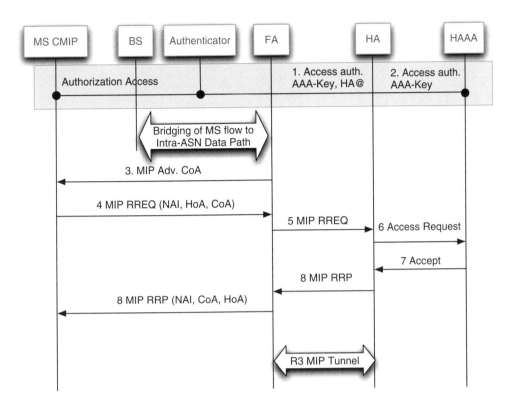

Figure 4.9 CMIP key generation and transfer – message sequence.

A FA is able to distinguish between PMIP and CMIP clients simply observing the way they start their connection setup. If a terminal is a PMIP client it uses a DHCP DISCOVER, in instead it is a CMIP it send a MIP RRQ.

4.4.8 CMIP Security Considerations

The security procedure is the same as used with PMIP but it includes the usage of a new key, a FA-RK to derive a MN-FA key to authenticate messages from the terminal to the FA. This key is generated and used into the MN and FA, the rest of the procedure is the same as the PMIP scenario.

The WiMAX Forum also suggests that for both CMIP and PMIP terminals the HA dynamic assignment should be supported. This procedure allows the user to configure the terminals without an HA IP address and to receive this parameter (and others) during the link layer authentication. More specifically during the EAP authentication phase the AAA inserts in the RADIUS Access-Accept packet some attributes about the authenticating MS configuration including HA IP address, and if needed a DHCP-Server address or a Framed-IP-Address[2]. The authenticator receives the RADIUS Access Accept packet and if the

---

[2] A Framed-IP-Address is an address statically bound to a MS. If this attribute is returned in the RADIUS Access Accept that address must be assigned to the MS without considering other policies.

terminal is a CMIP sends the HA configuration in the EAP success packet, otherwise if the authenticating MS is a PMIP the MS configuration attributes are shared with the PMIP Mobility Manager that handles the terminal MIP registration procedure (see Figure 4.9).

4.4.9 QoS

IEEE 802.16 defines a set of quality of service classes and the MAC layer parameters that define their quality. The WiMAX Forum addresses the management of QoS introducing the concept of *service flow*, each service flow is mapped to a set of quality parameters that are configurable from the network manager. When a node is authenticated it will be assigned a certain number of service flows of any type depending on its agreement with the operator. At least one service flow is always assigned to an MS, the *initial service flow*, that is used to move DHCP or MIP signalling. The creation and assignment of the following ones is dealt by logical entities that are described in the specification. Their interaction is quite complex and out of the scope of this chapter. The version 1.0 of the WiMAX Forum specification defines only fixed and pre-provisioned service flows. The final outcome is that the Access-Accept RADIUS packet will contain one more attribute for each service flow that must be activated for the MS. Each service flow is characterized by a set of parameters (i.e. tolerated jitter, maximum latency, etc.). The value of these fields can be moved using appropriate RADIUS attributes or they can be preconfigured in the ASN. In the most general case, the ASN will receive a set of RADIUS attributes that define the number of flows and the quality of each flow for each MS. Within the 1.0 specification, the ASN has a set of pre-configured service flows and it receives from the RADIUS server a directive to activate a subset for the current MS; service flows can not be activated or renegotiated once the MS has completed the authentication (this will be supported in the next revision).

Exactly as it happens with DHCP or MIP, the authenticator that receives the RADIUS attribute may be co-located with the BS or more likely included into the ASN-GW. In this second case it will use a custom protocol to move the QoS parameters into the BS to be associated to the corresponding MAC procedures.

4.4.10 A Complete Authentication Procedure

In Figure 4.10 is provided the complete key tree that is produced during an authentication, here we briefly summarize their functions and their usage.

After a successful EAP authentication two keys are generated into the MS and into the RADIUS server called MSK and EMSK. The first one is moved from the RADIUS server to the authenticator into the Access-Accept packet. It will be used by the MAC layer to generate the keys necessary to the cryptographic algorithms of IEEE 802.16. The second one is kept into the endpoints and will be used to generate MIP session keys: the MN-HA key and the FA-RK key. The MN-HA key is moved to the HA when the HA sends a RADIUS request to the server, the FA-RK is moved to the authenticator and will be used to generate the session key MN-FA in CMIP configuration[3] (see Figure 4.9).

---

[3] The MIP keys that are generated and reported in the figure are distinct for PMIP and CMIP and for IPv4 and IPv6. For simplicity we didn't focus on this detail in the text

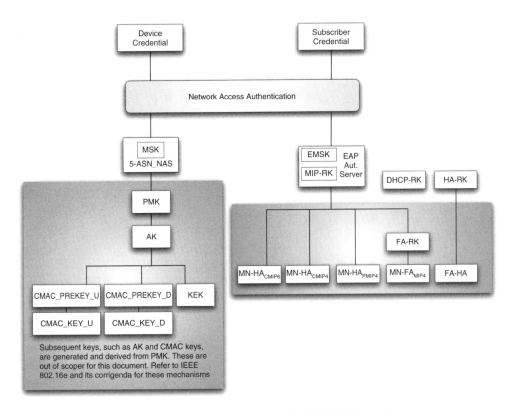

Figure 4.10 The global key tree of a WiMAX network.

The path from FA to HA and from DHCP relay to DHCP server does not depend on the authentication of a single MS, so that the keys used to secure those paths are independent by the MSK (Master Session Key) or EMSK. They are generated by the RADIUS server and later on moved into the ASN with the Access-Accept frame and to the HA and DHCP Server with custom requests generated directly by RADIUS clients embedded in their software.

In Figure 4.11 is shown the scheme of a complete EAP-TTLS MSCHAPv2 (Microsoft Challenge-Handshake Authentication Protocol version 2) authentication in the case of a MS that does not support MIP negotiations and triggers DHCP and PMIP procedures. The single steps have been illustrated so far and do not need further descriptions. What is evident is that in the case of CSN-Anchored mobility the procedure to perform an handover is very complex and consequently very time-consuming. Note that some of the packet exchanges are realized between hosts that do not reside in the same LAN network so that the whole procedure could need several seconds to be completed. This is the price to pay to have extreme flexibility and to support very dynamic configurations.

4.5   Further Reading

The WiMAX Forum website is a valuable resource for white papers and case studies (see [19]). In particular, industries participating in WiMAX Forum activities have released

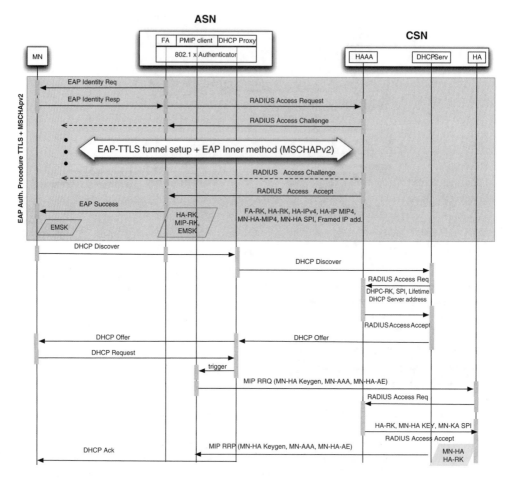

Figure 4.11   A complete WiMAX authentication and parameters exchange.

white papers on the security aspects of WiMAX networks. We suggest [20] and [21] for
AAA management. In [2] the WiMAX forum released a complete overview of WiMAX
deployment with a descriptive style and references to more detailed specifications. For a
more in-depth view of Mobile IP features including security we recommend [22] where
real-world use cases are explained and related to Cisco implementations as well as [23]
that is focused on the interactions of AAA protocols with mobility management.

4.6   Summary

WiMAX is considered, among the emerging standards, the one that is more likely to be
able to address all the bandwidth demands of the new-coming high-speed mobile voice
and data services. For this reason there are a lot of concerns about its effective security.
Even if most of the of security standards used for WiMAX are widely-trusted protocols a
collection of secure technologies does not, in itself, constitute a secure end-to-end network.

Consequently, WiMAX presents a range of security design and integration challenges that are worth to be discussed.

From a security point of view the central element of a WiMAX network is the AAA Server, a functional entity responsible of essentially three main functions: authorization, authentication and accounting. The *de facto* standard for AAA servers is the RADIUS protocol even if IEEE 802.16e foresees also the support for DIAMETER, a more advanced protocol that is expected to update RADIUS. Together with Privacy and Key Management Protocol Version 2 (PKMv2) as a key management protocol both device and user authentications can be performed relying on an AAA server that stores users and devices credentials.

Within the WiMAX architecture the AAA server is used extensively also to supply user related information through specific RADIUS attributes. The AAA server has the role of moving to the ASN the network configuration and QoS parameters related to the specific user. Those attributes are included in the last Access-Accept RADIUS message that is delivered to the 802.1x authenticator in the ASN.

The rationale behind this architecture is to concentrate all the connection related parameters in a reliable central storage point (the AAA server at the H-NSP) and to define primitives and procedures that allow to move those parameters in a secure way, in order to ensure seamless handovers and reliable QoS to mobile users. A MS can freely choose its point of attachment to the network whether the selected BS belongs to its provider or not and all the QoS and IP addressing parameters are moved to the ASN that handles that point of attachment during the EAP authentication phase.

References

[1] http://www.wimaxforum.org. Last visited 04/20/2009.

[2] WiMAX Forum, WiMAX Forum Network Architecture. Stage 2: Architecture Tenets, Reference Model and Reference Points. V. 1.2, WiMAX Forum Std., 2009.

[3] WiMAX Forum, WiMAX Forum Network Architecture. Stage 3: Detailed Protocols and Procedures, WiMAX Forum Std., 2009.

[4] IEEE, IEEE Standard for Local and metropolitan area networks Part 16: Air Interface for Fixed Broadband Wireless Access Systems, IEEE Computer Society and the IEEE Microwave Theory and Techniques Society Std., 2007.

[5] IEEE, Part 16: Air Interface for Fixed and Mobile Broadband Wireless Access Systems Amendment 2: Physical and Medium Access Control Layers for Combined Fixed and Mobile Operation in Licensed Bands, IEEE Computer Society and the IEEE Microwave Theory and Techniques Society Std., 2005.

[6] IEEE, Part 16: Air Interface for Fixed and Mobile Broadband Wireless Access Systems – Amendment 3: Management Plane Procedures and Services, IEEE Computer Society and the IEEE Microwave Theory and Techniques Society Std., 2007.

[7] IETF, *RFC 2865 – Remote Authentication Dial In User Service (RADIUS)*, IETF Internet Engineering Task Force – Network Working Group Std., 2000.

[8] IETF, *RFC 3588 – Diameter Base Protocol*, IETF Internet Engineering Task Force – Network Working Group Std., 2003.

[9] D. Johnston and J. Walker, 'Overview of ieee 802.16 security,' *Security and Privacy, IEEE* **2**(3): 40–8, May–June 2004.

[10] R. Fantacci, L. Maccari, T. Pecorella and F. Frosali, 'Analysis of secure handover for ieee 802.1x-based wireless ad hoc networks' *Wireless Communications, IEEE* **14**(5): 21–9, October 2007.

[11] IEEE, 802.1X – Port Based Network Access Control, IEEE Computer Society and the IEEE Microwave Theory and Techniques Society Std., 2001.

[12] IETF, RFC 3748 – Extensible Authentication Protocol (EAP), IETF Internet Engineering Task Force – Network Working Group Std., 2004.

[13] IETF, RFC 5246 – The Transport Layer Security (TLS) Protocol Version 1.2, IETF Internet Engineering Task Force – Network Working Group Std., 2008.

[14] IETF, RFC 5281 – Extensible Authentication Protocol Tunneled Transport Layer Security Authenticated Protocol Version 0 (EAP-TTLSv0), IETF Internet Engineering Task Force – Network Working Group Std., 2008.

[15] IETF, RFC 4187 – Extensible Authentication Protocol Method for 3rd Generation Authentication and Key Agreement (EAP-AKA), IETF Internet Engineering Task Force – Network Working Group Std., 2006.

[16] IETF, RFC 2560 – X.509 Internet Public Key Infrastructure – Online Certificate Status Protocol – OCSP, IETF Internet Engineering Task Force – Network Working Group Std., 1999.

[17] IETF, RFC 2131 – Dynamic Host Configuration Protocol (DHCP), IETF Internet Engineering Task Force – Network Working Group Std., 1997.

[18] IETF, *RFC 3344 – IP Mobility Support for IPv4 (Mobile IP)*, IETF Internet Engineering Task Force – Network Working Group Std., 2002.

[19] http://www.wimaxforum.org/resources/documents/marketing/whitepapers. Last visited 04/20/2009.

[20] B. Systems, 'Is your aaa up to the wimax challenge?', 2007.

[21] M. Inc., 'WiMAX security for real-world network service provider deployments', 2007.

[22] S. Raab and M. Chandra, *Mobile IP technology and applications* Cisco Press, 2005.

[23] M. N. Madjid Nakhjiri, *AAA and Network Security for Mobile Access: Radius, Diameter, EAP, PKI and IP Mobility* John Wiley & Sons, 2005.

Part C
Quality of Service

5

Cross-Layer End-to-End QoS Architecture: The Milestone of WiMAX

Floriano De Rango, Andrea Malfitano and Salvatore Marano
DEIS Department, University of Calabria, Italy

5.1 Introduction

WiMAX appears to be one of the most promising technologies of recent years and especially in its most recent version which specifies user mobility support and allows wireless multimedia services to be provided to a wide area. The term 'wide' has many advantages, both economic and practical. For example, consider the possibility of installing a WiMAX wireless infrastructure in a low density population area such as a small town or rural area, instead of creating a new fixed and wired infrastructure from scratch.

The real source of success will be to provide services that meet the user's needs, thus making the technology ever closer to the simplicity and quality that each generic user expects. To characterize the services provided with QoS (Quality of Service), the IEEE 802.16 protocol describes various mechanisms related to network topology. This chapter focuses on each aspect of these mechanisms in cross-layer QoS architecture, highlighting both PMP and Mesh topology aspects and their differences. Each kind of topology presents not only a different way to obtain QoS, but also other important aspects such as bandwidth allocation scheduling and call admission control algorithms, which are left to vendor implementation. These deficiencies, with reference to the MAC and PHY layers, and other important issues, are explained in this chapter, focusing on the importance of an end-to-end QoS concept. The chapter ends with a special section that outlines the challenges for WiMAX QoS. In particular, the future of QoS in the IP

WiMAX Security and Quality of Service: An End-to-End Perspective   Edited by Seok-Yee Tang,
Peter Müller and Hamid Sharif
© 2010 John Wiley & Sons, Ltd

world for multimedia applications is examined and the extension of interesting theories to QoS architecture is explored. A future WiMAX architecture with a mobile mesh node is illustrated in the final section of the work.

5.2   QoS Definitions

In order to understand the concept of QoS, we must examine it from different points of view. The user's point of view is the most abstract: a generic user tags a service as a qualitative satisfactory service if it meets his abstract qualitative expectations. For example, with regard to a video on demand service, the user will be satisfied if the video is displayed with no visible slowdown problems or distorted images. The user does not know the details about video transmission and network protocols but he is satisfied if the video is received in the correct way. The user and the requested application are the most obvious aspects of a communication scenario, but they are not the only aspects. The components that have a role are:

- user;
- application;
- network;
- protocol.

Each of these components provides different points of view and, excluding the user, each component is related to various technical aspects and provides a concrete definition of quality. The particular application defines its expectations in terms of well-defined constraints; the network affects the scenario with its particular architecture and physical constraints; and finally the protocol contributes with the definition of 'rules' and mechanisms available to ensure that the required quality levels can be achieved. An example of QoS constraints may be the following:

- end-to-end delay: the average packet delay from source to destination;
- delay jitter: end-to-end delay variation of packets;
- packet error rate (PER)/Bit error rate (BER): percentage value of packets/ bits lost;
- throughput: the percentage of sent packets correctly received at destination.

Other parameters can be defined according to the type of service considered and the network architecture.

5.3   QoS Mechanisms Offered by IEEE 802.16

Each protocol defines its particular mechanisms and algorithms so as to achieve high levels of QoS. It is important to note and to bear in mind that QoS is not related to a particular layer of the protocol stack. A protocol does not define only one layer but illustrates the behaviour of a series of layers. The coordination of all of these layers contributes to establishing the performance of the network based on a particular protocol. In this way QoS can be seen as a concept related to various layers which constitute the entire protocol stack.

WiMAX (Worldwide Interoperability for Microwave Access) is the commercial name which is used to indicate devices compatible with the IEEE 802.16 protocol. The IEEE 802.16 protocol [1] defines guidelines for providing wireless broadband services in a wide area. The protocol defines the physical layer (PHY), the medium access control (MAC) layer and also each management aspect; the PHY layer defines five air interfaces and the MAC layer allows itself to be interfaced with the IP (Internet Protocol) or ATM (Asynchronous Transfer Mode) upper layer protocol. In IEEE 802.16e [2], user mobility is also introduced.

The advent of the current state of protocol was developed by publication of a series of subsequent amendments [3]–[13]. This process has produced four different network architectures as specified by the IEEE 802.16 protocol; other new kinds of architecture are being considered.

In the rest of the chapter QoS mechanisms and algorithms provided by 802.16 are described and a series of protocol deficiencies will be discussed and analysed. For each kind of shortfall we are going to introduce some proposed solutions discussed in the literature. To describe the cross-layer end-to-end QoS issues related to WiMAX, we will consider a simple daily scenario: a user that tries to connect his personal computer with a remote server to obtain a particular service. This example will be taken up several times later in the chapter (see Figure 5.1). Using this example we are going to describe all of the QoS concepts related to an end-to-end communication based on the 802.16 protocol. The use of an example will allow us to illustrate complex concepts in a practical way. Obviously, the QoS issue will be seen in a cross-layer and end-to-end fashion.

5.3.1 Cross-Layer QoS Architecture

Figure 5.1 depicts a simple scenario in which a generic user tries to use a service offered by a remote server; this service may be a streaming video, VoIP call or other types of service. In Figure 5.1 the path from user to server is represented and the first step of the path is a WiMAX connection; that is the user device is connected to an SS (Subscriber Station), which in turn is connected to a BS (Base Station). In [1] protocols define an SS

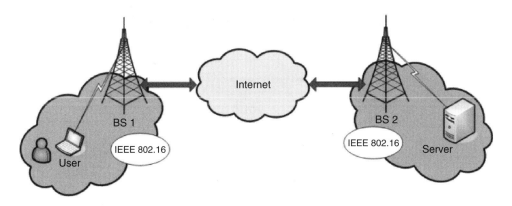

Figure 5.1 WiMAX scenario: user – server connection.

not only as a 'device' capable of providing service to only one user but also to a set of users in a building. In a user device, in a server and in each entity which plays a role in the communication the protocol stack defining the communication rules is implemented. For greater clarity and simplicity the user device and the server are considered as SSs. In our particular case, the MAC and PHY layers are defined by the IEEE 802.16 protocol, while the internet cloud can be constituted by each kind of technology. When a user application requires, for example, the transmission of a video, the higher protocol layer on the server side signals to the lower layer the need to send data with well-defined QoS constraints. Thus, the MAC and PHY layers of the 802.16 protocol begin a process to reach the common goal that is to send the real time data respecting the QoS constraints; PHY and MAC share a set of parameters passing information from one layer to the others, in this way they initialize a cross-layer paradigm that exploits inter-relations between network layers to improve efficiency and quality. The intrinsic nature of the algorithm which must guarantee a high QoS level can introduce a cross-layer aspect, this is because the QoS is not only related to one layer, but also involves all the stack protocol layers. Thus, to guarantee compliance with QoS constraints there is a need to create a collaboration between the various layers.

Considering the inherent characteristics of wireless communication and networking, the traditional layered network architecture can be considered to be inadequate to realise the full potential of wireless networks. Cross-layer design approaches can be used to improve and optimize the network performance by breaking the layer boundaries and passing information explicitly from one layer to the others. In Figure 5.2 a set of possible collaborations between pairs of protocol layers is illustrated. The black arrows of Figure 5.2 represent the information flows exchanged in the cross-layer architecture. The

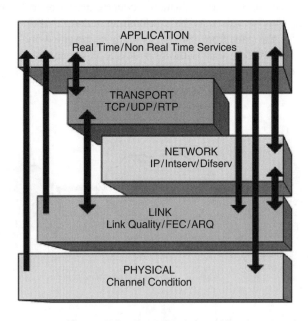

Figure 5.2 Cross-Layer example.

term cross-layer therefore does not refer only to a specific pair of layers but may also be associated with any level of the protocol stack.

5.3.2 MAC Layer Point of View

In this subsection, we will describe the MAC protocol layer, highlighting the mechanisms offered by the protocol to guarantee QoS. The MAC layer supports two different operative modes: point to multipoint (PMP) and mesh mode. In Figure 5.3 the two modes are depicted: the user segment operates in PMP mode and the BS server side operates in mesh mode.

These two operative modes provide different mechanisms that affect the MAC's behaviour in a different way. Overall, in a WiMAX network, we can identify three different entities: the base station (BS), the subscriber station (SS) and the mobile subscriber station (MSS). In PMP mode the BS has a central role and is the only entity that can manage the bandwidth allocation and schedule the bandwidth requests received from SSs. Only BS-SS links are admitted and the BS is the only station which can broadcast data and control messages without coordinating the transmissions or asking permission from other stations. In mesh mode there is a novelty: the capability to create direct links between SSs, thus an SS which goes behind the BS coverage area can also reach BS using a multi-hop route constituted by SS-SS links. Bandwidth scheduling can take place in a distributed or centralized manner; in the latter case the BS maintains a central role, while in distributed scheduling all entities are defined as 'mesh nodes'; we can distinguish the BS only because it is the gateway to reach 'the rest of the world'. MSS, finally, represents the mobile user and can only create a connection with BS in a PMP mode.

Figure 5.4 shows the protocol stack as defined by the IEEE 802.16 protocol. It is also possible to note the three sublayers which make up the MAC layer.

The Convergence Sublayer (CS), in PMP mode, performs the task of classification of SDU (Service Data Unit), mapping the various SDUs from higher layers in the proper

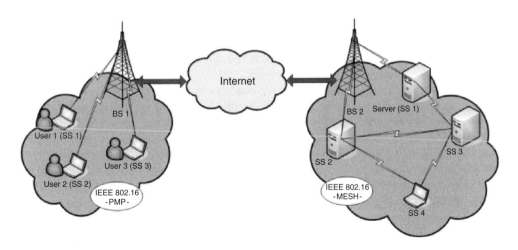

Figure 5.3   IEEE 802.16 PMP and mesh mode.

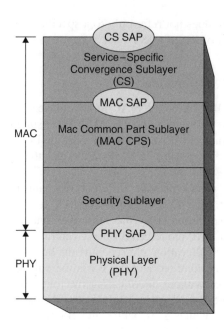

Figure 5.4 IEEE 802.16 protocol stack.

connection. To do this mapping in an effective way, a set of classifiers is defined and each SDU must be submitted to it. This task is related only to the PMP mode, because in PMP mode it is possible to create more than one connection between BS and SS. Each connection is related to a particular QoS level and this is true both in data and management connections.

The classifiers must respect an application order and if an SDU cannot be mapped in any type of connection it will be discarded. Another special feature of the CS sublayer is the suppression of parts of the PDU (Protocol Data Unit) header that are repeated in the packet, which can be rebuilt once the destination is reached. The central sublevel is the Common Part Sublayer (CPS). It performs the typical tasks of the medium access control layer, thus providing algorithms to ensure efficient coordination between the various entities that require transmission bandwidth allocation.

The task of the Privacy sublayer is to provide a strong protection to service providers against theft of service. Moreover, it protects the data flow from unauthorized access by strengthening the encryption of the flows passing through the network. The Privacy sublayer provides a client/server management protocol authentication key where the BS (server) monitors the key distribution to the clients. There are two main components in the Privacy sublayer: an encapsulation protocol for the encryption of data packets that are sent over the network and a key management protocol (Privacy Key Management: PKM).

The MAC PDU is presented in Figure 5.5. The MAC PDU consists of a fixed length header equal to 6 bytes, a payload that may contain one or more SDU or SDU fragments or may even be absent and, finally, a CRC (Cyclic Redundancy Check) field may be present. Also, the PDU payload can transport data or management messages. The fields of generic MAC header depicted in Figure 5.5, are the following:

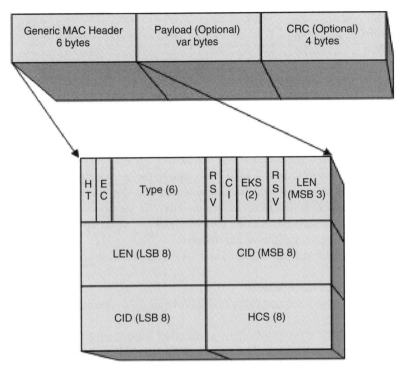

Figure 5.5   MAC PDU and generic MAC header.

- HT: header type, distinguishes between a generic header and bandwidth request header used in PMP mode;
- EC: encryption control, used to indicate if the payload is encrypted;
- Type: indicates if the payload contains one or more subheaders;
- Rsv: not used;
- CI: indicates if the payload ends with a CRC portion;
- EKS: indicates the payload encryption key;
- LEN: length of the PDU;
- CID: is the connection identifier;
- HCS: header check sequence.

In this section, we have described briefly the MAC layer and the structure of MAC PDU. In the following subsections we will introduce the QoS mechanisms used by protocol in PMP and mesh mode.

5.3.3 Offering QoS in PMP Mode

Considering the example introduced in section 5.2, if the network operates in PMP mode, when the server receives the request for a video on demand service, to guarantee a well defined QoS level, it may rely on three important concepts. These concepts are:

- connection;
- service class;
- service flow.

In Figure 5.6 these concepts are laid out in a simple way and it is possible to see how the basic mechanisms work together. In Figure 5.6 the four ellipses represent the components of protocol which are not defined in the IEEE 802.16 standard.

Compliance with the service constraints and consequent user satisfaction is related to these three important mechanisms which are not supported in mesh mode. In order to satisfy the client request, the server needs bandwidth. To simplify, we consider the server as an SS. The SS must send a bandwidth request for a particular connection to BS. The MAC protocol is strongly connection oriented and the connection is identified by a 16 bit CID (Connection Identifier). Consequently, the service in our example must be mapped on a well-defined connection. This connection can group each data flow which is characterized by the same QoS requirements.

In PMP mode the bandwidth request from an SS can be made in three different ways:

- using a bandwidth request header;
- by making a piggyback request using the Poll Me bit (PM) present in grant management subheader; or
- depending on service class in which the application is mapped, an SS can be polled periodically to verify any requests. The polling can be made in a broadcast or unicast way.

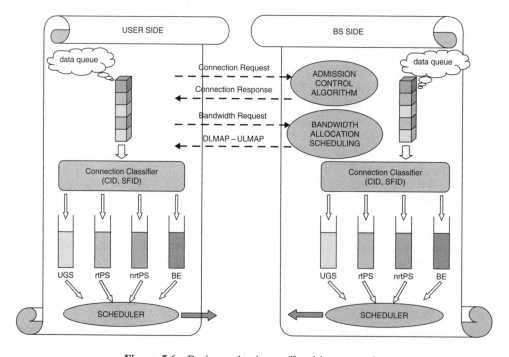

Figure 5.6 Basic mechanisms offered by protocol.

In the subsequent sections, we will see how the SS applications can be mapped in the various service classes. After the mapping process and request process the BS can assign bandwidth to SS in two different ways:

- The grant is assigned to the single connection; this mode is defined as grant per connection (GPC).
- The BS gives the SS an amount of bandwidth corresponding to all the connection requests of the SS; this mode is defined as grant per SS (GPSS).

5.3.3.1 Connection Oriented MAC

As already stated, the MAC IEEE 802.16 is strongly connection oriented. Everything happens in the context of a connection and each SS can activate more than one connection with the BS. The concept of connection is related to the service class to which the connection belongs, thus the SDU received from upper levels are mapped on the correct connection based on correspondence with the criteria of the classifiers. When MAC receives new SDU with qualitative constraints that do not correspond to any existing connection, the MAC of the SS can request the activation of a new connection and the allocation of bandwidth needed to meet the quality requirements. Hence a connection is a dynamic object which can be created, modified and deleted. In this way, the video on demand service of our example will be mapped on a service class and on the connection corresponding to it. The SDU of applications with specific quality constraints can only be mapped on a specific connection. This concept of well-defined connection-service class binding and consequently connection-QoS binding is evident not only in data flow but also in management message flow. The management messages are exchanged between BS and SS on three different management connections:

- **Basic management connection**: used to exchange short urgent messages.
- **Primary management connection**: carrying longer and delay tolerant messages.
- **Secondary management connection**: used to carry standard-based delay tolerant messages.

Each of these connections is characterized by different QoS levels.

5.3.3.2 Service Classes

In PMP mode a required service can be mapped onto one of the following service classes:

- **Unsolicited Grant Service (UGS)**: in this service class the real time applications generating fixed-size packets on a periodic basis are mapped. An example of these applications is the VoIP call; voice data flow generates periodic and fixed-size packets. Once the BS grants bandwidth for a connection belonging to this service class, the SS maintain this allocation for the lifetime of the connection. The SS can use the UGS connection to require bandwidth for other types of service using the *poll-me* bit of the grant management subheader. In this way SS activating the polling process allows for bandwidth optimization and in this way there is no waste of bandwidth to send a request.

- **real-time Polling Service (rtPS)**: in the real-time polling service class real-time applications which send variable-size packets on a periodic basis are mapped; an example of this service is the MPEG video. In this case, the SS verifying the existence of this connection; if false, can create a new connection for this service and then request bandwidth for it. The bandwidth request, for this service class, can be made with unicast polling mechanisms, that is, the BS grants the SS a lot of bandwidth in which the SS can send its request. Another way to request bandwidth is to insert the request in a data PDU; this is for optimized bandwidth management. This process is called a piggybacking request.
- **not-real-time Polling Service (nrtPS)**: this service class is suitable for non real-time applications such as FTP traffic. The services mapped on this class generate variable-size packets on a regular basis. Like the previous class, to request bandwidth for this service the SS can use piggybacking and polling mechanisms. In this case unicast and broadcast polling can be used, the difference being that broadcast polling can cause collisions and the collision can be resolved by an exponential backoff process.
- **Best Effort (BE)**: the Best Effort data service can map data flows without QoS constraints. To request bandwidth for this class, the SS can use all the previously described mechanisms, but normally the BS scheduler neglects the BE connections. In fact a BE connection receives the remaining amount of bandwidth after the BS serves all of the most important service classes.

In this way, the service of our example and any other service is classified qualitatively according to class membership and mapped on a connection. To complete the treatment we must introduce the quantitative aspect of QoS; the service flow is one of the quantitative aspects expressed in the 802.16 protocol.

5.3.3.3 Service Flow

The QoS concept in the IEEE 802.16 protocol is related to connection, service classes and service flow concepts. The latter is a bidirectional data flow which provides well-defined levels of QoS ensuring compliance with restrictions imposed by applications. Thus, considering our example where the server has to send a video to the user, the communication (in the first step of path: server-BS), is allowed if there is an active service flow associated with the relevant connection.

There is a set of attributes which characterize a service flow:

- Service Flow ID (SFID): identifies the service flow in a univocal way in the network;
- CID: identifies the associated connection;
- ProvisionedQoSParamSet: a set of QoS parameters which are used in network management scope;
- AdmittedQoSParamSet: a set of QoS parameters for which the BS has reserved resources;
- ActiveQoSParamSet: this set of QoS parameters defines the active state of current service flow;
- AuthorizationModule: a logical function of BS; it denies or authorizes each modification to the QoS parameters of service flow.

A service flow can be characterized by three different phases:

- Provisioned: a provisioned service flow is a created but not activated service flow; it has no bandwidth associated with it;
- Admitted: it has bandwidth reserved for it, this service flow does not transmit PDU until it becomes activated;
- Activated: it has bandwidth and a set of parameters associated with it; this service flow is ready to send data.

An admitted service flow is related to a well-defined CID; thus, each service flow is mapped onto a connection and each connection will belong to one of the scheduling data services offered by the protocol on the basis of the required QoS. Only a service flow tagged as activated may forward packets. A service flow, and consequently a connection CID, has a set of parameters associated with it:

- MSR: Maximum Sustained Rate;
- MRR: Minimum Reserved Rate;
- maximum-latency;
- maximum jitter;
- priority.

These parameters are useful to define the QoS for a particular service. Each connection, when it receives a grant, can try to transmit with a higher data rate, but the MSR serves to limit the connection and the MRR associated with it, acting as the 'guaranteed amount of bandwidth'. A service flow, as a connection, has a dynamic behaviour and can be created, modified or deleted using a set of primitive management functions provided by the protocol.

A number of service flows can share a set of parameters; thus, the protocol introduces the concept of service classes or service class names. A service class is an optional object that may be implemented at the BS and is a set of QoS parameters identified by a service class name (SCN).

Concluding the discussion about QoS in PMP mode, we can say that the service class classifies a service qualitatively and maps the service on a connection. The service flow provides a set of parameters that can guarantee respect for QoS constraints on the connection.

5.3.4 QoS Introduction in Mesh Mode

After the description of PMP mode, in this subsection we will introduce the mesh mode. In the subsequent subsection, QoS support in mesh mode will be described in detail. In this operating mode new terms unknown to the PMP mode are introduced:

- neighbour: a neighbour of a node is a node one step away from it;
- neighbourhood: the set of all neighbouring nodes;
- extended neighbourhood: it contains, in addition to the neighbourhood node, the neighbours of a neighbourhood.

The basic principle of transmission coordination in an 802.16 mesh network is that no one node, including the BS node, can transmit on its own initiative without coordinating its transmission within its extended neighbourhood. In a network operating in mesh mode there are two different ways to manage bandwidth allocation: centralized or distributed mode. The distributed scheduling, in turn, can be either coordinated or uncoordinated. In the distributed coordinated scheduling, all mesh nodes have to coordinate their transmissions in their extended neighbourhood and they use the same channel to transmit the scheduling information. The uncoordinated distributed scheduling allows fast setup communications between two nodes and does not cause collisions with the messages and the traffic of coordinated scheduling. Both modes of distributed scheduling, coordinated or not, use a three-way-handshake protocol. Considering the example introduced in previous sections, if the server is a mesh node of the network, in order to transmit data it has to send a *request* to the next hop mesh node, indicating the requested number of minislots it needs. The destination node replies with the *grant* message that is acknowledged with a *grant* copy by the server.

This is obviously true in the distributed mode; whereas in centralized scheduling the functionalities are different. In centralized scheduling a reachability tree must be considered; it is a logical structure built with a subset of all the network links. The server in the example, identified with a mesh node, has to collect the bandwidth requests received from the child mesh nodes and sends all the requests to its father mesh node in the tree. In this way all the requests will be received by BS and subsequently the BS can diffuse all the bandwidth grants along the reachability tree.

One of the advantages of mesh mode is the capability to reach the destination using a multi-hop path. If the server does not fall within the BS coverage area it can still reach the destination.

Other concepts related to the PHY protocol layer will be discussed here, because there are some mechanisms related to frame composition which influence the QoS protocol performances.

The mesh mode only supports the TDD (Time Division Duplexing) mode; thus, the frame is divided into two parts:

- control subframe;
- data subframe.

The data subframe is organized into a fixed number of minislots and each bandwidth grant can be constituted by a set of minislots. Two different types of control subframes are admitted by the protocol:

- Network control subframe: is exploited to transport network control messages used by nodes to acquire network synchronization and network configuration properties.
- Scheduling control subframe: is used to collect bandwidth requests and to send grant messages (centralized and/or distributed).

A frame does not contain both types of control subframes, but a well-defined number of frames containing scheduling control subframes are present between two frames with a network control subframe. The *Scheduling Frame* parameter, broadcast in the control messages, defines the periodicity of the two types of subframe.

The dispatch of the control and scheduling messages occurs in a collision free manner and to guarantee it, each mesh node inserts two parameters useful for calculating the next transmission in the messages:

- *xmt holdoff exponent*;
- *next xmt mx*.

Each mesh node, during message forwarding, has to calculate its next transmission and express it in the form of an interval using the two above-mentioned parameters. In practice, the node does not inform the neighbours about its next transmission instant, but it sends an interval into which the next transmission falls; this interval is defined by the following constraints:

$$\text{next xmt time} > 2^{\text{xmt holdoff exponent}} * \text{next xmt mx} \tag{5.1}$$

$$\text{next xmt time} <= 2^{\text{xmt holdoff exponent}} *(\text{next xmt mx} + 1) \tag{5.2}$$

Between one transmission and the next, a node must wait in silence for a time interval, defined as a *holdoff interval*, equal to:

$$\text{xmt holdoff time} = 2^{(\text{xmt holdoff exponent} + 4)} \tag{5.3}$$

5.3.5   QoS Application on Packet by Packet Basis

In PMP mode the IEEE 802.16 protocol defines various mechanisms useful for providing QoS. In mesh mode, the protocol does not have the following concepts: connection, service class and service flow. The protocol guidelines, inherently the QoS issues in mesh mode, are that the quality of service must be guaranteed, in the link context, *packet by packet*. The mesh node has the task of managing the received packets in such a way as to guarantee compliance with the application QoS constraints. In order to realize and satisfy the QoS constraints, a mesh node can use a set of PDU header fields defined by the protocol. The generic header of a MAC PDU contains a 16-bit CID field. In the mesh mode, in the case where the payload is constituted by a MAC management message, the CID field is split into two parts. The first portion, of eight bit length, is the logical network identifier and the second portion contains the link identifier. If the MAC PDU contains a data payload, the first 8-bit portion of the CID is redistributed over four fields used to implement the QoS policies. The four fields are:

- Type: indicates if PDU transports a management message or an IP datagram; it is two bits long. Two configurations of this field are reserved for future developments;
- Reliability: this field indicates the number of admitted retransmissions for the current MAC PDU. Two possible values are: no chance of retransmission or a maximum number of retransmissions equal to four;
- Priority/Class: indicates the priorities associated with the membership class of the message;
- Drop Precedence: a message with a high drop precedence value has a high probability of being eliminated in the case of network congestion.

The presence of these four fields, and especially the last two, provides the protocol with the capability to create service classes in which the various user applications can be mapped. Using these fields, when the server in the example receives SDU related to video application, a kind of classification giving higher priority to PDU with little end-to-end delay constraints can be created. This mapping mechanism, as well as other algorithms, does not fall within the scope of the protocol and this obviously can be an advantage because each WiMAX device implementer can search for the best optimized solution.

In mesh mode there are other parameters, related to frame structure, that affect the final QoS provided by networks. The influence of PHY and MAC parameters on provided QoS and the desire to optimize a performance means that researchers prefer a *cross-layer* architecture.

In the previous subsection we described the alternation between network control and scheduling control subframe; this being regulated by the *Scheduling Frame* parameter. Obviously there is no best choice for the parameter value although a compromise value is needed; this is because a low parameter value causes a small number of scheduling control subframes between two network control subframes and this leads to a slowdown in bandwidth request-grant procedure. Instead a high value for the *Scheduling Frame* parameter can lead to a slowdown in the network configuration process. Both of the two negative cases influence the guaranteed QoS levels.

Another 'interesting' parameter is the *holdoff exponent* of equation (5.3). A high setting value for this parameter tends to increment the silence time for a mesh node. On the contrary, a small value means that mesh nodes try to transmit with high frequency; thus, a node, which needs to submit its request, may have some difficulties in finding free transmission opportunities. The literature presents some solutions to optimize the setting parameters, such as those listed in [14]–[16].

5.3.6 PHY Layer Point of View

The IEEE 802.16 protocol defines guidelines for the MAC and PHY layers. The PHY protocol layer describes five different air interfaces which establish transmission guidelines in licensed and license-exempt bands. The air interfaces are listed here:

- WirelessMAN-SC: the 10–66 GHz band; supports TDD and FDD (Frequency Division Duplexing) techniques;
- WirelessMAN-SCa: defined as a licensed band below 11 GHz; supports TDD and FDD techniques;
- WirelessMAN-OFDM: defined as a licensed band below 11 GHz; supports TDD and FDD techniques;
- WirelessMAN-OFDMA: defined as a licensed band below 11 GHz; supports TDD and FDD techniques;
- WirelessMAN-HUMAN: defined as a license-exempt band below 11 GHz; supports TDD technique only.

Looking at this list, one may note that the protocol supports both single carrier modulation and multi carrier modulation techniques such as OFDM (Orthogonal Frequency Division Multiplexing). The modulation techniques used in the transmission chain start from the robust and simple BPSK (Binary Phase Shift Keying) to the more complex 256 QAM (Quadrature Amplitude Modulation). The OFDM is a multicarrier transmission technique which allows for good performance in particular scenarios affected by multipath fading or the Doppler effect. Another important concept provided by the protocol is MIMO (Multiple Input Multiple Output). This transmission technology uses more than one antenna in both transmission and reception devices. The MIMO functionality is based on the concept of spatial multiplexing. The data flow on the transmission side is divided into a series of subflows, each flow being modulated and transmitted in the same band using parallel transmission chains. On the receiver side the subflows are received on different antennas, elaborated and recomposed in a unique data flow. The MIMO technique, considering a fixed BER (Bit Error Rate) value is able to increment the throughput value; vice versa with a fixed value of throughput where it is possible to improve BER behaviour.

All of these mechanisms have been included in the PHY protocol in order to give the protocol the capability of providing a high QoS level. The encapsulation of mechanisms which influence the QoS, in both the PHY and MAC layers, confirms what we already anticipated: that the best QoS architecture is a *cross- layer* architecture in which both layers collaborate to achieve a common goal. In the following subsection, another instrument useful for achieving QoS is illustrated: adaptive coding and modulation.

5.3.7 ACM: Adaptive Coding and Modulation

All of the air interfaces of a PHY layer can use adaptive coding and modulation (ACM). The ACM concept is very simple: it is the capability to select, instant by instant, the modulation which has the higher efficiency, consistent with the condition of propagation and interference on the link between transmitter and receiver mesh node. The modulation efficiency is equal to the number of bits that can be transmitted in a modulation symbol. The ACM allows the modulation and the coding to be modified dynamically, according to the Signal to Noise ratio (S/N). In order to see this concept in a practical way we can consider our example of the server. If we identify, as we have already seen in previous section, the server with a mesh node or an SS, then the server can decide to modify *instant by instant* the modulation technique so as to obtain the best performance. In the case of a high value of S/N the server can choose a robust technique such as BPSK; instead; if the S/N value is small, then the most efficient modulation can be selected: for example the 64 or 256 QAM. In addition to modulation, encoding may also be changed dynamically from the server, adding more redundant bits in the case of high S/N values. With regard to a base station, the dependence of modulation and encoding on the S/N value creates the following situation:

- In transmission with the SS stations situated on the borders of the coverage area, the BS must use more robust modulation and coding;
- In transmission with SS stations situated near the center of the coverage area, the BS must select more efficient modulation and coding. These concepts are shown in Figure 5.7 where one may note that the BS uses a different modulation in each concentric area. The concentric areas are depicted in relation to *BS- SS* distance.
- The best selection of SS can be different from the best choice of BS, owing to the different transmission power.

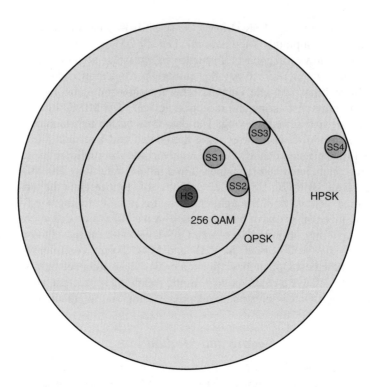

Figure 5.7   Selected modulation as function of BS-SS distance.

The adaptive modulation and coding combines, within a cross-layer architecture, to design an excellent solution in order to provide QoS in a wide range of scenarios.

5.3.8   Mobility Support in IEEE 802.16

The IEEE 802.16e version of the protocol introduces user mobility; it is the latest version of the IEEE 802.16 protocol and was devised at the end of 2005. This version is designed for the use of WiMAX in scenarios where users have mobility. The IEEE 802.16e document introduces the following features:

- support for adaptive antennae;
- handover management;
- roaming management.

The mobile device is identified as MSS and its admitted speed range is related to vehicular speed. With the addition of mobility capability, new problems are introduced. The advent of mobility allows interesting scenarios to be imagined and designed; each user, in this way, can rely on a broadband wireless connection everywhere and under every condition. Obviously, the addition of this capability can also have negative aspects and introduces new challenges; to guarantee respect for QoS constraints new considerations are needed. We start by considering the PHY layer of the protocol stack as user mobility introduces or increases the negative effects of various impairment phenomena. Considering a pair of transmitters and receivers, when there is a relative speed value greater than zero, the signal transmission is affected by the Doppler effect which causes a frequency shift in the modulated signal. The multipath fading effect is also amplified in a mobile scenario.

In the literature there are many works investigating these topics. For example, we can consider [17] where a performance evaluation of a vehicle to vehicle channel is given. The IEEE 802.16e protocol does not only work with fixed infrastructure, that is, with a fixed BS, but it can also be applied to a particular scenario in which an HAP (High Altitude Platform) [18] has the role of BS. HAPs are a new breed of airships or planes that will operate in the stratosphere at an altitude of 17–22 km. In recent years, the great potential of these network elements has captured the interest of academic and industrial bodies [19, 20, 21]. They have the potential of being deployed quickly and do not need a complex infrastructure, such as in the case of the terrestrial network. The applicability of the 802.16 protocol to this new scenario with an analysis of BER and PER performance can be found in [22], [23] and [24]. Another interesting case is the application to the railway [25] or airport [26] scenario. The focus of the researcher is to guarantee QoS everywhere by always using the best solution and by considering the integration of 802.16 with other technologies.

It is not only impairment effects that influence QoS, but also another phenomenon which appears in a simple scenario. Imagine that the user in our example has a device equipped with the 802.16e protocol and wants to see a video located in the server; the server starts to transmit the video and the data passes through the internet cloud until it reaches base station BS1 with which the user is connected. The user starts to move and reaches the border of the BS1 coverage area; thus the signal strength of BS1 is very low and the user device also receives the signal from another BS: BS2. Owing to the attenuation of the received signal, the protocol activates a process to leave BS1 and connect the device with BS2. This process, instigated in order to change the BS, is called handoff. The handoff process can be classified into two types:

- soft handoff: the connection with the old BS is closed only after the new connection is activated;
- hard handoff: in a first step the old connection is terminated and in a second step the new connection is activated.

In Figure 5.8 the deactivation of the connection with BS1 and the new connection for a mobile user are shown. QoS is influenced and diminished by the handoff process; it introduces latency and other new problems.

The IEEE 802.16 protocol only defines layers one and two of the protocol stack but the handoff process also involves layers three and four; this is increasingly evident in heterogeneous network architecture. When the user device has to change the BS and the new BS is in the same IP subnet of the previous BS, then setting changing occurs only in layers one and two; but if the new BS falls in a new IP subnet then an IP configuration process is needed. The trigger for the handoff process is not only the received signal strength but can also be QoS requests. If the old BS is not able to provide the requested QoS level then the device can decide to scan channels in order to find a new BS.

In this challenge the cross-layer characterization of the handoff process is clear but the need to consider an end-to-end approach to QoS is also increasingly clear. The handoff

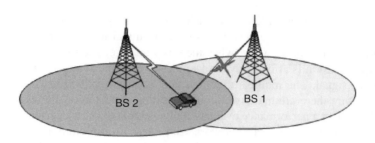

Figure 5.8 Handoff process.

problem in the IEEE 802.16e scenario is treated extensively in the literature. To enrich our knowledge of this issue, we recommend reading, for example, [27]–[31].

5.4 What is Missing in the WiMAX Features?

To use the term 'missing' when referring to a standardized and widespread protocol seems strange and misplaced, however the 'missing parts' of the protocol represent an advantage. The protocol defines the guidelines for PHY and MAC layers but various important aspects are neglected deliberately by protocol designers. With reference to the PHY layer, for example, the protocol illustrates all of the supported techniques but does not explain specific algorithms to optimize, for example, energy saving or algorithms to reach high throughput level using adaptive modulation or to perform other tasks. In the MAC layer there are other opportunities for improvement, for example, the definition of a Call Admission Control (CAC) algorithm is essential in order not to diminish the QoS of admitted calls, accepting all the received requests. Scheduling algorithms are also not present in the protocol. As we have already stated, the absence of these kinds of algorithms is voluntary, the purpose of protocol designers is to give a certain degree of freedom to device producers; in this way, it is possible to construct solutions characterized by cross-layer architecture or optimized algorithms which take into account the desired point of view.

In the following section, a set of algorithms for improving MAC and PHY behaviour will be introduced and commented on, in particular distinguishing between the PMP and mesh mode.

5.4.1 Absences in the MAC Layer

The absence of a scheduling algorithm and a call admission control algorithm can be observed in the MAC layer. Some proposals to improve the protocol can be found in the literature for both the PMP and mesh mode. For a practical approach to explore the two types of algorithms we introduce briefly some of the solutions provided by the literature.

When the server has data to send, it builds a message containing a request for a new connection and the amount of minislots that it needs. When the BS in PMP mode or a generic node in mesh mode receives the message it evaluates the admission of the new connection and then how much bandwidth to grant to the server. The first decision process is called call admission control and this is a long-term decision because the admission of a new request affects the QoS performance of existing connections for a period equal to the lifetime of the connection. The second decision (amount of bandwidth) influences the actual amount of available bandwidth and consequently is a short-term decision because bandwidth availability can change over the period.

5.4.2   Scheduling Algorithm

The scheduling algorithm decides when a station can forward data and how much bandwidth is granted from the BS or from another mesh node. The protocol defines QoS mechanisms, such as the service classes in PMP mode or the PDU classification by CID in mesh mode; the scheduler must 'exploit' these mechanisms in an efficient way to provide optimized bandwidth management.

5.4.2.1   PMP Mode

PMP has many mechanisms available to provide QoS, namely service flow, service class and connection concepts. Each PDU of a particular application is mapped on a well-defined service class characterized by the parameter of a service flow. A scheduling algorithm, in PMP mode as in mesh mode, has to guarantee to each service class the forwarding of PDU stored in data queues, respecting the *QoS constraints*. Another important concept is *fairness* among connections of the same class, however it is also important to avoid *starvation* of a service class with lower QoS constraints.

The protocol describes the concepts summarized previously but does not present a scheduling algorithm. Thus, in order to realize a realistic, complete and functioning BS, we can select a scheduling algorithm from amongst those presented in the literature. We mention some solutions as examples. To delineate BS behaviour it is necessary to establish how the BS makes the grants to the various service classes. In [32] there is a mathematical modelling of bandwidth allocation scheme. In this model, which also involves a mathematical queue model, the amount of grants is established as a function of network traffic. Usually, in each solution, as in this one, the UGS class is considered in a privileged manner, and rtPS and nrtPS receive grants in a dynamic way. In [33] BS behaviour is enriched with a mathematical analysis of the contention mechanism and from [34] we may understand how to handle voice traffic.

In this way, by collecting various ideas from the literature it is possible to improve the 802.16 protocol. It is important to note that the research is always in progress; the protocol was standardized five years ago but researchers are still stimulated by the possibility of constructing ever more efficient solutions. Evidence of this research activity is to be found in [35] and [36]. The former selects a set of scheduling algorithms such as:

- Earliest Deadline First (EDF);
- Weighted Round Robin (WRR);

and illustrates a performance comparison with the aim of proving that no scheduler is able to achieve the best performance under every condition. The conclusion is that the best ideal solution is a cross-layer scheme integrating scheduling, routing and call admission control algorithms. The latter, [36], analyzing an hybrid algorithm, appears to confirm this theory.

5.4.2.2 Mesh Mode

In the previous subsection we have seen the difficulties in creating an efficient BS scheduler in PMP mode; instead, in the following subsections we will analyze the mesh mode
case. The mesh mode introduces further complications owing to the nature of network
architecture. In this case the bandwidth allocation can be made in a centralized or distributed way, but in both cases the network topology is more complicated than with PMP.
In the case of PMP the network has a star topology, while in mesh mode, also using
centralized scheduling, the topology can be built randomly. The scheduling algorithm
therefore proves to be more complicated; it must also take into account transmission
coordination problems such as hidden terminal or exposed terminal. The scheduler has to
unite the concepts of coordination in the extended neighborhood and bandwidth allocation.

5.4.2.3 Distributed Algorithm

If a mesh network, in order to provide bandwidth allocation, utilizes distributed scheduling, it does not have a privileged entity taking the role of coordinator; the coordination
takes place in a distributed manner between mesh nodes belonging to the extended neighborhood. In Figure 5.9 the extended neighborhood for several mesh nodes is represented.
The scheduling algorithm, implemented in each mesh node, has to operate in order to
comply with the QoS constraints. An important aspect is that each node must provide a
classification of received PDU because the service class concept is not present. Summa-
rizing the focus of scheduling algorithm in distributed modes we have:

- to respect PDU QoS constraints applying a sort of PDU classification: QoS must be
  applied packet by packet;
- to decide the instance of transmission trying to avoid collision in the two-hop neighborhood.

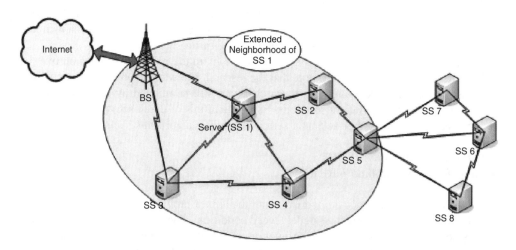

Figure 5.9 Extended neighborhood of server node.

In the literature there is little related to the distributed scheduler. For example, [15] examines the scheduler as delineated by the protocol and highlights that scheduler performance is influenced by network topology and by mesh node number. The authors create a stochastic model of the scheduler and prove the conclusions by the results of a simulation. In the literature, [37] is also interesting; it confirms the protocol gaps and focuses attention on the important problem of collided transmissions.

5.4.2.4 Centralized Algorithm

In centralized scheduling the focus of the scheduler is the same as in the distributed case, but its task is facilitated by the presence of a coordinator. In the previous case the scheduler can be imagined as distributed on each mesh node; in this case however, the only coordinator is the BS. Each mesh node has to send its request toward the BS and consequently the BS replies, spreading the bandwidth grants. To allow message forwarding, a coverage tree is considered. As we may understand from the previous concept, in the centralized way the coverage tree and the routing are important topics. In Figure 5.10 we see an example of a coverage tree for the server mesh node.

This is emphasized in [38], where the authors explain how the final performance is influenced by choice of routing. Interference also contributes to deterioration in network performance; as illustrated in [39], the interference is represented by concurrent transmissions in the two-hop neighborhood; the solution aims at minimizing this interference. Many works could be mentioned here, for example the scheduling issue related to VoIP traffic considered in [14] or multimedia traffic in [40]. A long list of related work can deflect attention from the primary task which is to understand the difference between the basic concepts of the scheduler in PMP and mesh mode. To conclude the section we summarize some ideas:

- Both PMP and mesh scheduler are related to bandwidth allocation.

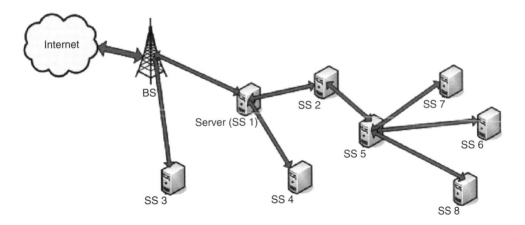

Figure 5.10 Example of coverage tree.

- The decisions in PMP mode and centralized mesh mode are taken only by BS, while in mesh distributed mode the decision algorithm is distributed.
- The topology in mesh mode influences scheduler performance and in this topic the major problem is the interference in the two-hop neighborhood.
- In mesh mode the optimized solution is a cross-layer solution integrating routing, scheduling and call admission control concepts.

5.4.3   Call Admission Control Algorithm

With regard to a WiMAX scenario: when a user makes a request to a remote server to obtain a particular service, the server must execute the following steps:

- to identify the next node in order to forward data;
- to forward to the next node a request for a new connection.

When the next node (BS in PMP mode or a generic WiMAX node in mesh mode) receives the request, it has to decide whether or not to admit the new call. The call admission control has to make this decision on the basis of network condition and traffic behaviour. This decision is very important because it influences not only the QoS for the new connection, but also the QoS of existing connections. In fact, in accordance with call admission policy the entity which takes the decision can opt to decrease the allocations already granted in order to leave room for the new call.

Any positive or negative decision, however, will cause a reduction in the overall network throughput.

5.4.3.1   PMP Mode

In PMP mode the entity which takes decisions is always the base station. The BS, in a centralized way, can organize and decide each new call admission or preemption of an old connection with lower priority. The BS can take its decision following a specially selected set of QoS parameters. Some of the parameters that can be considered are:

- end-to-end delay;
- throughput;
- number of refused calls.

In the literature the lack of a call admission control algorithm is met with various types of solutions. In [41] a call admission control algorithm is presented in which the concept of service preemption is introduced and the admission decision is based on traffic class and bandwidth utilization of each traffic class. Each traffic class has a bandwidth portion reserved for it and can also preempt the lower priority admitted services. In [42], however, the focus is on reducing the polling delay and a cost-based function for admission decision is proposed. Another simple idea to create a CAC algorithm is to exploit the service class concepts defined by the protocol and the mechanism of bandwidth reservation for traffic class, such as that realized in [43]. Another interesting solution, like a cross-layer scheme, can be found in [44], but it is important to keep in mind that in order to implement a new CAC algorithm, the algorithm task has to be established and, then, each QoS support mechanism provided by the protocol must be taken into account.

5.4.3.2 Mesh Mode

In mesh mode the call admission control process has the same tasks as the PMP mode. The main difference is that the entity which takes decisions is not the base station but a generic mesh node. Another important difference is that the PMP mode has a set of mechanisms to provide QoS that are not present in mesh mode. Thus, to realize the CAC algorithm the first step is to create a sort of classification of requested services; after this, it is possible to create, for example, a simple CAC scheme using the priority values of services.

The topics related to mesh mode are relatively recent and very little research has been published. We may note that a network operating in mesh mode is very similar to an ad-hoc network, thus it is possible to find inspiration in literature related to the ad-hoc topic. However, interesting work is present in the literature. For example in [45] the concept of preemption in mesh mode is presented; three different traffic classes with an assigned priority value are implemented and the decision regarding admission is based on the concept that all the bandwidth can be divided among the three classes. Also of interest is [46], where the admission problem for VoIP traffic is treated and the CAC is coordinated with an end-to-end bandwidth reservation process. Another idea, in [47], is based on the threshold mechanism. The requests with higher priority, if there are sufficient free minislots, are always admitted whereas low priority requests are refused in the case of congestion; the congestion is verified by a bandwidth utilization threshold. If this last value is less than a fixed threshold then the network is not congested. These mechanisms and the mapping of the service classes concept onto a set of values of the flags contained in the CID field are implemented exclusively so as to provide well-defined QoS levels.

Taking into account the server example, if we identify the server with a mesh node, the server must also implement the CAC algorithm in its protocol stack. This is false when considering a network in PMP mode.

5.4.4 PHY Layer Improvements

The PHY protocol layer, as introduced in the previous section, delineates the use of five different air interfaces with the support of different modulation techniques; it allows the use of a single carrier or multicarrier modulation such as OFDM. The possibility of using such different techniques allows different scenarios characterized by different impairment effects to be addressed. The protocol also defines guidelines to implement an adaptive coding and modulation algorithm. In the following subsection we introduce the topic of ACM algorithms.

5.4.5 QoS Based ACM Algorithm

The adaptive coding and modulation algorithm allows a change in modulation and coding in a dynamic way in order to obtain, under each condition, the best system performance.

The IEEE 802.16 protocol constructs the basis for this kind of algorithm to provide a differentiation in QoS mechanisms. The best way to exploit the ACM capabilities is to design a cross-layer scheme in which both the MAC and PHY layers can communicate and collaborate in order to allow high QoS levels. For example, a very interesting ACM algorithm can select modulation and coding, taking into account the packet error rate or the packet size. Thus, it is possible to play with a set of MAC and PHY parameters. Let us consider a mobile node with an actual speed value; the data transmission of this node will be affected by a well-defined value of Doppler effect and multipath fading and to contrast these effects, the node can select a robust modulation and low PDU size values. When the node becomes fixed and its speed value is equal to zero then the node choice falls in a less robust but more efficient modulation and can also use a high PDU size value. In this way MAC and PHY can 'collaborate' in order to estimate channel behaviour and consequently to modify not only the transmission parameters, but also bandwidth allocation. This last affirmation can be explained in this way: a node has a number of data minislots assigned to it; this amount of minislots is characterized by a well-defined bandwidth value, which in turn is related to modulation efficiency. Thus, a change in modulation causes a change in modulation efficiency and the amount of bandwidth.

In the literature there is much which lays the foundations for interesting cross-layer algorithms. In [48] and [49] the authors present channel behaviour modelling for 802.16 scenarios, in which Markov chain concepts are applied. Alternatively, in [50] the authors enrich the channel analysis with an algorithm useful for QoS tasks. Other interesting works are [51], [52] and [53].

All of these examples are introduced in order to illustrate the trends in research and to show in a practical manner how the problems are solved.

5.5 Future Challenges

We have considered the classic challenges related to QoS, explained how to build cross-layer solutions and highlighted the things to keep in mind in constructing cross-layer solutions. From this point, however, we are going to consider the most complex challenges which arise when looking to the future. One of these concerns end-to-end communication in an IP world, which represents the application of the IP protocol above 802.16 MAC. In the Wireless WAN (WWAN) context or other scenarios where WiMAX has to be integrated with other existing technologies, end-to-end QoS support will become an interesting challenge.

It is very interesting to take a look at new ways of addressing the cited problems, using theories and tools which belong to other branches of research.

5.5.1 End-to-End QoS in the IP World

In previous sections, we have cited the server example in which a user connected to a BS tries to get a service such as video on demand. A server connected to a BS, which is separated from the user by an internet cloud, has to provide the video. This internet cloud can be an IP world, and more specifically an IP based network. The communication that takes place between the user and the server is identified as point-to-point or as end-to-end communication. In communication of this type it is not only the two WiMAX segments (server and user) that are involved, but also everything that is encountered along the route from source to destination. The quality of the final service will be influenced by all of the mechanisms of the various technologies that come into play and also by the elements of discontinuity represented by the points where different technologies are applied.

Considering, for example, the IPv6 protocol which represents the future of the IP world, we can say that some problems arise in a world in which IP meets WiMAX. The neighbour discovery of IPv6 supports various functions for the interaction between nodes of a single subnet, such as address resolution. IPv6 was designed with no ties, independent from the underlying levels of protocol; however, to optimize the operations it requires the presence of multi-cast technology. Protocol 802.16 in PMP mode does not support bidirectional multicast and therefore appears inappropriate for the IPv6 features listed below:

- address resolution;
- router discovery;
- auto configuration;
- duplicated address detection.

The 802.16 protocol also enables encapsulation of an IP datagram in a MAC PDU, but does not define how it should be made. The PMP mode has a reluctance with regard to IPv6 features, unlike the mesh mode where the SS has the chance to distribute the messages in a multi-cast way; the problem remains with the MSS because it must connect itself with the BS in point-to-point mode. If we continue the analysis of the world built on the integration between IP and WiMAX, we can distinguish between two different access modes: fixed and mobile. The first is a valid alternative to XDSL connections, while the second creates support for new mobile data services, voice and multimedia traffic. Diversification can also be made in the mobile access; an IPv6 link can be defined as a shared IPv6 prefix link model or as a point-to-point. In the first case, with reference to Figure 5.11, a subnet consists of a single AR (Access Router) interface and multiple SS units.

In the second case, a subnet consists of only single AR, BS and MSS; so each connection is treated individually. Obviously, each scenario and each type of link introduced can influence the quality of service of an end-to-end connection. There are many problems to solve, for example, the need to make a mapping between the service classes of IEEE 802.16 and the IP concept of DiffServ. This is to answer the question: how can we continue to guarantee the QoS to a service that starts from a WiMAX node under a specific technology and is then mapped onto another service over a different technology? This is obviously true in the case of transition from different levels of the same protocol stack such as IP and MAC, as well as peer-level MAC 802.11 and MAC 802.16. It is necessary to note that every discontinuity that represents a transition from one protocol to another introduces the need to reconsider the quality of service. These concepts emphasize the importance of cross-layer solutions and this becomes more and more evident if we consider the possible handoff problems typical of mobile terminals. We recall, to this purpose, that, when a mobile terminal in its motion always remains within the same subnet then the handoff should not make an information update. However, when switching from one subnet to another, then reconfiguration at different levels of the protocol stack is required. This of course affects the quality of service offered, because these procedures lead to the introduction of delays. The concepts explained here are merely a brief introduction to

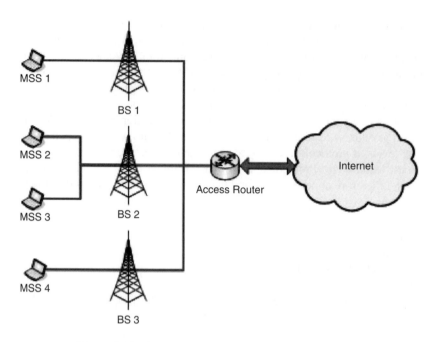

Figure 5.11   Example of shared IPv6 prefix link model.

those that influence the QoS in an IP-WiMAX world; in order to study these topics, we recommend reading [54]–[58].

5.5.2   New Ways to Resolve the WiMAX QoS Problem: Two Interesting Examples

We have already established the presence of voluntary gaps identifiable as algorithms of various kinds of integration functions, interworking and cross-layer capabilities. These gaps can be bridged by designing ad hoc solutions that always consider the ultimate goal of ensuring quality of service. In the literature there is a great number of works which approach these problems in an original way. To create even more optimized and original solutions, the researcher analyzes the classical telecommunication problems, with the help of interesting theories developed for applications in other disciplines. Many of these theories allow one to achieve fascinating but also elegant and efficient solutions. Two examples of these theories are the interesting game theory and fuzzy logic. In the following subsections we want to introduce some basic concepts of these two theories, with a brief description of related works found in the literature.

5.5.3   Game Theory in the WiMAX Scenario

The game theory was founded primarily for economic applications, dealing with situations of strategic interaction between decision makers which are intelligent and rational. The term 'intelligent' means that decision makers understand the situation they are faced by and are able to reason logically. 'Rational' means that preferences are consistent with the final outcomes of the decision-making process and are intended to maximize these preferences. The maximization is carried out by trying to achieve a certain gain, which is expressed through a utility function.

A game is therefore an iteration between multiple entities. An initial classification of games is as follows:

• Cooperative Games: studying the formation of coalitions with binding agreements that may be of benefit to the individual components.
• Non-cooperative Games: the theory of non-cooperative games is concerned with mechanisms of individual decisions, based on individual reasoning, in the absence of mandatory alliances.

Game theory deals with situations where there are at least two entities that interact according to the rules of the game. As with roulette, it deals with situations where there are at least two entities that interact according to the rules of the game. A game is classifiable as a game with complete information if the rules of the game and the utility functions of all players are common knowledge amongst all players. Also, a game is over when each player has a finite number of moves available and the game ends after a finite number of moves. The basic assumption of game theory is that all players behave rationally, that is, no player chooses an action if he has at his disposal another choice that allows him to get better results, regardless of the behaviour of his adversaries.

The formal description of a non-cooperative game takes two forms:

• extended form: the description of the game is made with a tree structure;
• normal form or strategic form: specifies the number of players, the space of strategies and the utility function (payoff) of each player.

A strategy is a complete plan of actions related to the individual player. There are several ways to identify solutions to a game, one is the identification of the Nash equilibrium. The Nash equilibrium is a pair of strategies, and this pair is a point of equilibrium if each player, assuming the strategy set of the other, has no interest in deviating from the selected strategy. Therefore, the Nash equilibrium is the point where if the player departs from it, the utility obtained will certainly be lower. The application of this theory to networks issues is becoming more and more widespread.

The game theory is used successfully for protocol design and optimization of radio resource management. In a resource management game, each player acts rationally to achieve his goals. Now we will see how to implement an algorithm for allocation management and call admission control using game theory. Consider a BS operating in point to multipoint mode (PMP), when a new connection arrives at the BS, the BS will check the type of class membership of the connection and invoke the algorithms for call admission control and bandwidth allocation. Suppose you want to regulate access connections for services such rtPS and nrtPS. When one of these new connections reaches the BS, there is a conflict because each of the existing connections will maximize their quality of service, but at the same time, the BS will steal a portion of their bandwidth to be allocated to the new connection. The concept of utility of each connection that is defined in terms of throughput and average delay then comes into play. The game can be identified as a non-cooperative game, the players are the various connections, the strategy is the choice of the amount of bandwidth to be offered to the new connection and the payoff is the total utility of existing connections and the new connection. We can determine a solution of the game through the identification of the Nash equilibrium. Consequently, the resolution indicates the amount of bandwidth to be offered, which is the solution identified by the allocation algorithm. From here it is easy to implement the call admission control algorithm: if the solution meets all of the quality constraints of the old and new connections then the new connection can be accepted or alternatively, should be rejected. An example of the procedure described here can be found in [59]. There it is possible to find the procedure described previously in detail. The solutions derived using game theory are interesting and are compared, by some authors, to results obtained by static and adaptive allocation and call admission control schemes. Using the adaptive scheme, the obtained call block probability is minimal, but the delay and throughput requirements cannot be satisfied when the traffic load becomes higher. Instead, the bandwidth grants obtained with the static scheme can meet the requirements of throughput and delay. As expected, when the call arrival rate of new connections has an increasing trend, the call block probability increases for both the static scheme and game theory solution. This means that in order to provide QoS guarantees to the links that already exist, some new connections can be blocked. However, it is very important to note that the framework for the bandwidth grants, based on game theory, may provide a slighter delay for the rtPS connections and a greater throughput for nrtPS connections, and that block probability is similar to that of the static scheme.

There are many other works [60–63] which apply game theory to network issues; the presence of these works proofs that the application of game theory to introduced issue is valid.

5.5.4   Fuzzy Logic: What Idea to Guarantee QoS?

Fuzzy logic allows us to resolve the issues we have introduced in a different way. Fuzzy logic was born in computer science and more specifically, arises in the application of artificial intelligence. It certainly looks different from the classical Boolean logic in which the only allowed values are true or false, identified as 0 and 1. Fuzzy logic introduces the concept of degree of belonging to a set. In fact, while for classical logic an element can belong to a given set or its complement in an exclusive way, in fuzzy logic an element can belong to both sets, and the concept of membership is accompanied by a degree of ownership that can take values between 0 and 1. This level of membership can be interpreted as the degree of truth of 'the element belongs to'. For clarification, here is an example. To define the state of congestion of a network we consider a threshold that represents the current use of the bandwidth and we say: for a utilization value greater than 70 % the network is congested, while if the current bandwidth utilization is less than or equal to 70 % the network is not congested. At this point, according to classical logic, a network with utilization value equal to 71 % is defined as congested while a network that has utilization of 70 % is not congested even if the true condition of the two networks is almost identical. The advantage of fuzzy logic is the ability to express a concept in shades and express the degree of membership. To use fuzzy logic in the mechanism of a

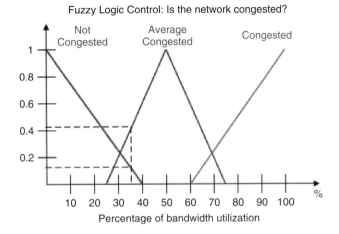

Figure 5.12 Example of fuzzy logic based control.

congestion threshold it is necessary to divide the range of values into multiple functions. These functions can have the most disparate shapes (triangles, trapezoids, Gaussian, etc.). In this way you can associate a real number with each membership function by a process called *fuzzyfication* and therefore you can have predicates that are partially true. As an example, according to possible selected functions, we could say that a network with utilization coefficient equal to 71 % can be defined as congested to the degree of 0.4 and not congested to a degree equal to 0.6. A more sophisticated representation, by fuzzy logic, for the congested concepts, can be seen in Figure 5.12, where three different cases are considered:

- not congested;
- averagely congested;
- congested.

It is interesting to note that we can obtain a degree of membership for each case. Thus, a network with bandwidth utilization value equal to 35 % is not congested with a degree of truth equal to 0.12 and is averagely congested with a degree of truth equal to 0.415.

Fuzzy logic is therefore another aid in dealing with network problems; it has an expressive power greater than classical logic, or rather the opportunity to express predicates in a way that is very close to the real world. The possibility of using fuzzy logic for the construction of logical control seems clear. A practical use similar to the mechanism described in the congestion example can be found in [64], or in [65] and [66].

For example, we can consider the new ideas presented in [64]. Here is introduced fuzzy logic to design handoff mechanisms which take into account the provided quality of service. The main issues are the vertical and horizontal handoff: the horizontal handoff is the classic process of migration from one base station to another within the same segment of the protocol; for clarity, this is the transition from one BS to another BS. The vertical handoff, however, happens when the mobile terminal switches from one protocol to another. In the case under consideration, the authors address the transition from a WiMAX BS to a WiFi (Wireless Fidelity) access point. This type of handoff is not triggered by the received signal strength, but it is based on other metrics. The authors propose, in addition to a classical handoff scheme, a bandwidth scheme based on the adaptive fuzzy logic algorithm. The considered scenario is the following: there are a number of WiMAX cells in which there are WiFi cells. When the mobile user is in a WMAN cell, it tries to verify through threshold mechanisms whether it is possible to migrate to the WLAN cell, otherwise it checks the option to switch to another WMAN cell. If the mobile unit is in a WLAN cell it checks first whether it is possible to switch to another WLAN cell, otherwise it verifies the possibility of transiting to a WMAN cell.

At this point, how does fuzzy logic come into play? The authors enrich a classical handoff algorithm with the presence of a fuzzy logic module that receives two variables as input: the user speed and the traffic of the most promising WLAN. This module is characterized by two membership functions of fuzzy variable speed and traffic. The output of the fuzzy logic rules are the thresholds that are used in the handoff algorithm.

This logic is then used to implement the additional module; the authors, using classical logic, were not able to describe the different speed and traffic levels. In addition, to obtain the migration of the mobile user toward the most promising network, the algorithm can reduce the number of unnecessary handoffs. An unnecessary handoff happens if the received signal strength (RSS) is evaluated in an incorrect window size. The proposed algorithm considers a dynamic window size, depending on the user speed. Thus, the unnecessary handoff is eliminated; it was eliminated in cases when the user is slow and the size of the window is great.

5.5.5 *Designing* Mobility – Mesh *WiMAX*

The IEEE 802.16 protocol presents very interesting features in both PMP and mesh mode. These two operating modes offer the opportunity to create a broad spectrum of scenarios that can meet the most disparate needs. Despite this, the research seeks to go beyond the limits imposed by the protocol and analyze, from different points of view, the opportunity to further enrich the WiMAX by having mobile stations that support the mesh mode. Currently, MSS stations are restricted by the presence of a BS, the MSS is able to establish a connection only with a BS and it is not possible to create links with other MSS or SS.

5.5.6 *How to Extend QoS Mechanisms*

The introduction of the mesh mode in mobile WiMAX makes it necessary to introduce changes at every level of the protocol stack in order to ensure a well-defined QoS and the proper working of the network. WiMAX has still not become part of everyday communication and in the literature there are many proposals to change the protocol so as to support the mesh mode in the mobile version of WiMAX. Without doubt, to ensure high throughput it is appropriate to introduce a physical level using the SOFDMA-1024 (Scalable Orthogonal Frequency Division Multiple Access with 1024 sub carriers) technique that would increase the efficiency of available bandwidth. Interesting changes also affect the MAC level; to support mobile mesh it is proposed to limit the scheduling to the distributed manner only, in order to remove the bottleneck formed by the presence of BS. The use of the distributed mode prevents the introduction of delays typical of the centralized mode; in this latter mode each bandwidth request and each bandwidth grant should always pass through the BS. The mechanism of registration of a new station and especially the mechanism of mesh election as presented in the mesh protocol should be changed. Currently the mechanism of mesh election is based on coordination in the two-hop neighborhood of a node, where the neighborhood is a fixed set of nodes. This constraint must be removed in the mobile mesh. A cross-layer solution for updating the routes and the neighbours of a node would be interesting at this point. We must also mention the possible changes to the handoff process which at this point would not only be engaged with a BS but with any SS or other MSS. In the literature there is research that deals with the interesting possibility of the introduction of mobile mesh [67] and which may also consider new WiMAX network architectures implemented in clusters [68].

5.6   Summary

In this chapter we have described briefly the IEEE 802.16 protocol, known as WiMAX. Besides introducing the basic mechanisms of the protocol, we have discussed how the various mechanisms of the protocol ensure well-defined quality of service levels. This has been done by highlighting the differences between the two modes supported by the protocol: PMP and mesh. Subsequently, we discussed all of the various gaps neglected deliberately by the developers of the protocol and which are relevant in the absence of algorithms for scheduling, call admission control and adaptive modulation and coding. These shortcomings, as mentioned above, have been neglected deliberately so as to enable the implementers of WiMAX devices to create ad hoc solutions which are optimized according to certain objective functions. In discussing the various issues the importance of cross-layer solutions has been emphasized.

We concluded the discussion by introducing exciting new challenges in the research, such as the introduction of mobile mesh. Another interesting aspect of the chapter is the introduction of games theory and fuzzy logic: two theories developed in other disciplines but used recently to resolve network problems such as bandwidth resource allocation or call admission control. This is very interesting from a didactic point of view because it is an example of the integration of different disciplines through an interdisciplinary way of thinking.

References

[1] IEEE 802.16-2004. IEEE Standard for Local and Metropolitan Area Networks, Part 16: Air Interface for Fixed Broadband Wireless Access Systems.

[2] IEEE 802.16e-2005. IEEE Standard for Local and Metropolitan Area Networks, Part 16: Air Interface for Fixed and Mobile Broadband Wireless Access Systems – Amendment 2: Physical and Medium Access Control Layers for Combined Fixed and Mobile Operation in Licensed Bands and Corrigendum 1.

[3] IEEE 802.16-2001. IEEE Standard for Local and Metropolitan Area Networks, Part 16: Air Interface for Fixed Broadband Wireless Access Systems.

[4] IEEE 802.16c-2002. IEEE Standard for Local and Metropolitan Area Networks, Part 16: Air Interface for Fixed Broadband Wireless Access Systems – Amendment 1: detailed system profiles for 10-66 GHz.

[5] IEEE 802.16a-2003. IEEE Standard for Local and Metropolitan Area Networks, Part 16: Air Interface for Fixed Broadband Wireless Access Systems – amendment 2: medium access control modification and additional physical layer specifications for 2-11 GHz.

[6] IEEE 802.16.2-2004. IEEE Recommended Practice for Local and Metropolitan Area Networks, Coexistence of fixed broadband wireless access systems.

[7] IEEE 802.16f-2005. IEEE Standard for Local and Metropolitan Area Networks, Part 16: Air Interface for Fixed Broadband Wireless Access Systems – Amendment 1: Management Information Base.

[8] IEEE 802.16k-2007. IEEE Standard for Local and Metropolitan Area Networks: Media Access Control (MAC) Bridges, Amendment 2: Bridging of IEEE 802.16.

[9] IEEE 802.16g-2007. IEEE Standard for Local and Metropolitan Area Networks, Part 16: Air Interface for Fixed Broadband Wireless Access Systems – Amendment 3:Management Plane Procedures and Services.

[10] IEEE 802.16 Conformance01-2003. IEEE Standard for Conformance to IEEE 802.16, Part 1: Protocol Implementation Conformance Statement (PICS) Proforma for 10-66 GHz WirelessMan-SC air interface.

[11] IEEE 802.16 Conformance02-2003. IEEE Standard for Conformance to IEEE 802.16, Part 2: Test Suite Structure and Test Purpose for 10-66 GHz wirelessMan-SC air interface.

[12] IEEE 802.16 Conformance03-2004. IEEE Standard for Conformance to IEEE 802.16, Part 3: Radio Conformance Tests (RCT) for 10-66 GHz WirelessMAN-SC Air interface.

[13] IEEE 802.16 Conformance04-2006. IEEE Standard for Conformance to IEEE 802.16, Part 4: Protocol Implementation Conformance Statement (PICS) Proforma for Frequencies below 11 GHz.

[14] V. Schwingenschlogl, V. Dastis, P.S. Mogre, M. Hollick and R. Steinmetz, 'Performance Analysis of the Real-time Capabilities of Coordinated Centralized Scheduling in 802.16 Mesh Mode', IEEE 63rd Vehicular Technology Conference. VTC 2006-Spring. May 7–10, 2006, pp. 1241–5, Melbourne, Vic.

[15] M. Cao, W. Ma, Q. Zhang and X. Wang, 'Analysis of IEEE 802.16 Mesh Mode Scheduler Performance' IEEE Transactions on Wireless Communications 6(4) April 2007: 1455–64.

[16] M. Cao, W. Ma, X. Wang, Q. Zhang and W. Zhu, 'Modelling and Performance Analysis of the Distributed Scheduler in IEEE 802.16 Mesh Mode', MobiHoc '05, Urbana-Champaign, Illinois, USA.

[17] B. Wang, I. Sen and D.W. Matolak, 'Performance Evaluation of 802.16e in Vehicle to Vehicle Channels', IEEE 66th Vehicular Technology Conference, VTC-2007 Fall. Sept. 30 2007–Oct. 3 2007, pp. 1406–10.

[18] F. De Rango, M. Tropea and S. Marano, 'Integrated Services on High Altitude Platform: Receiver Driven Smart Selection of HAP-Geo Satellite Wireless Access Segment and Performance Evaluation' International Journal Of Wireless Information Networks 13(1): 77–94, Jan. 2006.

[19] G. Avdikos, G. Papadakis. and N. Dimitriou, 'Overview of the Application of High Altitude Platform (HAP) Systems in Future Telecommunication Networks', 10th International Workshop on Signal Processing for Space Communications, 2008. SPSC 2008. 6–8 Oct. 2008, pp. 1–6.

[20] S.K. Agrawal and P. Garg, 'Effect of Urban-Site and Vegetation on Channel Capacity in Higher Altitude Platform Communication System' Microwaves, Antennas & Propagation, IET 3(4), June 2009: 703–13.

[21] Z. Yang and A. Mohammed, 'High Altitude Platforms for Wireless Sensor Network applications', IEEE International Symposium on Wireless Communication Systems. 2008. ISWCS '08. 21–24 Oct. 2008, pp. 613–17.

[22] F. De Rango, A. Malfitano and S. Marano, 'PER Evaluation for IEEE 802.16 - SC and 802.16e Protocol in HAP Architecture with User Mobility under Different Modulation Schemes', IEEE Global Telecommunications Conference, Globecom'06, 28 Nov–2 Dec. Alaska, 2006.

[23] F. De Rango, A. Malfitano and S. Marano, 'BER and PER Evaluation for IEEE 802.16e Protocol in HAP Architecture with User Mobility', Wireless Telecommunication Symposium, WTS 2006, Apr.27–29, 2006. Pomona, CA, USA.

[24] F. De Rango, A. Malfitano and S. Marano, 'Wireless Channel Evaluation of IEEE 802.16e Protocol in HAP Architecture with Mobility Scenario under Different Modulation Schemes', 13th International Conference on Telecommunications, ICT 2006, May 9–12, 2006. Madeira Island, Portugal.

[25] W. Wang, H. Sharif, M. Hempel, T. Zhou and P. Mahasukhon, 'Throughput vs. Distance Tradeoffs and Deployment Considerations for a Multi-Hop IEEE 802.16e Railroad Test Bed', IEEE Vehicular Technology Conference, VTC Spring 2008, 11–14 May 2008, pp. 2596–2600.

[26] I. Sen, B. Wang and D.W. Matolak, 'Performance of IEEE 802.16 OFDMA Standard Systems in Airport Surface Area Channels', Integrated Communications, Navigation and Surveillance Conference, ICNS '07, April 30 2007–May 3 2007, pp. 1–12.

[27] J. Park, D.H. Kwon and Y.J. Suh, 'An Integrated Handover Scheme for Fast Mobile IPv6 Over IEEE 802.16e Systems', IEEE 64th Vehicular Technology Conference, VTC-2006 Fall. 25–28 Sept. 2006, 1–5.

[28] Y. Choi and S. Choi, 'Service Charge and Energy-Aware Vertical Handoff in Integrated IEEE 802.16e/802.11 Networks', INFOCOM 2007. 26th IEEE International Conference on Computer Communications. May 6–12, 2007, pp. 589–97.

[29] H.J. Lee, J.H. Kim, J.H. Kwun and O.S. Lee., 'A handover time negotiation mechanism for seamless service in IEEE 802.16E', IEEE Military Communications Conference, MILCOM 2008. 16–19 Nov. 2008, pp. 1–7.

[30] J. Hur, H. Shim, P. Kim, H. Yoon. and N.O. Song, 'Security Considerations for Handover Schemes in Mobile WiMAX Networks', IEEE Wireless Communications and Networking Conference, WCNC 2008. March 31 2008–April 3 2008, pp. 2531–6.

[31] H.J. Yao and G.S. Kuo, 'An Integrated QoS-Aware Mobility Architecture for Seamless Handover in IEEE 802.16e Mobile BWA Networks', IEEE Military Communications Conference, MILCOM 2006, 23–25 Oct. 2006, pp. 1–7.

[32] E. Hossain and D. Niyato, 'Queue-Aware Uplink Bandwidth Allocation and Rate Control for Polling Service in IEEE 802.16 Broadband Wireless Networks' *IEEE Transactions on Mobile Computing* **5**(6) (June 2006): 668–79.

[33] Y.P. Fallah, F. Agharebparast, M.R. Minhas, H.M. Alnuweiri and V.C.M. Leung, 'Analytical Modeling of Contention-Based Bandwidth Request Mechanism in IEEE 802.16 Wireless Networks' *IEEE Transactions on Vehicular Technology* 57(5) Sept. 2008: 3094–107.

[34] B.P. Tsankov, P.H. Koleva and K.M. Kassev, 'Scheduling Algorithms for Carrier Grade Voice over IEEE 802.16 Systems', The 14th IEEE Mediterranean Electrotechnical Conference. MELECON 2008. May 5–7 2008, pp. 126–31.

[35] P. Dhrona, N. Abu Ali and H. Hassanein, 'A Performance Study of Scheduling Algorithms in Point-to-Multipoint WiMAX Networks', 33rd IEEE Conference on Local Computer Networks. LCN 2008. October 14–17 2008, pp. 843–50.

[36] S.C. Lo and Y.Y. Hong, 'A Novel QoS Scheduling Approach for IEEE 802.16 BWA Systems', 11th IEEE International Conference on Communication Technology. ICCT 2008. November 10–12 2008, pp. 46–9.

[37] H. Zhu, Y. Tang and I. Chlamtac, 'Unified Collision-Free Coordinated Distributed Scheduling (CF-CDS) in IEEE 802.16 Mesh Networks' *IEEE Transactions on Wireless Communications* **7**(10) October 2008: 3889–903.

[38] L.W. Chen, Y.C. Tseng, D.W. Wang and J.J. Wu, 'Exploiting Spectral Reuse in Resource Allocation, Scheduling, and Routing for IEEE 802.16 Mesh Networks', IEEE 66th Vehicular Technology Conference. VTC-2007 Fall. Sept. 30 2007–Oct. 3 2007, pp. 1608–12.

[39] J. El-Najjar, B. Jaumard and C. Assi, 'Minimizing Interference in WiMAX/802.16 Based Mesh Networks with Centralized Scheduling', IEEE Global Telecommunications Conference. GLOBECOM 2008. Nov. 30 2008–Dec. 4 2008, pp. 1–6. New Orleans, LO.

[40] S. Xergias, N. Passas and A.K. Salkintzis, 'Centralized Resource Allocation for Multimedia Traffic in IEEE 802.16 Mesh Networks', in Proceedings of the IEEE, Jan. 2008, 96(1): 54–63.

[41] H. Wang, W. Li and D.P. Agrawal, 'Dynamic admission control and QoS for 802.16 wireless MAN', Wireless Telecommunications Symposium 2005. April 28–30, 2005, pp. 60–6.

[42] B.J. Chang, Y.L. Chen and C.M. Chou, 'Adaptive Hierarchical Polling and Cost-Based Call Admission Control in IEEE 802.16 WiMAX Networks', IEEE Wireless Communications and Networking Conference. WCNC 2007. March 11–15, 2007, pp. 1954–8. Kowloon.

[43] F. Hou, P.H. Ho and X. Shen, 'Performance Analysis of a Reservation Based Connection Admission Scheme in 802.16 Networks', IEEE Global Telecommunications Conference. GLOBECOM 2006. Nov. 27 2006–Dec. 1 2006, pp. 1–5.

[44] C. Tian and D. Yuan, 'A Novel Cross-Layer Scheduling Algorithm for IEEE 802.16 WMAN', International Workshop on Cross Layer Design. IWCLD07. 20–21 Sept. 2007, pp. 70–3.

[45] T.C. Tsai and C.Y. Wang, 'Routing and Admission Control in IEEE 802.16 Distributed Mesh Networks', IFIP International Conference on Wireless and Optical Communications Networks. WOCN '07. July 2–4, 2007, pp. 1–5. Singapore.

[46] C. Cicconetti, V. Gardellin, L. Lenzini, E. Mingozzi and A. Erta, 'End-to-End Bandwidth Reservation in IEEE 802.16 Mesh Networks', IEEE International Conference on Mobile Adhoc and Sensor Systems. MASS 2007. Oct.8–11, 2007, pp. 1–6. Pisa.

[47] F. Liu, Z. Zeng, J. Tao, Q. Li and Z. Lin, 'Achieving QoS for IEEE 802.16 in Mesh Mode', 8th International Conference on Computer Science and Informatics. Salt Lake City, USA.

[48] F. De Rango, A. Malfitano and S. Marano, 'Markov Chain Based Models Comparison in IEEE 802.16e Scenario', WINSYS. July 26–29, 2008. Porto, Portugal.

[49] F. De Rango, A. Malfitano and S. Marano, 'Markov Chain Based Models Comparison and Hybrid Model Design in IEEE 802.16e Scenario', MILCOM. November 17–19, 2008. San Diego, CA.

[50] F. De Rango, A. Malfitano and S. Marano, 'Instant Weighed Probability Model to Guarantee QoS in IEEE 802.16e Scenario', WCNC. 5–8 April, 2009. Budapest, Hungary.

[51] F. Qiao and P. Lin, 'A Particle Swarm Approach for IEEE 802.16m System Scheduling', International Conference on Computer Science and Software Engineering, 2008. Dec.12–14, 2008, pp. 1234–7.

[52] M. Settembre, M. Puleri, S. Garritano, P. Testa, R. Albanese, M. Mancini and V. Lo Curto, 'Performance Analysis of an Efficient Packet-Based IEEE 802.16 MAC Supporting Adaptive Modulation and Coding', International Symposium on Computer Networks, 2006, pp. 11–16.

[53] S.A. Filin, S.N. Moiseev, M.S. Kondakov, A.V. Garmonov, D.Y. Yim and J. Lee, 'QoS-Guaranteed Cross-Layer Adaptive Transmission Algorithms for the IEEE 802.16 OFDMA System'. IEEE Wireless Communications and Networking Conference. WCNC 2006. April 3–6, 2006, pp. 964–71.

[54] J. Park, D.H. Kwon and Y.J. Suh, 'An Integrated Handover Scheme for Fast Mobile IPv6 Over IEEE 802.16e Systems', IEEE 64th Vehicular Technology Conference. VTC-2006 Fall. Sept., 25–28 2006, pp. 1–5. Montreal, Que.

[55] Y.W. Chen and F.Y. Hsieh, 'A Cross Layer Design for Handoff in 802.16e Network with IPv6 Mobility', IEEE Wireless Communications and Networking Conference. WCNC 2007. March 11–15, 2007, pp. 3844–9.

[56] M.K. Shin, Y.H. Han and H.J. Kim, 'IPv6 Operations and Deployment Scenarios over IEEE 802.16 Networks', 9th International Conference on Advanced Communication Technology. Feb. 12–14, 2007, pp. 1015–20.

[57] Y.H. Han, H. Jang, J.KH. Choi, B. Park and J. McNair, 'A Cross-Layering Design for IPv6 Fast Handover Support in an IEEE 802.16e Wireless MAN', IEEE Network, November-December 2007 21(6): 54–62.

[58] E. Barka, K. Shuaib and H. Chamas, 'Impact of IPSec on the Performance of the IEEE 802.16 Wireless Networks', New Technologies, Mobility and Security. NTMS '08. Nov. 5–7, 2008, pp. 1–6. Tangier.

[59] D. Niyato and E. Hossain, 'Radio Resource Management Games in Wireless Networks: an approach to bandwidth allocation and admission control for polling service in IEEE 802.16 [Radio Resource Management and Protocol Engineering for IEEE 802.16]', IEEE Wireless Communications, Feb. 2007, 14(1): 27–35.

[60] M.P. Anastasopoulos, P.D.M. Arapoglou, R. Kannan and P.G. Cottis, 'Adaptive Routing Strategies in IEEE 802.16 Multi-Hop Wireless Backhaul Networks Based On Evolutionary Game Theory' IEEE Journal on Selected Areas in Communications September 2008, 26(7): 1218–25.

[61] S. Chowdhury, S. Dutta, K. Mitra, D.K. Sanyal, M. Chattopadhyay and S. Chattopadhyay, 'Game-Theoretic Modeling and Optimization of Contention-Prone Medium Access Phase in IEEE 802.16/WiMAX Networks', Third International Conference on Broadband Communications, Information Technology & Biomedical Applications, 2008. Nov. 23–26, 2008, pp. 335–42. Gauteng.

[62] D. Niyato and E. Hossain, 'A Game Theoretic Analysis of Service Competition and Pricing in Heterogeneous Wireless Access Networks' IEEE Transactions on Wireless Communications 7(12), Part 2. December 2008: 5150–5.

[63] D. Niyato and E. Hossain, 'Integration of IEEE 802.11 WLANs with IEEE 802.16-based Multihop Infrastructure Mesh/Relay Networks: A game-theoretic approach to radio resource management'. IEEE Network. May–June 2007, 21(3): 6–14.

[64] J. Nie, L. Zeng and J. Wen, 'A Bandwidth Based Adaptive Fuzzy Logic Handoff in IEEE 802.16 and IEEE 802.11 Hybrid Networks', International Conference on Convergence Information Technology, 2007. Nov.21–23, 2007, pp. 24–9.

[65] Y. Tang, F. Yang and Y. Yao, 'Dynamic Data Path Strategy Based on Fuzzy Logic Control for Mobile WiMAX', 2nd International Conference on Anti-counterfeiting, Security and Identification. ASID 2008. Aug. 20–23, 2008, pp. 194–8. Guiyang.

[66] D. Niyato and E. Hossain, 'Delay-Based Admission Control Using Fuzzy Logic for OFDMA Broadband Wireless Networks', IEEE International Conference on Communications. ICC '06. June 2006, pp. 5511–16. Istanbul.

[67] V.D. Hoang, M. Ma, R. Miura and M. Fujise, 'A Novel way for Handover in Maritime WiMAX Mesh Network', 7th International Conference on Telecommunications. ITST '07. June 6–8, 2007, pp. 1–4. Sophia Antipolis.

[68] M.B.K. Hartzog and T.X. Brown, 'WiMAX – Potential Commercial Off-The-Shelf Solution for Tactical Mobile Mesh Communications', IEEE Military Communications Conference. MILCOM 2006. Oct. 23–25, 2006, pp. 1–7.

6

QoS in Mobile WiMAX

Neila Krichene and Noureddine Boudriga
Communication Networks and Security Research Laboratory (CNAS),
University of the 7th November at Carthage, Tunisia

6.1 Introduction

Recent years have been marked by a growing need for the provision of advanced applications and Internet-related services at high throughput and low costs while guaranteeing the required QoS and continuous and open access to such services. In particular, subscribers wish to enjoy ubiquitous access to their preferred services in a transparent way while on the move or driving. To provide wireless access to Internet services and multimedia applications on a metropolitan scale even for underdeveloped regions, mobile WiMAX marries mobility support to high bandwidth and enhanced availability, reliability and flexibility. For example, mobile WiMAX amendments implement service differentiation and adopt a connection-oriented philosophy in order to reserve the required resources for the required level of QoS while minimizing any possible waste of the network assets.

Mobile WiMAX needs to address multiple constraints in order to guarantee QoS. First, the wireless nature of the channels implies constant changes in links capacity; thus changing the state of available resources and causing the re-negotiation of service level agreements. Second, the wireless nature of the channels induces important transmission error rates, which may violate QoS requirements. Third, the mobility of users implies constant changes in established routes and unavoidable handover delays that may negatively affect the provision of real-time services. Fourth, mobile WiMAX needs to guarantee a significant level of security in order to become deployed widely. However, security guarantee and QoS provision are always seen as contradictory as security guarantee is resource and time consuming while QoS provision is based on restricting time and resources constraints. Last but not least, support of the mesh mode renders the provision of QoS more challenging as multihop communication introduces significant delay and

WiMAX Security and Quality of Service: An End-to-End Perspective   Edited by Seok-Yee Tang,
Peter Müller and Hamid Sharif
© 2010 John Wiley & Sons, Ltd

is achieved by untrustworthy nodes that may easily violate QoS requirements without being detected.

Mobile WiMAX amendments defined standardized mechanisms for supporting QoS but left many QoS functions unspecified so that researchers and constructors could design and adopt the mechanisms best suited to fulfilling particular requirements. For example, connection-oriented MAC along with the request-grant mechanisms and the usage of service flows with five defined service classes have been defined by mobile WiMAX designers in order to pave the way for advanced provision of QoS. However, additional intelligence needs to be implemented within Subscribers Stations (SSs) in order to distribute their allocated bandwidth among connections with different QoS requirements. Besides, the IEEE 802.16 amendment did not specify particular mechanisms in order to perform resources management, traffic shaping, admission control and scheduling. QoS signalling mechanisms have been defined by the standard but the definition of algorithms that may use such mechanisms for bandwidth allocation was left to the vendors.

In this chapter, we address QoS management in WiMAX networks; we aim especially at demonstrating how mobile WiMAX technology offers continuity of services while providing enhanced QoS guarantees in order to meet subscribers' demands. This chapter will be organized as follows: first, we survey the architectural QoS requirements that have to be fulfilled with regard to subscribers' mobility. We then describe in detail the mechanisms built by the Mobile WiMAX network to provide QoS. We focus on the defined service flows, the 'connection-oriented' nature of the MAC layer, the bandwidth request and allocation procedures and the scheduling service. After that, we present the mechanisms designed to maintain QoS during handover. Finally, we survey some research studies that analyze the limits of the standardized Mobile WiMAX QoS procedures and propose new mechanisms aiming at enhancing QoS provision.

6.2   Architectural QoS Requirements

In this section, we will describe the QoS-related challenges introduced by the wireless nature of communications coupled to subscribers' mobility on a metropolitan scale. We will then define the architectural requirements that need to be fulfilled in order to provide the required QoS in such environments.

6.2.1   QoS-Related Challenges

QoS is determined by different parameters in the network indicating the types of traffic that can be supported and the type of experience a user will have. Generally speaking, typical parameters that determine the QoS level are bit error rate, jitter, latency, average data throughput and minimum throughput, [1]. IEEE 802.16 was designed to support various types of applications with different QoS requirements. For instance, mobile subscribers can have access to data transfer and web browsing, which may be considered as best effort applications although they can also support Voice over IP (VoIP), Video on Demand (VoD) and IPTV services which have significant needs in terms of bandwidth, latency and jitter. To support multiple applications with different QoS requirements, mobile WiMAX defines different scheduling services optimized for different requirements and adopts a connection-oriented philosophy to handle subscribers' traffic.

Nevertheless, mobile WiMAX faces different constraints that may prevent the delivery of the required QoS. First, the wireless nature of the communication can be highly affected by the environmental conditions such as multi-path, fast fading, refraction and weather conditions. The latter greatly affects the delivered QoS and particularly the bit errors rate. Note that bit errors and the resulting packet loss have a negative impact on voice communications since retransmission is not an option, [2]. Maintaining a constant bit error rate without inducing significant complexity of processing and delays is a challenging issue for QoS provision over mobile WiMAX networks.

Moreover, transmission over the air introduces significant delay resulting from channel propagation delays, serialization delays, channel coding delays and delays caused by MAC processing. At the network layer, the forwarding and buffering delays at the network layer and the packetization, coding/decoding at the application layer increase the total latency. Latency greatly affects voice communications as it is very boring for listeners to have long and inconsistent pauses. Audio and video traffic are tolerant of bit errors as they are characterized by an inherent redundancy, although they are greatly affected by jitter which disturbs the intra-frame or inter-frame synchronization required for decoding the video signal, [2].

Other external factors may also be considered as constraints facing the provision of QoS by mobile WiMAX networks. For example, the distance of the terminals from the transmitter, the speed at which the subscriber is moving and the amount of traffic on both the uplinks and the downlinks may affect the delivered QoS. Users' mobility at high speed may also result in a loss of the level of QoS required by the ongoing connection unless advanced mobility management mechanisms are implemented. In more detail, mobile subscribers are exposed to multi-path, fast fading and refraction. Multi-path signals prevent higher sustained throughput as signal reflections are received by the mobile station at different times. Such multi-path interferences have a serious effect on the quality of communication when the delay spread and the time span separating the reflection is in the order of the transmitted symbol time [3]. Users' mobility becomes more challenging when Base Stations (BSs) and mobile users have wireless interfaces from different manufacturers since the mobile stations have to authenticate and associate with a new BS each time a move is made.

Meanwhile, intelligent policy, call admission control, traffic shaping and scheduling functions need to be implemented at both the BS and the SS in order to provide the required QoS level and maintain it during the connections. For example, it is not acceptable to let an SS get a higher or lower QoS level than that contracted. Mobile WiMAX needs to maintain the level of QoS of the existing and new connections despite an increased load scenario. The network assets need to be exploited in an optimized fashion so that the resources are not overcommitted, the users follow their correspondent policy rules and only authorized users can use the resources. Note also that the supplied resources should be maintained despite changing channel conditions and that the mobile WiMAX needs to ensure that committed resources can be supplied.

Packets scheduling in mobile WiMAX networks is a hot research issue as the packet scheduler handles different flows of complex applications with different QoS requirements. The most challenging aspect is that the wireless nature of the WiMAX channels may highly impair the scheduler's QoS-support capability and invalidate theoretical fairness in the assignment of available resources [3]. The supported mesh mode also has several constraints related to the multi-hop nature of the communications, thus preventing the guarantee of the required QoS level, especially during handover.

6.2.2 Architectural Requirements

To satisfy QoS guarantees, mobile WiMAX needs to implement mechanisms at both the physical and the MAC layers. At the physical layer, self-interference should be minimized in order to minimize errors in the received signals. Different signal bandwidths should also be supported in order to enable transmission over longer range in different multi-path environments [3]. Symbol duration and guard-band value need to be studied carefully in order to minimize multi-path impact. Advanced multi-antenna signal processing techniques also needs to be implemented in order to minimize the impact of multi-path and reflections while maximizing the signal to noise ratio. Robust error correction techniques need to be implemented in order to be immunized against corrupted packets resulting from fading.

Mobile WiMAX should be able to recover signal noises and errors resulting from bad channel conditions or high mobility by implementing adaptive modulation scheme and coding. Bandwidth resources at the radio interface should be managed in an efficient manner and monitored in order to ensure that the Service Level Agreement established for the connection is always met. Concatenation, fragmentation and packing of MAC PDUs and MAC SDU should be performed in order to achieve better bandwidth utilization.

On the other hand, optimal scheduling of space, frequency and time resources over the air interface on a frame-by-frame basis should be implemented. Resources allocation and scheduling need to adapt dynamically to the bursty and unforeseeable nature of the traffic while providing a large dynamic range of throughput to specific users based on their demand without degrading overall network performances or causing starvation to particular users or traffic flows. In more detail, the MAC layer needs to be more deterministic than contention-based in order to satisfy the required QoS guarantees in a more reliable fashion [3]. Scheduling algorithms should adopt a flexible strategy in allocating the timeslots to subscribers' stations (SSs) according to their needs; and the required resources should remain assigned to the correspondent SS for the entire duration of the communication. Second, the implemented scheduling scheme should be priority-based in order to distribute the available resources correctly among the various classes of service depending on their QoS requirements. However, inter-class and intra-class fairness must be guaranteed. The implemented scheduling algorithm should manage the network assets efficiently despite overload and over-subscription. It should be simple, efficient and fair with a low computational complexity. It needs also to guarantee throughput and delay performance while guarding against misbehaving. It should also be channel-aware and take opportunistic decisions by serving subscribers with good channel quality without violating the QoS requirements of unlucky subscribers experiencing poor channel quality [3]. Currently, much research is being conducted in order to propose optimized scheduling algorithms for both the point-to-multi point mode and the mesh mode. Nevertheless, designing efficient scheduling methods for the mesh mode remains a little bit harder owing to the distributed nature of the mesh mode and the constraints introduced by multi-hop communication.

Call Admission Control (CAC) at the MAC layer should guarantee the required QoS to the new and the existing connections by preventing oversubscription for a particular bandwidth. QoS based CAC may be implemented according to need so that new connection requests are classified into particular queues with regard to their associated service class types and are then processed. Policy functions should include general and application-dependent rules that prevent subscribers from experiencing a higher or a lower QoS level than the one specified in their QoS profile. Moreover, end-to-end QoS control needs to be implemented in order to guarantee the required level of QoS and act appropriately in the case of QoS violations. QoS parameters have to be monitored continuously and adjusted dynamically with regard to dynamic service demand. On the other hand, QoS mechanisms should be applied to both uplink and downlink directions in order to improve the provision of QoS. QoS management functions such as a configuration and registration function enabling the pre-configuration of subscribers' QoS-enabled flows and traffic parameters, as well as a signalling function enabling the dynamic establishment of such parameters, need to be implemented. Mobile WiMAX needs also to adopt common definitions of global service class names and their associated authorized QoS parameters in order to facilitate operations across a distributed topology by agreeing on a baseline convention for communication.

Optimized handover schemes that achieve latencies of less than 50 milliseconds should be designed in order to guarantee the required level of QoS for real-time applications such as VoIP. Flexible key management schemes also need to be implemented in order to maintain the required security during handover without causing degradation of services. The MAC layer should ensure smooth handover without violating the QoS requirements of the ongoing connection. The handover decision should not be based only on the received signal strength, but also on the QoS that may be provided by neighbouring BSs and load conditions. Soft handover schemes should be implemented in order to ensure the continuity of the service. Fast scheduling on a per-frame basis needs to be adopted in order to minimize handover latency. Scheduling also needs to be able to be readjusted quickly depending on the new available assets and channel conditions when handover occurs.

Last but not least, QoS-aware routing protocols should be implemented in WiMAX so that the best route to choose is the one guaranteeing QoS requirements. Redundant routes should be addressed in order to recover route breaks caused by subscriber's mobility, especially in the context of mesh mode.

To conclude, a cross-layer approach involving the physical, MAC and network layers should be adopted in order to guarantee the delivery of the required QoS.

6.3   Mobile WiMAX Service Flows

The mobile WiMAX MAC layer defines service flows and their associated QoS parameters in order to facilitate operation across a distributed topology. Data are transported through connections. Each connection is associated with a scheduling service and each scheduling service is associated with a set of QoS parameters that dictate its behaviour, [4, 5].

In this section, we will present the Unsolicited Grant Service (UGS), the Real Time Polling service (rtPS), the Extended Real Time Polling Service (ertPS), the non-real Time Polling Service (nrtPS) and the Best Effort (BE) service flows adopted by Mobile WiMAX.

More precisely, we will state the set of QoS parameters associated to each service flow while explaining the principle of QoS provision.

6.3.1 Service Flows

A service flow is a unidirectional flow guaranteeing a particular QoS described by its QoS Parameters Set, [4]. More precisely, a service flow is either defined for the uplink or the downlink direction; it may exist even if it is not activated to transport traffic. A service flow is identified by the following attributes:

- *Service Flow ID (SFID)*: which identifies the service flow between a BS and an SS.
- *CID*: the connection ID of the transport connection. It exists only when the connection has an admitted or active service flow. An SFID is associated with a unique transport CID and a transport CID should be associated with a unique SFID.
- *ProvisionedQoSParamSet*: a QoS parameter set provisioned by a network management system, for example.
- *AdmittedQoSParamSet*: which defines the set of QoS parameters for which the BS or the SS are reserving resources such as bandwidth or memory.
- *ActiveQoSParamSet*: defines the set of QoS parameters defining the service being provided to the service flow. Note that only an active service flow may forward traffic.
- *Authorization Module*: a logical function implemented in the BS in order to approve or deny changes in the QoS parameters and classifiers associated with a service flow.

It is valuable to note that the *ActiveQoSParamSet* is always a subset of the *AdmittedQoSParamSet* which is in turn a subset of the authorized 'envelope' determined by the *ProvisionedQoSParamSet* [4].

The service flows can be further classified into three types: the 'provisioned', the 'admitted' and the 'active'. The provisioned service flow is known via provisioning by the network management system. Its *ActiveQoSParamSet* and its *AdmittedQoSParamSet* are both null and its activation and admission are deferred. The MS may request the activation of a provisioned service flow by passing the SFID and the associated QoS parameter set to the BS. The latter may respond by mapping the service flow to a CID if the authorization is verified and there are enough available resources. Besides, the BS may choose to activate a service flow by passing the SFID, the CID, and the associated parameter set to the MS [4].

The admitted service flow has resources reserved for its *AdmittedQoSParamSet*, but these parameters are not active. The adopted activation model is usually used in telephony applications as it presents two phases of activation. First, the resources for a call are admitted. After performing an end-to-end negotiation, those resources are activated. The service flow that has resources assigned to its *AdmittedQoSParamSet* but whose resources are not yet completely activated is in transient state. Note that an activation request of a service flow presenting an *ActiveQoSParamSet*, which is a subset of the *AdmittedQoSParamSet*, will be allowed. Meanwhile, an admission request where the *AdmittedQoSParamSet* is a subset of the previous *AdmittedQoSParamSet* and where the *ActiveQoSParamSet* remains a subset of the *AdmittedQoSParamSet* should be served [4].

Finally, the active service flow has resources committed by the BS for its *Active-QoSParamSet* which is non-null. An admitted service flow may be activated by providing an *ActiveQoSParamSet* and signalling the resources required at the current time in order to complete the second stage of the two-phase activation model. A service may be provisioned and activated immediately. It may also be created dynamically and then activated immediately so that the two-phase activation is skipped and the service flow is operational immediately upon authorization.

Service flows may be created, admitted or activated in response to triggers other than network entry. A service flow is first instantiated then provided with a flow ID and a 'provisioned' type. The service flow is enabled after the transfer of operational parameters and it may change its type to 'admitted' or 'active'. When the service flow type becomes active, it will be mapped onto a particular transport connection. Service flow encodings include either the full definition of the service attributes or a service class name. The service class name is a particular string known to the BS; it defines indirectly the set of QoS parameters [4].

6.3.2 Scheduling Services Supporting Service Flows

Note that each transport connection is associated with a data scheduling service. Let us provide an overview of the Unsolicited Grant Service (UGS), the Real Time Polling service (rtPS), the Extended Real Time Polling Service (ertPS), the non-real Time Polling Service (nrtPS) and the Best Effort (BE) scheduling services adopted by Mobile WiMAX.

6.3.2.1 The Unsolicited Grant Service (UGS)

The UGS service is intended to support real-time uplink service flows transporting periodically fixed-size data packets such as Voice over IP traffic without silence suppression. The service guarantees fixed size grants on a real-time basis in order to minimize the overhead and latency that may result in MS requests. The BS provides the MS periodically by 'Data Grant Burst IE' based upon the Maximum Sustained Traffic Rate of the service flow. Meanwhile, the MS is banned from using contention request opportunities for a transport connection associated with UGS. The required QoS parameters are the Maximum Sustained Traffic Rate, the Maximum Latency, the Tolerated Jitter, the Uplink Grant Scheduling Type and the Request/Transmission Policy [5, 6]. If the Minimum Reserved Traffic Rate parameter is present, it should be set to the value of Maximum Sustained Traffic Rate. The BS should not allocate more bandwidth than the Maximum Sustained Traffic Rate of the Active QoS Parameter Set except when a clock rate mismatch compensation is mandatory.

6.3.2.2 The Real Time Polling Service (rtPS)

The rtPS service is intended to support real-time uplink service flows transporting periodically variable-size data packets such as Moving Picture Expert Group (MPEG) video traffic. The service flow provides real-time periodic unicast request opportunities to enable the MSs in indicating the size of the required grant. The adopted request/grant policy should meet the flow's real-time constraint. The rtPS service achieves a better data transport efficiency comparing to the UGS as only the required resources are granted; however, it adds more request overhead. The BS issues unicast request opportunities even if prior requests are unfulfilled. Therefore, the MS will use only the unicast request opportunities and the data transmission opportunities to gain uplink transmission opportunities. Meanwhile, the MS is banned from using contention request opportunities for a transport connection associated with rtPS. The required QoS parameters are the Minimum Reserved Traffic Rate, the Maximum Sustained Traffic Rate, the Maximum Latency, the Traffic Priority, the Uplink Grant Scheduling Type and the Request/Transmission Policy [5, 6].

6.3.2.3   The Extended Real Time Polling Service (ertPS)

The ertPS is a scheduling service added recently by the mobile WiMAX amendments; it is intended to support real-time uplink service flows transporting periodically variable-size data packets such as Voice over IP (VoIP) traffic with silence suppression [5]. The ertPS combines the advantages of both the UGS and the rtPS services. The BS provides unicast grants in an unsolicited manner as in UGS but the allocations are dynamic as in rtPS. Therefore, we may save the latency of a bandwidth request while optimizing the utilization of resources. The BS offers periodic uplink allocations that may be used to request bandwidth as well as transferring data. The default size of allocations is the value of the Maximum Sustained Traffic Rate at the connection. Then, the MS may request the change of the allocation's size by using an extended piggyback request field of the Grant Management subheader, using the Bandwidth Request (BR) field of the MAC signalling headers or sending a codeword over Channel Quality Information Channel (CQICH). The BS keeps unchanged the size of the allocation until it receives a bandwidth change request from the MS. If the MS does not find available unicast bandwidth request opportunities, it may use the contention request opportunities for that connection or send a CQICH codeword to inform the BS of its needs. When the BS receives the CQICH codeword, it should start allocating the UL grant. The required QoS parameters are the Maximum Sustained Traffic Rate, the Minimum Reserved Traffic Rate, the Maximum Latency, the Tolerated Jitter, the Traffic Priority and the Request/Transmission Policy [5, 6].

6.3.2.4   The non Real Time Polling Service (nrtPS)

The nrtPS provides unicast polls on a regular basis allowing uplink service flow receiving request opportunities despite network congestion. It is intended to support delay-tolerant applications with variable-rate data streams such as File Transfer Protocol (FTP) [5, 6]. Generally speaking, the BS polls nrtPS CIDs every one second or less. The MS should be able to use the contention request opportunities for a transport connection associated with nrtPS. Therefore, the MS may use the contention request opportunities, unicast request opportunities and data transmission opportunities. The required QoS parameters are the Minimum Reserved Traffic Rate, the Maximum Sustained Traffic Rate, the Traffic Priority, the Uplink Grant Scheduling Type and the Request/Transmission Policy [5, 6].

6.3.2.5   Best Effort (BE)

The BE service is intended to offer timely unicast request opportunities for supporting best effort applications such as data transfer and web browsing [5, 6]. The MS should be able to use the contention request opportunities. Therefore, the MS may use contention request opportunities, unicast request opportunities and data transmission opportunities. The required QoS parameters are the Maximum Sustained Traffic Rate and Traffic Priority [6].

6.3.3 QoS Parameters

Mobile networks need to share a common definition of service class names linked with *AuthorizedQoSParamSets* in order to facilitate operation across a distributed topology [5]. In the following, we give an overview of the global service flow class name QoS parameters.

6.3.3.1 The Maximum Sustained Traffic Rate

The *Maximum Sustained Traffic Rate* represents the peak information rate of the service. This rate is stated in bits per second and pertains to the Service Data Units (SDUs) at the input of the system [5]. More precisely, the *Maximum Sustained Traffic Rate* parameter does not encompass the transport, protocol or network overhead such as MAC headers or Cyclic Redundancy Check (CRC) fields or the non-payload session maintenance such as Session Initiation Protocol (SIP) administration [7]. It is useful to note that this parameter does not limit the instantaneous rate of the service as it is determined by the physical attributes of the ingress port. Nevertheless, the service should be policed so as to cope with this parameter on the average over time at the destination network interface at the uplink direction. On the network in the downlink direction, it may be assumed that the service was already policed at the ingress to the network [5]. If the *Maximum Sustained Traffic Rate* parameter is set to zero, it means that there is no explicitly mandated maximum rate. It should be noted that the maximum sustained traffic rate field specifies an upper bound only and not the guarantee of provision.

6.3.3.2 The Minimum Reserved Traffic Rate

The *Minimum Reserved Traffic Rate* represents the minimum rate expressed in bits per second that should be provided to the service flow. If the MS requests less than the bandwidth value specified by that parameter, the BS may use the excess of the reserved bandwidth for other use. However, the BS has to fulfil the bandwidth requests of a connection up to its minimum reserved traffic rate. The value of this parameter should be determined after excluding the MAC overhead. If the value of this parameter is set to zero, no minimum traffic rate will be reserved [5].

6.3.3.3 The Maximum Latency and the Tolerated Jitter

The *Maximum Latency* indicates the maximum time value between the reception of a packet at the Convergence Sublayer of the BS or the MS and the arrival of that packet at the peer device. When this parameter is set to a non null value, it represents a service commitment and should be guaranteed. A null value of this parameter is the synonym of no commitment. The *Tolerated Jitter* parameter indicates the maximum delay variation (i.e. commonly known as jitter) for the connection.

6.3.3.4 The Request/Transmission Policy

The *Request/Transmission Policy* reflects the ability to specify some attributes for the associated service flow. Examples of such attributes include, for uplink service flows, restrictions on the types of bandwidth request options that may be used. Another example is the options for Protocol Data Unit (PDU) formation. An attribute is enabled when its correspondent bit position is set to one. A null value of the attributes affecting the uplink bandwidth request types refers to the default actions stated in the scheduling service description while a value of one indicates that the action associated with the attribute bit overrides the default action.

6.3.3.5 The Traffic Priority

The value of this parameter represents the priority given to a service flow. When two service flows share the same values in all QoS parameters except that which has priority, the service flow with higher priority should be attributed a lower delay and higher buffering preference. However, when two service flows have different QoS parameters values, the priority parameter should not take precedence over any conflicting service flow QoS parameter. The BS uses the *Traffic Priority* parameter when arranging the request service and grants generation while the MS should preferably select contention Request opportunities for Priority Request CIDs based on this priority and its *Request/Transmission Policy*. The Traffic priority values range from 0 to 7 where higher numbers indicate higher priority and the default value is 0 [5].

6.4 Admission Control

In this section we explain the 'connection-oriented' nature of the WiMAX MAC layer and show how to associate a connection with a required QoS. We also survey the operations realized by the MAC Common-Part Sublayer (CPS) and Service Specific Convergence Sublayers (SSCS) in order to process the admission control of bandwidth and service flow requests and the bandwidth allocation of the accepted requests.

6.4.1 MAC Layer Connections

The IEEE 802.16e MAC layer is divided into three sublayers which are: the Service Specific Convergence Sublayer (*SSCS*), the MAC Common-Part Sublayer (*CPS*) and the MAC Privacy Sublayer (*PS*), as illustrated by Figure 6.1. The *SSCS*'s role is to assign Service Data Units (*SDUs*) to the corresponding MAC connection, and enable bandwidth allocation. Classification of *SDUs* is implemented by transforming or mapping the external network data received through the *SSCS* Service Access Points (*SCSS SAPs*) to MAC SDUs and then transmitting them to the MAC *CPS* through MAC *SAPs*. To perform mapping services to and from MAC connections correctly, SSCS is divided into two convergence sublayers which are the Asynchronous Transfer Mode (*ATM*) convergence sublayer (*ATM CS*) and the Packet convergence sublayer (*Packet CS*). The *MAC CPS* implements all the MAC functions and mechanisms allowing system access, bandwidth allocation and connection maintenance.

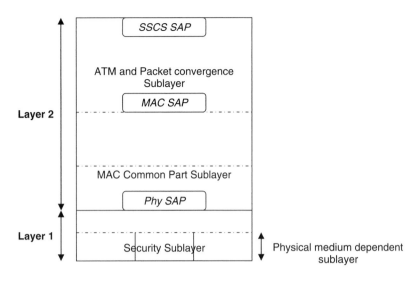

Figure 6.1   IEEE 802.16 reference model and protocol stack.

The IEEE 802.16 MAC layer is 'connection oriented' as all data communication is associated with a connection. A connection may be identified as a unidirectional mapping between BS and MS MAC peers. We distinguish two types of connection, the management connection and the transport connection [5]. IEEE 802.16e amendments specify that each air interface of the MS should have a 48-bits unique MAC address. In the Point-to-Multipoint mode, this address is used at the initial ranging process and during the authentication process. During initial ranging, the MAC address is used to establish the appropriate connections for the MS. During authentication, the BS and the MS may mutually verify their identities. Each connection is identified by a 16-bits Connection IDentifier (*CID*). When the MS enters the initialization process, two pairs of management connections (uplink and downlink) are established between the MS and the BS while a third pair of management connections may be generated optionally. We may distinguish the mandatory **basic connection** which is used by both the MS MAC and the BS MAC layers in order to exchange short and time-urgent MAC management messages. The mandatory **primary management connection** is used by both the MS MAC and the BS MAC layers in order to exchange longer and more delay-tolerant MAC management messages. Finally, the optional **secondary management connection** serves for transferring delay tolerant and standards-based messages as the Dynamic Host Configuration Protocol (*DHCP*) and Trivial File Transfer Protocol (*TFTP*) messages [5]. Note that a MS can have a static configuration and may not require the establishment of the secondary management connection. For example, an MS may have a static IP address; thus it does not need the exchange of *DHCP* messages, [8].

Besides the management connections, the IEEE 802.16 amendments define *transport connections* which are used to transport the user data and even the traffic related to connectionless protocols, such as Internet Protocol (IP). More precisely, the transport connections carry the Protocol Data Units (*PDUs*) received by the MAC layer from the upper layers and associated with a scheduling service corresponding to the required

QoS level. Transport connections are associated with service flows and new transport connections may be established when a MS's service needs change. A transport connection defines both the mapping between peer convergence processes that use the MAC and a service flow where the service flow defines the QoS parameters guaranteed for the PDUs exchanged on the connection. Note that the BS initiates the set-up of service flows for bearer services based on the collected provisioning information. Moreover, the registration of a MS or the modification of the contracted service at the MS level triggers the higher layers of the BS to initiate the setup of the correspondent service flows. Admitted or active service flows are associated uniquely with transport connections while MAC management messages should never be exchanged over transport connections. Besides, bearer or data services are never transferred on the basic, primary or secondary management connections [5]. The BS should send *DSA-REQ* (Dynamic Service Addition request) messages to the managed MSs after the transfer of operational parameters and to unmanaged MSs after registration in order to set up connections for preprovisioned service flows belonging to these MSs. The concerned MS responds with *DSA-RSP* messages. Once established, transport connections may require active maintenance owing to stimulus from either the MS or the network side of the connection. Finally, transport connections may be terminated by the BS or the MS generally after a change in the MS's service requirements [5].

When operating in mesh mode, every node should have a 48-bits unique MAC address that is used during network entry and as part of the authorization processes. An authorized node should receive a 16-bit Node Identifier (*Node ID*) from its mesh BS. That Node ID is used to identify the mesh node during normal operation. Besides, each mesh node uses an 8-bit Link Identifier (*Link ID*) to address other nodes in its local neighbourhood. Link IDs are assigned and communicated during the Link Establishment process when neighbouring nodes establish new links. The Link ID is transmitted as part of the *CID*; it should be used in distributed scheduling in order to identify resource requests and grants. Note that the *CID* in the mesh mode is specified to convey broadcast/unicast, service parameters and the link identification.

We may deduce that connections are the lower level of data transport services and are associated with a higher level service flow. More precisely, a connection and its associated QoS parameters form a service flow [5]. All the requests for transmissions are based on the *CIDs*. In fact, the requested bandwidth may differ for different connections even if these connections belong to the same service type. Besides, higher-layer sessions may use the same wireless *CIDs*. The type of service and some other parameters of a service are implicit in the *CID*; they may be accessed by a lookup indexed by the *CID* [4].

6.4.2   Bandwidth Request Procedures

Bandwidth requests are used by the MSs to inform their managing BSs that they demand uplink bandwidth allocation. A MS may choose one of two procedures in order to request bandwidth. In fact, it can send explicitly a *Bandwidth Request* message or it may send an optional *Piggyback Request* in a Grant Management subheader [8]. Requests may be sent in any uplink burst but they should not be sent during initial ranging. We distinguish two types of bandwidth request, namely incremental or aggregate requests. An incremental request adds the requested bandwidth to the bandwidth already allocated by the BS while an aggregate request replaces the current bandwidth value with the value in the request. The *Type* field in the Bandwidth Request header specifies whether the request is incremental or aggregate. Piggyback requests are always incremental as they do not have a particular *Type* field to specify the type of the request. Besides, requests which are done during bursts where collisions may happen should be aggregate. The self correcting nature of the request/grant protocol allows the MSs to use aggregate Bandwidth Requests periodically as a function of the QoS of a service and of the quality of a link [5]. The BSs should support incremental and aggregate bandwidth requests while the MSs must support aggregate bandwidth requests. Support of incremental bandwidth requests is optional for the MSs.

As stated earlier, bandwidth requests may come in the form of a standalone Bandwidth Request header. In this case, the Bandwidth Request PDU should be composed by only a Bandwidth Request header and should not contain a payload [5]. Every MS that receives a Bandwidth Request header on the downlink has to discard the PDU. The *CID* field of the Bandwidth Request header specifies the connection for which the uplink bandwidth is requested while the Bandwidth Request (*BR*) field indicates the number of requested bytes. As the uplink burst profile may change dynamically, the BR value should indicate the number of bytes required to carry the MAC header and the payload but not the physical layer overhead. A MS should not request bandwidth for a connection while it does not have any PDU to transmit on that connection [4, 5]. A MS requests bandwidth for a particular connection. Nevertheless, the BS allocates bandwidth addressed to the MS basic management connection and not individual connections; thus rending the MS unaware of which request the grant refers to [8]. The BS may also assign bandwidth for the explicit purpose of MS bandwidth request messages using specific burst types in the Uplink Access Definition (*UL-MAP*) management message.

Polling may be defined as the process by which the BS allocates bandwidth for the MSs so that they can issue their bandwidth requests. Polling can be unicast, multicast or broadcast. Unicast polling means that the MS is polled individually; it is done using bursts directed at the MS's Basic CID. In this case, no explicit message is transmitted to poll the MS. However, the BS allocates in the UL-MAP message sufficient bandwidth to enable the MS to respond with a Bandwidth Request. If the MS does not require bandwidth, the previous allocation is padded. MSs which have an active UGS connection and which do not set the poll-me (*PM*) bit in the header of the packets corresponding to that connection or MSs which have sufficient bandwidth should not be polled individually in order to save the resources.

When the BS lacks available bandwidth to poll many inactive MSs individually, some MSs may be polled in multicast groups or a broadcast poll may be adopted. In this case, no explicit message is transmitted in order to poll the MSs and the bandwidth is allocated in the UL-MAP. However, that allocation is associated to some reserved multicast or broadcast CIDs and not to a particular MS's Basic CID. When multicast or broadcast polling is adopted, any MS belonging to the polled group may request bandwidth during any request interval allocated to the corresponding CID in the UL-MAP by a Request IE. Consequently, collisions may occur. To minimize collision probability, only the MSs which require bandwidth will answer; they will also implement a contention resolution algorithm to select the slot in which they should transmit the initial bandwidth request. A MS deduces an unsuccessful transmission when it does not receive any grant in the

Interval Description	UL-MAP IE Fields		
	CID on 16 bits	UIUC on 4 bits	Offset on 12 bits
Initial Ranging	0000	2	0
Multicast group 0xFFC5 Bandwidth Request	0xFFC5	1	405
Multicast group 0xFFDA Bandwidth Request	0xFFDA	1	605
Broadcast Bandwidth Request	0xFFFF	1	805
SS { XE "SS"} 5 Uplink Grant	0x007B	4	961
SS { XE "SS"} 21 Uplink Grant	0x01C9	7	1126
-	-	-	-
-	-	-	-
-	-	-	-

Figure 6.2   Sample UL-MAP with multicast and broadcast IE.

number of subsequent UL-MAP messages indicated by the Contention-based reservation timeout parameter.

An example of bandwidth allocation using multicast and broadcast polling is provided by Figure 6.2 which illustrates the Information Elements (*IEs*) of an *UL-MAP* message, [4, 5]. The *Offset* field specifies where the burst starts while the *UIUC* field indicates the encoding and the modulation type that the MS should adopt while transmitting during this interval. On the other hand, the *CID* field specifies how the bursts are addressed while *Uplink Grants* are intervals where the MSs can send uplink data. It is valuable to note that bandwidth request regions are used to send bandwidth requests only [8].

6.4.3   Bandwidth Allocation Procedures

Contrary to the bandwidth requests which are done on the basis of individual *CIDs*, each bandwidth grant is addressed to the MS's Basic *CID*. More precisely, the MS cannot determine which request is being served as the BS simply grants the bandwidth for the MS without specifying its actual use. Therefore, when the MS receives a shorter transmission opportunity than expected, no explicit reason is given. To address this issue, the MS refers to the latest information received by the BS and the status of the request and may decide to execute backoff and request again bandwidth or drop the SDU. A MS always transmits in an interval defined by a Data Grant IE directed at its Basic *CID*. Therefore, unicast polling of a MS should be done by allocating a Data Grant IE directed at its basic *CID*. The procedure implemented by the MS to obtain the required bandwidth is illustrated by Figure 6.3 [4, p. 143].

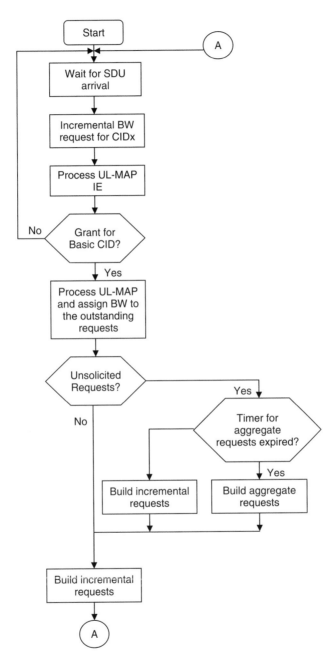

Figure 6.3 MS Request/Grant flow chart.

The IEEE 802.16e amendments do not specify particular algorithms for bandwidth allocation. Nevertheless, the BS has to implement an advanced bandwidth allocation algorithm in order to provide the required QoS level for the managed SSs. In particular, the maximum latency parameter should be taken into account in order be able to support time-sensitive applications. The authors of [9] consider the head-of-line waiting time of packets as the scheduling metric for the real time traffic but the QoS service classes are not involved. The Two-Phase Proportionating (TPP) scheme proposed in [10] utilizes the dynamic adjustment of the DL and UL in order to maximize the bandwidth utilization but does not address the physical layer characteristics such as the adaptive Modulation and Coding Scheme (MCS). The authors of [11] apply strict priority while taking into account the employed MCS but they do not consider latency; thus causing starvation for the low-level service classes.

To address such issues, the authors of [12] propose a Highest Urgency First (HUF) algorithm which may be adopted with the physical-layer OFDMA-TDD. That algorithm begins by translating the data bytes of requests to slots reflecting the MCS of every SS then calculating the number of frames to guarantee the required maximum latency for every request of service flows. Then, the algorithm pre-calculates the number of the slots that will be used by the DL/UL requests which should be transmitted in that scheduling frame. After that, the algorithm determines the portion of DL/UL subframe. The third phase consists in allocating the slots for every flow using U-factor which considers the urgency of every bandwidth request. Finally, the slots will be allocated for every queue to the SSs.

6.5 Scheduling Service

In this section, we discuss the Mobile WiMAX scheduling service by examining the properties that enable it to fulfil broadband data service provision while offering the required QoS over time-varying broadband wireless channels.

6.5.1 Scheduling Architecture in Mobile WiMAX

Mobile WiMAX has been designed with QoS issues in mind. Providing the required QoS level depends on four operations, which are the Connection Admission Control, packets scheduling, traffic policing and traffic shaping. Packets scheduling consists of serving the resource requests which have been sent by the mobile users and queued at the BS level. Scheduling is based on two orthogonal components which are deciding the order of serving the users' requests and managing the service queues [13]. Schedulers implemented at the mobile WiMAX BSs determine how much bandwidth should be granted to a particular SS (Subscriber Station) without distinguishing between the traffic destined for that SS. Therefore, the SSs need to schedule their different traffic at their own level [14]. Consequently, data scheduling is anticipated at both the BS and the SSs in order to select the packets that should be transmitted on the downlink and the uplink.

The general QoS architecture defined by the IEEE 802.16 standard is depicted by Figure 6.4. The dotted blocks indicate the parts that are not defined by the standard. More precisely, the IEEE 802.16 amendments did not specify the uplink scheduling for the rtPS, nrtPS and BE services that should be implemented at the BS level; they also omitted to

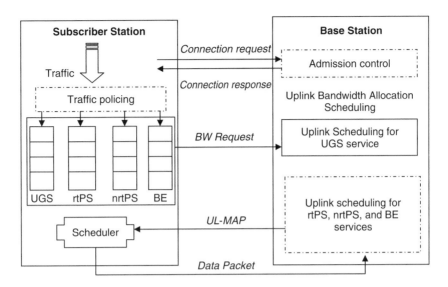

Figure 6.4 General QoS architecture defined by IEEE 802.16 standard.

define the admission control and the traffic policing processes [15]. According to the
IEEE 802.16 amendments, the BS schedules the transmission of the SSs' downlink data
packets and their bandwidth requests. That downlink scheduling is processed by the BS at
the beginning of each frame in order to construct the DL-MAP and the UL-MAP control
information messages required for the management of the downlink and uplink subframes.
The scheduler in the downlink direction has a complete knowledge of the queues' status;
therefore, it can use classical scheduling algorithms as Weighted Round Robin (WRR)
and Weighted Fair Queuing (WFQ) [16]. It is useful to note that additional downlink
flows cannot be served till the end of the subsequent uplink subframe after downlink
flows are served in their corresponding subframes. The uplink scheduling is much more
complex than downlink since the input queues are located in the SSs. More precisely, the
uplink scheduling implemented at the BS level needs to coordinate its decisions with all
the managed SSs.

   The scheduling architecture for an IEEE 802.16 system should guarantee efficient bandwidth utilization while providing fairness between users and responding correctly to the
QoS constraints of real time applications [17]. Two types of scheduling architecture arise:
traditional scheduling architecture and new scheduling architecture. Traditional scheduling architectures are based on classical and simple scheduling algorithms that were used
initially in wired networks. Examples of such algorithms are the First In First Out (FIFO)
algorithm and Round Robin (RR). Some traditional architectures may use these algorithm
in their initial form while other architectures may implement additional mechanisms in
order to extend these algorithms so that they become more suited to the IEEE 802.16
QoS context. Although the additional mechanisms respond better to the class structure of
the standard and guarantee fairness and traffic differentiation, they are more complex to
implement. New scheduling architectures implement new scheduling techniques especially
designed for the IEEE 802.16 standard.

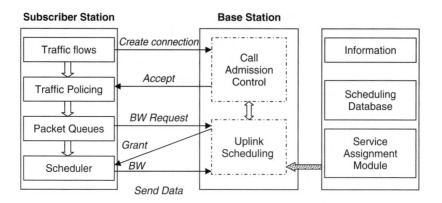

Figure 6.5 New QoS architecture.

The authors of [18] proposed a hierarchical model based on the implementation of enhanced mechanisms for a traditional architecture with a modification of the IEEE 802.16 standard architecture. That new architecture is depicted by Figure 6.5; it is based on the hierarchical distribution of the total bandwidth. More precisely, the total bandwidth is distributed among the four classes of the IEEE 802.16d standard. Each class distributes its bandwidth according to the state of its queues and may implement the scheduling algorithm that fits its constraints. The *Information* module depicted in Figure 6.5 collects, at the beginning of each time frame, the queue size information from the received bandwidth requests during the previous time frame. After that, the *Information* module updates the *Scheduling Database* module. The *Service Assignment* module refers to the *Scheduling Database* module in order to generate the adequate UL-MAP messages. The BS broadcasts the UL-MAP to the managed SSs in the downlink subframe while each SS's scheduler transmits its data packets according to the received UL-MAP.

6.5.2 Packet Schedulers Overview

Several scheduling algorithms exist for both wired and wirelesss networks. Nevertheless, adopting wired scheduling algorithms for the wireless context does not achieve good performance owing to the variable state of the wireless channel. As stated earlier, the mobile WiMAX amendments did not specify a particular data scheduling algorithm but left it vendor-specific. Consequently, many researchers tried to design simple, scalable and optimized scheduling algorithms able to accommodate different QoS requirements while fulfilling efficient bandwidth utilization and being fair in both the long run and the short run. It is worth to note that scheduling algorithms may adopt different approaches. For instance, some of them are priority based as they schedule all the connections by one centralized scheduling. The second approach selects a different scheduling algorithm for each different class of traffic. Other algorithms are just designed for a particular class of traffic [15]. In the following, we present an overview of some of the scheduling algorithms that may be adopted in the wireless context.

6.5.2.1 Proportionate Fair (PF) Scheduling

This scheduling algorithm was proposed by the Qualcomm Company, which was realized in the IS-856 standard for the downlink traffic scheduling (also known as High Data Rate (HDR)) [19, 20]. The designers of this scheduling algorithm intended to enhance system throughput while guaranteeing fairness for the active queues. This is achieved by managing the compromise between the system throughput and the starvation of low priority users, [15]. The PF scheduling is based on the priority function $Ui(t) = ri(t)/Ri(t)$ where ri(t) represents the current data rate and Ri(t) is an exponentially smoothing average of the service rate received by the SS i up to slot t. The queue having the highest value of Ui(t) is served at the time slot t. The average throughput of the different queues is then updated as follows:

$$Ri(t + 1) = (1 - 1/Tc)Ri(t) + (1/Tc)ri(t) \text{ if the connection i is served at the time slot t}$$
$$Ri(t + 1) = (1 - 1/Tc)Ri(t) \text{ if the connection i is not served at the time slot t}$$

where Tc is a time-constant assumed to be 1000 slots (1.66 second) in the CDMA-HDR system [19]. Adopting a larger Tc value makes the perceived throughput less sensitive to the short-time starvation in the queue. Therefore, the scheduler may wait for a longer period of time for a particular user switching from a bad channel condition to a good one. When the PF scheduling algorithm is used to manage a great number of users, an additional throughput gain may be obtained by scheduling them to use the characteristics of fast fading channels, called multi-user diversity gain. The PF scheduling algorithm is simple but it does not guarantee specific QoS requirements such as delay and jitter as it was designed originally for saturated queues with non real-time data service [21].

6.5.2.2 Integrated Cross-Layer Scheduling

The authors of [22] and [23] have proposed scheduling schemes integrating different algorithms in order to handle different classes of service with different QoS constraints. The scheduling algorithm proposed in [22] is based on a priority function relative to each queue. The priority metric of each queue is updated with respect to its service status and the channel condition. First, the algorithm allocates a fixed number of time slots for the UGS queues as indicated in the IEEE 802.16 amendments. The queues of the real-time Polling Service are managed with the Earliest Deadline First (EDF) [24] an algorithm which is sensitive to delay. Note that with EDF, the priority of scheduling a packet increases with the time period it spends in the queue.

A scheme similar to the PF algorithm is deployed for the queues of the non real-time Polling service and the Best-Rate discipline is adopted for the management of Best Effort queues. In order to distinguish the priority of the four types of service, the class coefficients are associated to the queues of each service type. The authors of [22] based their algorithm on the following: assuming that the total time slots allocated to the UGS data streams is N_{ugs} per frame, then the residual time slots assigned to the other QoS classes is $N_r = N_d - N_{ug}$ where N_d is total time slots in one frame. The priority function for a connection i at time slot t is defined as follows:

$$\varphi_i(t) = \beta_{class}(R_i(t)/R_N)^*(1/F_i(t)) \text{ if } F_i(t) \geq 1 \text{ and } \varphi_i(t) = \beta_{class} \text{ if } F_i(t) < 1.$$

The different parameters of the priority function are set as follows:

- $\beta_{class}(\in [0, 1])$ is the coefficient of the service classes. As the priorities of the different classes may be ordered as

$$priority \; (rtPS) > priority \; (nrtPS) > priority \; (BE),$$

  then the coefficients of those classes may be classified as β rtps $> \beta$ nrtPS $> \beta$ BE.
- Ri (t) is the current number of bits that may be carried by one symbol on frame t via Adaptive Modulation and Coding (AMC), which is determined by the channel condition.
- RN refers to the maximum number of bits in one symbol.
- $F_i(t)$ is an indicator of the delay satisfaction function that can be defined as follows:
  - For real-time connections, it is given by
    $F_i(t) = T_b \text{-} W_i(t)$ with T_b is the delay bound specified for the connection and $W_i(t)(\in [0, T_i])$ is the longest packet waiting time.
  - For non real-time connections, $F_i(t)$ is the ratio of the average transmission rate to the minimum reserved rate.

The priority function for the BE connections is $\varphi_i(t) = \beta_{BE}(R_i(t)/R_N)$. $\varphi_i(t)$ depends on the normalized channel quality regardless of the delay or rate performance. Note that this scheme provides a diverse QoS support for multiple connections. Nevertheless, authors of [22] did not determined how to choose the upper bounds of the β_{rtPS}, β_{BE}, N_r and N_{ug} parameters values in order to obtain the optimal performances. Besides, it is hard to implement the scheduler owing to its high complexity.

6.5.2.3 TCP-Aware Uplink Scheduling Algorithm for IEEE 802.16

The TCP-Aware Uplink Scheduling Algorithm for IEEE 802.16 works was particularly designed for the BE service class [25]. The authors of [25] based their reasoning on two arguments. First, it is not advantageous to request bandwidth for BE connections as they do not have particular QoS requirements. Second, it is not wise to allocate the remaining bandwidth equally to all remaining BE connections as some connections would not use the allocated bandwidth while others would not be satisfied with the amount of allocated resources.

To address these issues, the authors of [25] propose calculating the bandwidth for a particular connection according to the sending rate of that connection. They avoid allocating a fixed amount of bandwidth for each connection as the sending rate is changing continuously; they also define the demand of a flow as the amount of access link bandwidth requested for achieving its maximum throughput so as not to be limited by the access link bandwidth. The first step of the proposed algorithm consists of computing the sending rate. If that rate is smaller than the allocated bandwidth, then the demand will be equal to the sending rate. However, if the sending rate is equal to the allocated bandwidth, then the allocated bandwidth will be slightly higher than the current sending rate. Finally, if the sending rate is larger than the allocated bandwidth, then the allocated bandwidth will increase until the stabilization of the sending rate. We can deduce that the allocated bandwidth is always higher than the sending rate of connection so that the sending rate is correctly estimated at any given time. Changes in the sending rate are detected as the maximum and the minimum values are maintained during a time period and changed accordingly. When these values need to be updated, the algorithm described previously is used for demand estimation. After estimating the demand for each connection, maximum-minimum fair scheduling is used for allocating the total bandwidth among all connections.

6.5.2.4 Slots Allocation Scheme

The authors of [26] proposed a scheduling architecture that allocates slots based on the QoS requirements, bandwidth requests sizes and the network parameters. More precisely, the BS translates the QoS requirements of the managed SSs into a number of allocated slots. The proposed algorithm consists of three stages. In the first stage, the BS computes the minimum number of slots that ensures the basic QoS requirements for each connection. In the second stage, the remaining unused slots are allocated to other connections. Note that if these unused slots are not allocated, they will be lost because the BS is not able to serve additional downlink flows until the end of the subsequent uplink frame. To address this issue, the authors of [26] propose allocating unused slots to the rtPS, nrtPS and BE connections. The allocated slots for each connection are then interleaved in order to decrease the maximum jitter and delay values.

6.6 Maintaining QoS During Handover

In this section, we will address the supported Mobile WiAMX handover schemes in order to determine the advantages and drawbacks of each scheme and then determine in which situations it should be adopted. We will then provide an overview of the adopted handover monitoring functions and the different parameters that should be taken into account in order to optimize the handover operation. For instance, the handover decision should not be based uniquely on signal strength but should also consider the provided QoS, the available resources and the cost.

6.6.1 WiMAX Handover Schemes

The IEEE 802.16e specifications support full mobility access which provides service continuity at high speeds up to 160 km/h and achieves seamless handover with less than 50 ms latency and less than 1 % packets loss ratio. Three handover methods are defined: the mandatory Hard Handover (HHO), the optional Fast Base Station Switching handover (FBSS) and the Macro Diversity Handover (MDHO). Initially, HHO is the only type that is required to be implemented by certified Mobile WiMAX equipment.

6.6.1.1 Hard Handover (HHO)

Hard handover adopts the 'break before make' philosophy. More precisely, it results in a sudden connection transfer from one managing BS to a second since the MS can communicate with only one BS at the same time. Consequently, all connections with

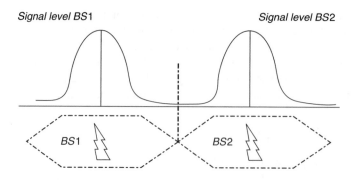

Figure 6.6   The hard handover process.

the serving BSs are broken before a new connection with the target BS is established. In Figure 6.6, the thick line at the border of the cells labelled 'BS1' indicates the place where HHO is executed . Generally speaking, HHO is the less complex handover type: nevertheless it induces high latency. Consequently, HHO is used mainly for data as it is not suited to real time latency-sensitive applications such as VoIP [27].

6.6.1.2   Macro-Diversity Handover (MDHO)

Macro Diversity Handover (MDHO) allows the MS to maintain a valid connection simul-taneously with more than one BS. A Diversity Set referred to as the Active Set is also maintained. The active set consists of a list of BSs that may be involved in the handover procedure. That list is updated through the exchange of MAC management messages based on the long-term Carrier to Noise plus Interface Ratio (CINR) of BSs. In more detail, the list update is performed with regard to two threshold values broadcast in the Downlink Channel Descriptor (DCD) which are the Add Threshold H_Add_Threshold and the Delete Threshold H_Delete_Threshold. A serving BS is dropped from the diver-sity set when the longterm CINR is less then H_Delete_Threshold while a neighbour BS is added to the diversity set when its long-term CINR is higher than H_Add_Threshold.

The MS needs to monitor the BSs permanently in the Diversity Set then choose an Anchor BS from among them. Moreover, the MS synchronizes and registers to the anchor BS then performs ranging while monitoring the downlink channel for control information. The MS communicates with the anchor BS and the BSs of the diversity set as depicted by Figure 6.7. Two or more BSs transmit data on the downlink so that multiple copies are received. Therefore, the MS needs to combine the received information using any of the well-known diversity-combining techniques. At the uplink, the MS transmission is received by multiple BSs and selection diversity of the received information is performed to pick the best uplink. Note that the BSs referred to as 'Neighbor BSs' in Figure 6.7 receive communication from the MS and the other BSs but the received signal level is not sufficient to add them into the Diversity Set.

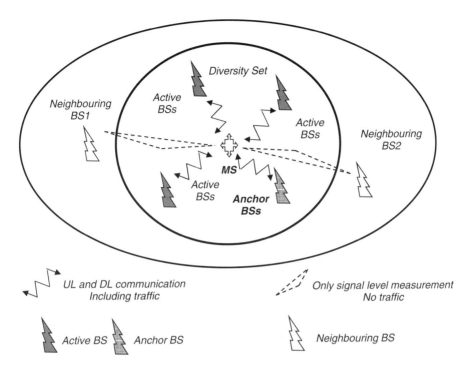

Figure 6.7   The macro diversity handover process.

6.6.1.3   Fast Base Station Switching (FBSS)

When the Fast Base Station handover is supported by both the MS and the BS, the MS maintains a list of BSs referred to as the Active Set and then monitors it continuously. The MS may perform ranging and maintain a valid connection ID with the BSs of the Active Set but it is allowed to communicate with only one BS called the anchor BS as depicted by Figure 6.8. In more detail, the MS is registered and synchronized with the anchor BS; both entities exchange uplink and downlink traffic including management messages. That anchor BS may be changed from frame to frame with respect to the BS selection scheme. In that case, the connection is switched to the new anchor BS without performing explicit handover signalling as the MS simply reports the ID of the newly selected BS on the Channel Quality Indicator Channel (CQICH). Every frame can then be sent via a different BS belonging to the Active Set.

The anchor BS may be updated by either implementing the 'Handover MAC Management Method' or the 'Fast Anchor BS Selection Mechanism'. The first updating method is based on the exchange of five types of MAC management messages while the second updating method transmits anchor BS selection information on the Fast Feedback channel. The new anchor BS should belong to the current diversity set; its selection is based on

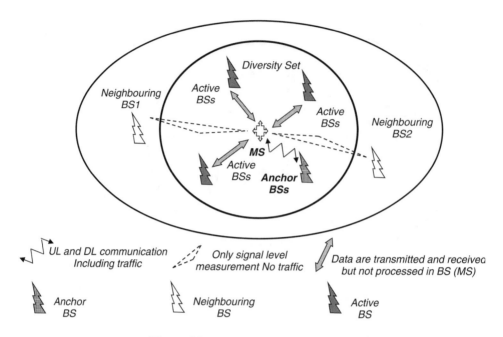

Figure 6.8   Fast base station switching.

the signal strength reported by the MS. Adding BSs to the diversity set and removing others is done on the basis of the comparison of their long-term Carrier to Noise plus Interface Ratio (CINR) to H_Add_Threshold and H_Delete_Threshold. To conclude, the FBSS and the MDHO have many similarities and achieve better performance compared to HHO. Nevertheless, they are more complex as they require the BSs of the active set or the diversity set to be synchronized, to use the same carrier frequency and to share the network entry information.

6.6.2   Optimizing Handover to Maintain the Required QoS

Handover optimization is a key challenge for network management as it results in the enhancement of the network performances by optimizing throughput, routing, delay profiles, delivered QoS and communication costs. The handover process consists of one MS migrating from the air-interface provided by one BS to the air interface provided by a second BS [5]. First, the MS performs a cell reselection by using the neighbour BS information or making a request to schedule scanning intervals or sleep-intervals to scan. The handover begins with a decision for the MS to handover from a serving BS to a target BS. Such decision is taken either by the MS or the serving BS. The MS needs then to synchronize to downlink transmissions of target BS and obtain DL and UL transmission parameters. IEEE 802.16e amendments implement some optimization mechanisms related to handover as shortening the handover is a key issue for minimizing the impact of such process on the provided QoS. For example, the synchronization to target BS downlink may be shortened by a previous reception of the MOB_NBR-ADV message which indicates to the MS the target BS IDentifier, the physical frequency, the Down-link Channel Descriptor (DCD) and the Uplink Channel Descriptor (UCD). To further establish synchronization with the target BS downlink process, the target BS may allo-cate non-contention-based initial ranging opportunities to the handing over MS if it has previously received handover notification from the serving BS. BS-to-BS communication which includes the allocation is processed over the backbone network. The MS and the target BS should then perform ranging. The network re-entry stage may be also shortened if the target BS possesses information from the serving BS. More precisely, the target BS may skip one or several of the following network entry phases depending on the nature of the provided information:

- negotiation of basic capabilities;
- PKM authentication phase;
- TEK establishment phase;
- sending the REG-REQ;
- sending an unsolicited REG-RSP message with updated capabilities information. The BS may choose to skip this stage.

The IEEE 802.16e amendments define the HO Process Optimization Type-Length-Value (TLV) which should be included in the ranging response (RNG-RSP) message when the MS is trying to perform network re-entry or handover and the target BS needs to identify re-entry process management messages that may be omitted during the handover attempt. In particular, the HO Process Optimization TLV defines the re-entry process manage-ment messages that may be skipped during the current handover owing to the availability of the MS service and further operational context information and the MS service and operational status post handover completion. In this case, the target BS may choose to use the MS service and operational information obtained over the backbone in order to send unsolicited SS Basic Capabilitiy response (SBC-RSP) and/or REG-RSP manage-ment messages to MS operational information. However, the MS will not enter Normal Operation with target BS until it has received all messages related to network re-entry, and MAC management as indicated in HO process Optimization. HO Process Optimiza-tion TLV is 8-bits long where each bit's value equalling 0 indicates that the associated re-entry management message is required and each bit's value equalling 1 indicates that the associated re-entry management message may be omitted. More precisely, we have: bit 0 specifies whether the SS Basic Capability request or response messages are required during re-entry, bit 1 specifies whether the PKM authentication phase except TEK phase may be omitted during current re-entry, bit 2 specifies whether the PKM TEK creation phase during re-entry processing is required, bit 3 specifies whether the REG-REQ/RSP management may be omitted, bit 4 indicates whether the network address acquisition management is required, bit 5 indicates whether we may omit the day acquisition man-agement, bit 6 indicates whether we may omit TFTP management messages and finally bit 7 indicates whether we adopt a full service and operational state transfer or a sharing between the serving BS and the target BS.

Several authors have focused on shortening the network entry process which is greatly responsible for causing handover delay [28, 29, 30, 31, 1]. Nevertheless, they mainly addressed the ranging process and the reduction of redundancies in the exchanged sig-nalling messages. For example, the authors of [28] proposed a fast handover algorithm that reduces the waste of wireless channel resources while reducing the handover latency. More precisely, a target BS estimation using mean CINR and arrival time difference was designed in order to reduce unnecessary neighboring BSs scanning and association process. Also, some redundant processes existing in the IEEE 802.16e draft standard and related to network topology acquisition and scanning were analyzed and abbreviated. Nevertheless, the research was based on the IEEE 802.16e draft standard and the analysis was not detailed [29]. Besides, the ongoing service may be interrupted as the transport connection is released by the serving BS then reassigned by the target BS after handover, [31].

The authors of [29] proposed several fast handover schemes in order to reduce the redundancies in the IEEE 802.16e MAC layer handover process and optimize the network re-entering process. More precisely, they proposed a target BS estimation algorithm allowing the mobile station to select the neighboring BS with the largest CINR value as the single target BS for scanning; thus avoiding unnecessary synchronization or association processes. These authors also proposed adopting fast ranging by instructing the target BS to allocate a dedicated uplink ranging opportunity to the mobile station; thus avoiding the time wasted in contention-based ranging. Finally, they recommended adopting pre-registration so that the target BS can obtain the service flow and the authentication information related to the mobile station through backbone networks before handover without being obliged to communicate with the authorization server.

The authors of [30] proposed two new scanning strategies to reduce the handover delays. The first strategy intends to reduce the number of frequencies checked during each scanning operation. The mobile station maintains information on the frequencies being used and may adopt one of the 'Most Recently Used' or the 'Most Frequently Used' approaches in order to make its choice. The second strategy uses the history of the mobile station handovers along with the information provided by the currently serving BS in order to improve the choice of a handover target neighbour BS to begin scanning. The authors modelled and simulated an area of Worldwide Interoperability for Microwave Access (WiMAX) coverage using real-world mobility trace data in order to show that their proposed strategies reduced the time required for scanning and handover.

The authors of [1] proposed adding a message called FastDL_MAP_IE to receive the real-time traffic from the target BS even if the mobile station was not synchronized with uplink from target BS. However, the mobile station will maintain old Connection IDentifiers (CIDs) until they are updated by the Registration Response (REG-RSP) message; thus leading to possible collisions with the existing CIDs in the target cell. If they occur, such collisions may harm the real-time services. To address this issue, the authors of [31] proposed a transport CIDs mapping scheme for real-time applications that guarantees continuous communications between the handover related BSs and the mobile station using former CIDs (unique among neighbouring cells); thus avoiding CIDs of handing over services conflicting with those of ongoing services in the target BS. They also intended to accelerate the HHO by adopting an enhanced link-layer QoS aware handover scheme known as Passport Handover.

6.7   Enhancing WiMAX QoS Issues: Research Work

In this section, we discuss the limits of the standardised Mobile WiMAX QoS procedures and we survey some research studies that propose new mechanisms aiming at addressing such limits and enhancing QoS provision. In particular, we present some new QoS mechanisms and a proposition for WiFi and WiAMX QoS integration. We also present the WEIRD (WiMAX Extension to Isolated Research Data networks) project that proposes a solution for an end-to-end WiMAX network architecture offering a support for end-to-end QoS through a full integration between the resource allocation in the WiMAX wireless link and the signalling for resource reservation in the wired segment of the access network and towards the core network.

6.7.1 New QoS Mechanisms

The IEEE 802.16 standard has been designed with QoS issues in mind. Its MAC connection-oriented nature along with the request-grant mechanisms and the service flows usage with five defined service classes enable subscribers to enjoy real-time and complex applications. However, additional intelligence needs to be implemented within SSs in order to distribute their allocated bandwidth among their different connections with different QoS requirements. Besides, the IEEE 802.16 amendment did not specify particular mechanisms to perform resources management, traffic shaping, admission control and scheduling. QoS signalling mechanisms have been defined by the standard but the definition of algorithms that may use such mechanisms for bandwidth allocation was left to the vendors [32].

Many researchers have focused on developing scheduling algorithms for uplink and downlink bandwidth allocation. For example, many have proposed a new QoS architecture integrating its own packet scheduler. In particular, the authors of [33] propose a scheduling algorithm and an admission control policy along with system parameters and traffic characteristics for which the network can provide QoS. Authors in [34] detail a new architecture and propose a priority-based scheduler that manages the priority according to channel and service quality. Some scheduling algorithms were discussed in the preceding section. Nevertheless, the majority of the proposed algorithms were simulated and not tested in practice so that we do not have a clear idea of performance achieved. Moreover, Admission control procedures should be combined with the proposed scheduling algorithms in order to build a complete QoS architecture and evaluate its performance as a whole. Finally, the mesh mode should be addressed in order to provide QoS-based applications for mobile and Non Line of Sight Nodes.

6.7.2 The WEIRD Project

The WiMAX Extension to Isolated Research Data networks (WEIRD) project is a European project established with the goal of implementing research testbeds based on WiMAX technology in order to enable isolated and remote areas to connect to the GEANT2 research backbone network. WEIRD members participate in the standardization of the WiMAX integration into next generation networks through building four European testbeds connected via GEANT2. Three application groups are adopted: Volcano and seismic activities monitoring, fire prevention and tele-medicine. The resulting application scenarios are deployed within the implemented testbeds while enhanced Network Control and Management entities and an improved version of WiMAX are prototyped and then validated. In particular, the WEIRD project aims to enhance the WiMAX technology

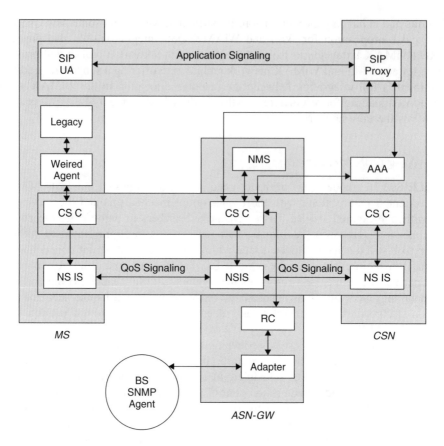

Figure 6.9   General architecture defined by the WEIRD project.

through enhancing the handover and access control mechanisms at the convergence layer while combining interoperability and mobility management and studying Radio-over-fiber techniques for massive and cost-effective WiMAX deployment. The WEIRD project also aims to enhance the IP control plane by studying the Advanced Authorization, Authentication and Accounting mechanisms, enhancing QoS support for real time and mission critical applications and enhancing resource management mechanisms [35].

The application scenarios considered by the WEIRD project are supported by the WEIRD architecture depicted by the Figure 6.9 [36]. That architecture is built upon the WiMAX architecture and we may distinguish the *Access Service Network* (ASN) and the *Connectivity Service Network* (CSN). The ASN provides radio access to the *Mobile Station* (MS) and comprises the *Base Stations* (BSs) and the *ASN Gateways* (ASN-GW). The CSN provides IP connectivity to the subscribers. Vertically, the WEIRD architecture is made up of the *Application and Service Stratum* and the *Transport Stratum*. The *Application and Service Stratum* implements the management and control of the supported applications while the *Transport Stratum* manages the available resources and guarantees the data exchange through the network architecture. Horizontally, the WEIRD architecture is made up of the *Management Plane*, the *Control Plane* and the *Data Plane*.

The *Management Plane* supervises medium- and long-term tasks such as QoS provisioning through the *Network Management System* (NMS) and the traffic management while the *Data Plane* supervises the user/application data. The *Control Plane* monitors short-term tasks such as QoS reservation through the introduction of new modules aimed at supporting real time applications with QoS differentiation. In particular, both SIP and non SIP (legacy) applications are supported. A *SIP User Agent* (SIP UA) module implemented at the MS level communicates directly with the SIP Proxy in the CSN. A *WEIRD Agent* module is implemented at the MS level in order to configure and monitor the required QoS parameters for the legacy applications. The *Connectivity Service Controllers* (CSC) modules placed in every element of the architecture play the most important role in terms of QoS monitoring. For instance, the CSC module of the ASN (CSC_ASN) coordinates the QoS functionalities such as resources allocation and admission control in the ASN and WiMAX segments. The CSC module of the MS (CSC_MS) communicates with the *WEIRD Agent* in order to collect the QoS parameters required by the legacy applications and then forwards them to the CSC_ASN. The latter triggers the Resource Controller (RC) module on the ASN gateway which acts as a generic layer between the upper modules of the WEIRD architecture and the lower ones. Note that the RC manages the WiMAX links and the QoS on these links by implementing the creation, modification and deletion of the service flows in these links as well as the associated *Service Classes* (CSs) and *Convergence Sublayer Classifiers* (CSCl). Both static and dynamic models have been defined and implemented in order to manage service flows and their status. Finally, the CSC module placed at the CSN level (CSC_CSN) gets the QoS reservation requests sent by the CSC_ASN then establishes the QoS paths in the core network. The CSC_CSN also implements medium- and long-term management functions including QoS provisioning, as stated earlier.

   The WEIRD supported applications can be grouped into two classes. The first class of application, called 'WEIRD aware' applications, can be updated to use the WEIRD defined services. Examples in this class include the SIP-based modules. Session based applications may follow the '*QoS assured*' model or the '*QoS enabled*' model. The QoS assured model establishes a session/flow only if the requested QoS can be guaranteed. To the contrary, the availability of the QoS resources does not influence the success of the session setup but may influence the effective level of QoS associated with it when adopting the QoS enabled model. In both cases, the SIP proxy always performs the reservation process. The second class of applications (i.e., legacy applications) cannot be updated but they may use the WEIRD services through an agent. Dedicated software modules have been developed and implemented at the MS level in order to support such applications. WEIRD supports environmental monitoring applications designed for fire and volcano monitoring, telemedicine applications based on high resolution video and data streaming from medical equipments and general purpose applications with mobility support including voice over IP (VoIP), video conference over IP, and video streaming for generic usage in monitoring and content diffusion [36].

6.7.3   WiFi and WiMAX QoS Integration

IEEE 802.11 Medium Access Control is based on either the Distributed Coordination Function (DCF) implementing the Carrier Sense Multiple Access with Collision

Avoidance or the Point Coordination Function (PCF) implementing a centrally controlled access. The IEEE 802.11e standard extends the basic IEEE 802.11 in order to provide QoS. More precisely, IEEE 802.11e implements two coordination functions which are the Enhanced Distributed Coordination Function (EDCF) which is a QoS enabled version of the basic DCF and the Hybrid Coordination Function (HCF) Controlled Channel Access (HCCA) which is similar to the PCF.

EDCF classifies the traffic into one of four queues known as Access Categories (ACs). Each AC is associated with three parameters, say the minimum and maximum Contention Window (CW) and the Arbitration Interframe Space (AIFS). The values of these parameters can be adjusted in order to assign a different priority to the AC. For example, an AC with higher priority will have smaller CW values and shorter AIFSs. After sensing an idle medium, each AC should wait for its AIFS then start the backoff timer. If an AC is the only one whose timer has expired, it will transmit a frame. Otherwise, the AC with the highest priority will transmit. A possible mapping between ACs and application types is the following: AC number 0 is assigned to background applications, AC number 1 is assigned to Best Effort applications, AC number 2 is assigned to video applications and AC number 3 is assigned to voice applications [37].

HCCA defines a superframe with a contention-free period followed by a contention period. A mobile subscriber needs to use the MAC QoS signalling in order to set up a Traffic Stream with particular QoS requirements. Up to eight traffic streams may exist with HCCA and packets assigned to a particular traffic stream are served with the same QoS requirements.

The authors of [38] propose an integrated architecture combining the WiMAX and the WiFi technologies and introducing a WiMAX/WiFi Access Point called W^2-AP. In the system under consideration, each WiMAX BS manages multiple SSs and multiple W^2-APs located at its coverage area. The WiFi clients are connected to the WiMAX network via the W^2-AP devices and share the WiMAX connection between the BS and the W^2-AP while each SS possesses a dedicated connection between it and the BS. The proposed MAC layer module aiming at integrating the WiMAX and WiFi technologies is based on two MAC frameworks which are the MultiMAC and the SoftMAC frameworks as shown by Figure 6.10. The MultiMAC lies between the physical device and the network layer in order to switch between different MAC protocols. The idea behind this is to enable the most appropriate MAC variant to claim then decode the incoming frames and encode the outgoing ones depending on the current network conditions. The SoftMAC may be seen as a convergence sublayer that encapsulates the WiMAX Packet Data Units (PDUs) into a single WiFi PDU over 802.11a OFDM physical layer or decapsulates a single WiFi PDU into its WiMAX PDU.

In more detail, the proposed MAC module is made up of three components which are the W^2-AP, the WiFi node and the WiMAX BS. The W^2-AP module encompasses the MultiMAC framework, the proposed convergence 802.16 MAC and the 802.16 OFDM physical layer embedded within a conventional 802.11 access point. As it uses the Multi-MAC and the SoftMAC frameworks embedded in the proposed 802.16 MAC, the W^2-AP acts as a bridge in translating frames between a WiFi and an Ethernet interface or as a relay transferring frames between a WiFi and a WiMAX network. Moreover, the W^2-AP plays the role of a WiMAX subBS which forwards bandwidth requests issued by WiFi nodes to the WiMAX BS then allocates the granted bandwidth from the WiMAX BS to

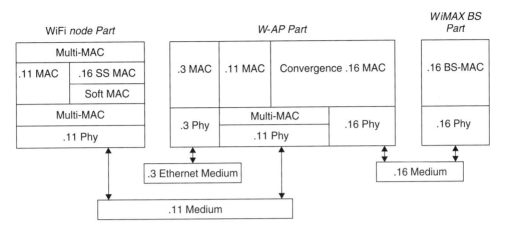

Figure 6.10   MAC layer module for WiMAX/WiFi integration.

that WiFi nodes. A WiFi node belonging to the proposed architecture adds the MultiMAC, SoftMC and 802.16 SS-MAC functions to the IEEE 802.11 MAC layer thanks to software upgrade techniques so that the upgraded 802.11 Network Interface Card allows the users located within the coverage area of a W^2-AP to access the backhaul services provided by the local WiMAX network. Finally, the MAC module in the WiMAX BS implements the initial IEEE 802.16 standard amendments. Since WiFi and WiMAX implement different protocols for the QoS management and the bandwidth access mechanisms, it is wise to implement additional functionalities in the WiFi MAC layer and Access Points so that they may support the connection-oriented services as WiMAX does [38]. The authors of [38] propose a Two-level Hierarchical Bandwidth Allocation (THBA) scheme in order to request and grant bandwidth within the proposed integrated network so that WiFi and WiMAX subscribers are controlled via a single protocol based on the IEEE 802.16 standard. The proposed scheme has two major advantages. First, it does not induce major modifications so that the time and expense required for the implementation of the integrated network are reduced. Second, WiFi is upgraded to support a fine level of QoS as in WiMAX without the need of implementing QoS mapping mechanisms; thus reducing the complexity of the implementation of the integrated network and guaranteeing continuity of QoS and consistency of bandwidth management throughout the network.

6.8   Further Reading

Mobile WiMAX has been deployed in the US, Russia and other countries around the world [39]. It is interesting to consider the degree of customer appreciations and satisfaction regarding this technology. It is also interesting to compare the mobile WiMAX technology with other wireless networking technologies such as Wibro, UMTS and HSPA/HSPA+ in terms of QoS provision for mobile subscribers [40, 41, 42, 43]. Moreover, providing seamless connectivity, handoff and mobility between WiMAX and 3G/3.5G/4G/WiFi networks requires the implementation of particular QoS mechanisms such as mapping and re-negotiation; it therefore represents a hot topic for research that needs to be investigated further.

Call admission control and bandwidth allocation for mobile WiMAX has been investigated by the authors of [44] who propose a dynamic scheme that supports voice, data and multimedia services with differentiated QoS. The authors of [45] developed a transmission algorithm for Mobile WiMAX system that selects the position of the boundary between the downlink and uplink subframes, the positions of the service flows within the frame, the coding and modulation schemes and the transmission power values in order to better satisfy the QoS requirements.

6.9 Summary

Mobile WiMAX has been designed to support advanced applications with specific QoS requirements while addressing mobility on the metropolitan scale. Therefore, mobile WiMAX subscribers will connect to their preferred services continuously with nearly the same QoS level while moving or driving from 75 to 93 miles per hour. This chapter intends to address QoS management in WiMAX networks; it aims at demonstrating how the mobile WiMAX technology will offer continuity of services while providing enhanced QoS guarantees in order to meet subscribers' demands. In particular, we present the architectural QoS requirements that have to be fulfilled during subscribers' mobility and then we discuss the mechanisms supported by Mobile WiMAX network to provide the required QoS. After that, we discuss the mechanisms adopted in order to maintain QoS during handover. Finally, we survey some research studies that analyze the limits of the standardized Mobile WiMAX QoS procedures and propose new mechanisms aimed at enhancing the provision of QoS.

References

[1] S. Choi, G. Hwang and T. Kwon, 'Fast Handover Scheme for Real-Time Downlink Services in IEEE 802.16e BWA System', Vol. 3 (2005), pp. 2028–32, IEEE VTC 2005 Spring, Sweden, May 2005.

[2] http://www.wimax.com/commentary/spotlight/what-every-company-needs-to-know-about-mobile-wimax-and-qos, accessed on 5 June 2009.

[3] 'Handoff Management in WiMAX networks', in WiMAX Network Planning and Optimization, ISBN: 9781420066623, Y. Zhang (Ed), Auerbach Publications, Taylor & Francis Group, April 2009.

[4] IEEE Standard for Local and Metropolitan Area Networks, Part 16: Air Interface for Fixed Broadband Wireless Access Systems, 2004.

[5] EEE 802.16e-2005 'IEEE Standard for Local and Metropolitan Area Networks Part 16: Air Interface for Fixed and Mobile Broadband Wireless Access Systems Amendment for Physical and Medium Access Control Layers for Combined Fixed and Mobile Operation in Licensed Bands', IEEE, 2006.

[6] Mobile WiMAX – Part 1: A Technical Overview and Performance Evaluation, WiMAX Forum, 2006, available at http://www.wimaxforum.org.

[7] M. Handley, Eve Schooler, H. Schulzrinne and J. Rosenberg, 'Session Initiation Protocol', RFC 2543.

[8] M. Carlberg Lax and A. Dammander, WiMAX – A Study of Mobility and a MAC-layer Implementation in GloMoSim, Master's Thesis in Computing Science, April 6, 2006.

[9] M. Andrews et al., 'Providing Quality of Services over a Shared Wireless Link' IEEE Communication Magazine, Feb. 2001, pp. 150–154.

[10] Y.N. Lin, S.H. Chien, Y.D. Lin, Y.C. Lai and M. Liu, 'Dynamic Bandwidth Allocation for 802.16e-2005 MAC', Chapter in Current Technology Developments of WiMax Systems, edited by M. Ma, Springer, 2009.

[11] A. Sayenko, O. Alanen, J. Karhula and T. Hamalainen, 'Ensuring the QoS Requirements in 802.16 scheduling', ACM MSWiM '06, Oct. 2006.

[12] Y. Lin, C. Wu, Y. Lin and Y. Lai, 'A Latency and Modulation Aware Bandwidth Allocation Algorithm for WiMAX Base Stations', in Proceedings of Wireless Communications and Networking Conference, 2008. WCNC 2008.

[13] S. Keshav, An Engineering Approach to Computer Networks, ATM networks, the Internet and the Telephone Network, Addison-Wesley, September 1997.

[14] D. Tarchi, R. Fantacci and M. Bardazzi, 'Quality of Service Management in IEEE 802.16 Wireless Metropolitan Area Networks', in Proceedings of IEEE ICC '06, Jun. 2006, Istanbul, Turkey

[15] A. Jain and A.K. Verma, Comparative Study of Scheduling Algorithms for WiMAX, available at http://tifac.velammal.org/CoMPC/articles/2.pdf.

[16] M. Gidlund and G. Wang, 'Uplink Scheduling Algorithms for QoS Support in Broadband Wireless Access Networks' Journal of Communications 4(2), March 2009.

[17] A. Khalil and A. Ksentini, 'Classification of the Uplink Scheduling Algorithms in IEEE 802.16', available at http://www.irisa.fr/dionysos/pages_perso/ksentini/aymen_iwdyn07.pdf.

[18] K. Wongthavarawat and A. Ganz, 'IEEE 802.16 Based Last Mile Broadband Wireless Militarily Networks with Quality of Service Support', in Proceedings of IEEE Milcom, 2003.

[19] A. Jalali, R. Padovani and R. Pankaj, 'Data throughput of CDMA-HDR a high efficiency-high data rate personal communication wireless system', in Proceedings IEEEE VTC pp.1854-1858, 2000, Tokyo, Japan.

[20] P. Bender, P. Black, M. Grob, R. Padovani, M. Sindhushayana and A. Viterbi, 'CDMA/HDR: a bandwidth-efficient high-speed wireless data service for nomadic users' IEEE Communication Magazine 38(7): 70–7, 2000.

[21] Xiaojing Meng, 'An Efficient Scheduling For Diverse QoS Requirements in WiMAX', a master degree thesis, University of Waterloo, available at https://uwspace.uwaterloo.ca/bitstream/10012/2736/1/thesis.pdf.

[22] Q. Liu, S. Zhou and G.B. Giannakis, 'A Cross-Layer Scheduling Algorithm with QoS Support in Wireless Networks' IEEE Transactions on Vehicular Technology 55(3), May. 2006.

[23] D. Tarchi, R. Fantacci and M. Bardazzi, 'Quality of Service Management in IEEE 802.16 Wireless Metropolitan Area Networks', in Proceedings of IEEE ICC'06, Jun. 2006, Istanbul, Turkey.

[24] L. Georgiadis, R. Guerin and A. Parekh, 'Optimal Multiplexing on a Single Link: delay and buffer requirements' IEEE Transactions on Information Theory 43(5): 1518–35, 1997.

[25] S. Kim and I. Yeom, 'TCP-aware Uplimk Scheduling for IEEE 802.16', IEEE Communication Letter, Feb., 2007.

[26] A. Sayenko, O. Alanen, J. Karhula and T. Hamalainen, 'Ensuring the QoS Requirements in 802.16 Scheduling', in Proceedings of ACM MSWiM, Montreal, 2006.

[27] Z. Becvar and J. Zelenka, Handovers in the Mobile WiMAX, In Research in Telecommunication Technology 2006 – Proceedings [CD-ROM]. Brno: Vysoké učené technické v Brně, 2006, Vol. I, pp. 147–50..

[28] D.H. Lee, K. Kyamakya and J.P. Umondi, 'Fast Handover Algorithm for IEEE 802.16e Broadband Wireless Access System', in Proceedings of the 1st International Symposium on Wireless Pervasive Computing, pp. 1–6, 2006.

[29] L. Wang, F. Liu and Y. Ji, 'Performance Analysis of Fast Handover Schemes in IEEE 802.16e Broadband Wireless Networks', in Proceedings of the Asia-Pacific Advanced Network (APAN) Network Research Workshop 2007, Xian, China, 2007.

[30] P. Boone, M. Barbeau and E. Kranakis, 'Strategies for Fast Scanning and Handovers in WiMAX/ 802.16', International Journal of Communication Networks and Distributed Systems 1, Issue 4/5/6: 414–32, 2008.

[31] W. Jiao, P. Jiang and Y. Ma, 'Fast Handover Scheme for Real-Time Applications in Mobile WiMAX', in Proceedings of the IEEE International Conference on Communications (ICC), pp. 6038, 6042, 2007.

[32] M.C. Wood, 'An Analysis of the Design and Implementation of QoS over IEEE 802.16', available at http://cec.wustl.edu/~mcw2/QoS_over_802_16/QoS_over_802_16.html .

[33] K. Wongthavarawat and A. Ganz, Packet Scheduling for QoS support in IEEE 802.16 Broadband Wireless Access Systems, John Wiley & Sons, 2003.

[34] Qingwen Liu, Xin Wang and G.B. Giannakis, 'Cross-Layer Scheduler Design with QoS Support for Wireless Access Networks', Second International Conference on Quality of Service in Heterogeneous Wired/Wireless Networks, p. 21, August 22–24, 2005.

[35] The WEIRD project home page, available at http://www.ist-weird.eu/.

[36] S. Mignanti, M. Castellano, M. Spada, P. Simoes, G. Tamea, A. Cimmino, P.M. Neves, I. Marchetti, F. Andreotti, G. Landi and K. Pentikousis, 'WEIRD Testbeds with Fixed and Mobile WiMAX Technology

for User Applications, Telemedicine and Monitoring of Impervious Areas', 4th International Conference on Testbeds and Research Infrastructures for the Development of Networks & Communities. TridentCom, 2008.

[37] G. Boggia, P. Camarda, L.A. Grieco and S. Mascolo, 'Feedback-Based Bandwidth Allocation with Call Admission Control for Providing Delay Guarantees in IEEE 802.11e Networks', November 2004, available at http://www.sciencedirect.com/science.

[38] H. Lin, Y. Lin, W. Chang and R. Cheng, 'An Integrated WiMAX/WiFi Architecture with QoS Consistency over Broadband Wireless Networks', IEEE, 2009.

[39] http://www.findarticles.com/p/articles/mi_m0EIN/is_2008_Oct_1/ai_n29468857/.

[40] Telecommunications Technology Association, 'Specifications for 2.3GHz Band Portable Internet (WiBroTM) Service', TTA Standard TTAS.KO-06.0082/R1, December 2005.

[41] S.N.P. Van Cauwenberge, 'Study of soft handover in UMTS', Technical University of Denmark, University of Gent (July 2003).

[42] HSPA, Release 5 (Downlink) and Release 6 (Uplink), 3GPP specifications.

[43] HSPA+, Release 7, 3GPP specifications.

[44] H.Y. Tung, K.F. Tsang, L.T. Lee and K.T. Ko, QoS for Mobile WiMAX Networks: call admission control and bandwidth allocation, in Proceedings of the IEEE Consumer Communications and Networking Conference, 2008.

[45] S.A. Filin, S.N. Moiseev and M.S. Kondakov, 'Fast and Efficient QoS-Guaranteed Adaptive Transmission Algorithm in the Mobile WiMAX System' IEEE Transactions on Vehicular Technology Journal 57(6): 3477–87, 2008.

7

Mobility Management in WiMAX Networks

Ikbal Chammakhi Msadaa, Daniel Câmara and Fethi Filali
EURECOM Mobile Communications Department, Sophia-Antipolis, France

Worldwide Interoperability for Microwave Access (WiMAX) is one of the most promising technologies for the next generation networks as it provides high data rates at medium and long range with full support of mobility. The technology is based on IEEE 802.16 standards and amendments specifying the MAC and PHY layers for fixed, nomadic, portable and mobile access.

Figure 7.1 illustrates different possibilities of WiMAX technology deployment. Fixed access is one of the primary applications for which WiMAX technology has been developed mainly for customers residing in rural areas. This has been addressed by the IEEE 802.16d-2004 standard [1] with support of non-line-of-sight (NLOS) access between a base station (BS) and a consumer premises equipment (CPE). Figure 7.1 shows another application of 802.16 BWA which is backhauling; a BS is linked to another BS in direct LOS or to WiFi hotspots (WiFi hotspot backhaul) to provide access to the Internet backbone. The technology has then been extended to nomadic and mobile environment through the publication of IEEE 802.16e standard [2].

This book chapter focuses on the latter standard – an amendment of the IEEE 802.16d-2004 standard – which provides enhancements related mainly to mobility management. The book chapter is organized as follows. Section 7.1 describes the logical architecture of a mobile WiMAX network. This architecture has been defined by the Network Working Group[1] (NWG) of the WiMAX Forum.[2] Section 7.2 describes the horizontal handoff

---

[1] A working group from the WiMAX forum. It is responsible for creating higher level networking specifications for fixed, nomad, portable and mobile WiMAX systems.

[2] An industry group created in June 2001 to promote the conformity and interoperability of the IEEE 802.16 products.

WiMAX Security and Quality of Service: An End-to-End Perspective   Edited by Seok-Yee Tang, Peter Müller and Hamid Sharif
© 2010 John Wiley & Sons, Ltd

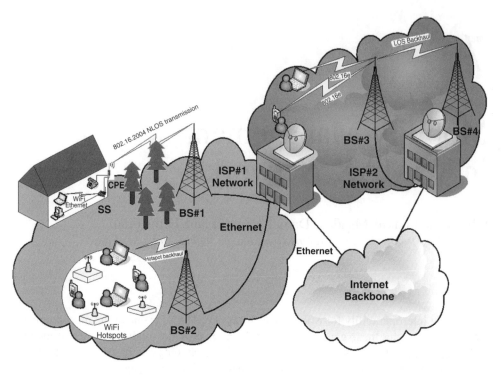

Figure 7.1   Examples of WiMAX deployments.

procedure proposed by the IEEE 802.16e standard. Section 7.3 presents some procedures
presented in the literature and aiming at improving the handover mechanism.

   Moreover, because this technology is more likely to co-exist with other access technologies in future networks, we dedicate section 7.4 to study the vertical handover mechanisms
in heterogeneous environment involving mobile WiMAX systems. Roaming, which has
been referred to as 'the missing piece of the WiMAX puzzle', is addressed in section 7.5.
In section 7.6, we describe the mobility support for the particular case of WiMAX mesh
networks and explain the difference between mesh and point-to-multipoint (PMP) networks from a mobility perspective. Section 7.7 concludes the chapter by highlighting the
main conclusions.

7.1   Mobile WiMAX Architecture

A Network Reference Model (NRM), presenting the logical architecture of a WiMAX
network, has been proposed by the NWG[3] [3]. It has been developed with an objective
of supporting many architectural profiles and addressing many deployment scenarios of
mobile WiMAX networks. In this section, we first describe the different entities of the
NRM and then discuss the technical and business merits of each profile.

---

[3] A working group from the WiMAX forum. It is responsible for creating higher level networking specifications
for fixed, nomad, portable and mobile WiMAX systems.

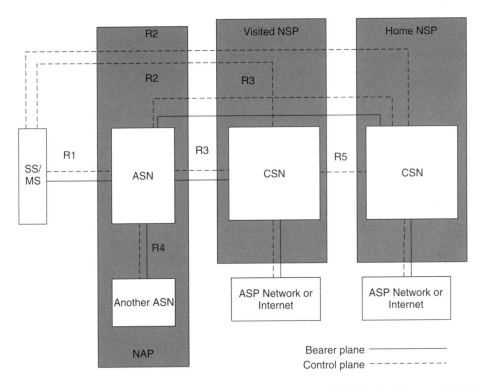

Figure 7.2   Network reference model. Copyright © 2007–2008 WiMAX Forum. All rights reserved.

As shown in Figure 7.2, the WiMAX NRM consists of three logical entities (Mobile Station MS, Access Service Network ASN, and Connectivity Service Network CSN) interconnected by R1-R5 reference points. These reference points insure multi-vendors interoperability between the different logical entities belonging to the network. Each of the MS, ASN, and CSN represents a grouping of functional entities (within an ASN, between an ASN and a MS, between an ASN and a CSN, etc.) that may be realized by a single or multiple physical devices:

1. Mobile Station (MS) is a generalized mobile equipment set which provides connectivity between a WiMAX subscriber equipment and a base station (BS).
2. Access Service Network (ASN) refers to a set of network functions providing radio access to the WiMAX MS. The mandatory functions that need to be provided by the ASN are: L2 and L3 connectivity with WiMAX subscriber, radio resource management (RRM), relay of AAA (Authentication, Authorization and Accounting) messages, network discovery and selection, mobility management, etc. An ASN consists of one or more BS and one or more ASN-Gateway (ASN-GW):
   (a) Base Station (BS) is a logical entities that incorporates a full instance of MAC and PHY layers compliant with the IEEE 802.16 suite of applicable standards.
   (b) ASN-Gateway (ASN-GW) is a logical entity that represents an aggregation of control plane functions. It may also perform bearer plane routing or bridging function.

3. Connectivity service network (CSN) refers to a set of network functions that provide IP connectivity functions to the WiMAX subscribers. Among the functions that the CSN may provide, we find: Internet access, inter-ASN mobility, admission control based on user profiles, etc. The CSN may include network elements such as routers, AAA proxy/servers, user databases, etc.

The distribution of the different functions within the ASN (between the BS(s) and the ASN-GW(s)) is an implementation choice. Nevertheless, to guarantee network interoperability requirements, the NWG Release 1.0.0 [3] defines three different implementations of the ASN. These implementations, whose respective reference models are depicted in Figure 7.3, are called interoperability profiles A, B, and C. Each of them corresponds to a specific distribution of ASN functions between the two entities composing the ASN: the ASN-GW(s) and the BS(s). As we can see it from Figure 7.3, in Profile A, for instance, the radio resource control RRC (which is given here as example for function mapping) is in the ASN-GW while in Profile C it is accomplished by the ASN-GW. In Profile B, however, all the functions are located within a single ASN entity, which includes the case where all the functions are grouped in the same physical device.

As discussed in [4], each profile has its own technical and business merits and selecting one or combining two or more of these profiles may seriously impact the handoff support in WiMAX networks. In [4], Hu *et al.* have investigated both the hierarchical and flat

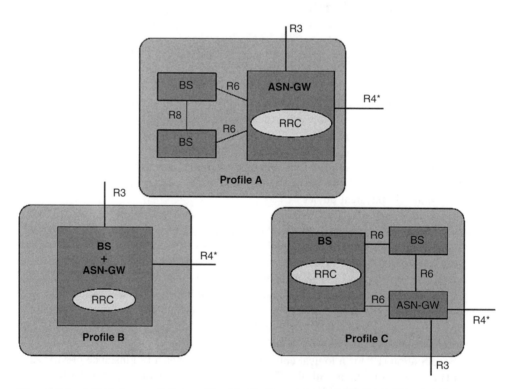

Figure 7.3 ASN interoperability profiles [4, 3]. Copyright © 2007–2008 WiMAX Forum. All rights reserved.

network architectures and their respective impacts on the performance of handoff in terms of latency, scalability, complexity, financial cost, etc. The authors have then mapped the different interoperability profiles to a hierarchical, flat or hybrid design which could help to choose the most appropriate architecture when deploying a technological solution.

7.2 Horizontal Handover in 802.16e

The IEEE 802.16e standard [2] defines three handover schemes:

- A mandatory hard HO mode also known as break-before-make HO. In this mode, the air interface link between the MS and the Serving BS is broken at all layers before being established again at the target BS. The HO process may be initiated either by the MS or by the BS.
- Two optional soft, known as make-before-break, HO modes:
  - Macro diversity HO (MDHO): this mode is defined in [2] as the process in which the MS migrates from an air interface provided by one or more other BSs to the air interface provided by one or more BSs. In the DL (respectively UL), this is achieved by having two or more BSs transmitting (respectively receiving) the same PDU to (respectively from) the MS.
  - Fast BS switching (FBSS): in this mode, an active set is maintained. It consists in a set of candidate BSs to which the MS is likely to handoff in near future. At any given frame, the MS is exchanging data only with one BS – anchor BS – of this active set [2].

More details about these three modes are provided in this section. Nevertheless more insight is given on the hard HO mode which is the only mandatory mode.

7.2.1 Network Topology Acquisition

1. The BS broadcasts periodically the network topology information using the MOB_NBR-ADV message. The message includes the BSIDs of the neighbouring BSs along with their respective channel characteristics normally provided by each BS own Downlink/Uplink Channel Descriptor (DCD/UCD) message transmission. This information is intended to enable the MS to perform fast synchronization with the advertised BSs by removing the need to monitor the DCD/UCD broadcasts from each neighbouring BS.
2. Based on the information provided by the MOB_NBR-ADV, the MS becomes aware of the neighbouring BSs and triggers the scanning and synchronization phase. Indeed, to handoff, the MS needs to seek available BSs and check if they are suitable as possible target BSs.
   Therefore, the MS sends MOB_SCN-REQ message to the serving BS indicating a group of neighbouring BSs for which a group of scanning intervals is requested. The MOB_SCN-REQ message includes the requested scanning interval duration, the duration of the interleaving interval, and the requested number of scanning iterations. In the example illustrated in Figure 7.4, these parameters correspond to P frames,

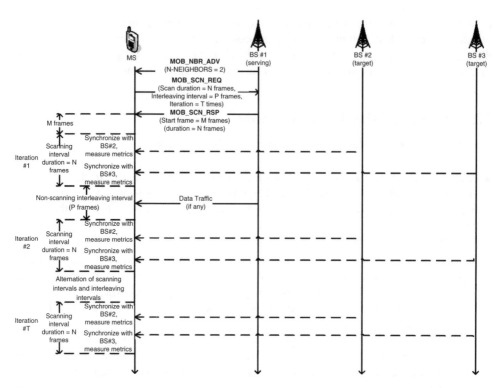

Figure 7.4 Example of neighbour BS advertisement and scanning (without association) by MS request [2]. Copyright © 2006 Institute of Electrical and Electronics Engineers, Inc. All rights reserved.

N frames, and T iterations, respectively. Note that the scanning phase could be triggered by the serving BS. If it is the case, the serving BS shall send to the MS a MOB_SCN-RSP message indicating a list of recommended neighbouring BSs.

3. Upon reception of the MOB_SCN-REQ message, the serving BS responds with a MOB_SCN-RSP message. In this message, the serving BS either grants a scanning interval at least as long as the one requested by the MS (which is the case in our example of Figure 7.4) or rejects the request.

4. After receiving the MOB_SCN-RSP message granting the request, the MS may scan – beginning at *Start frame* – one or more BSs during the time allocated by the serving BS. Each time a neighbouring BS is detected through scanning, the MS may attempt to synchronize with its downlink transmissions and estimate the quality of the PHY channel to evaluate its suitability as a potential target BS in the future. The serving BS may ask (by setting the report mode field to 0b10 in the MOB_SCN-RSP) the MS to report the scanning results by transmitting a MOB_SCN-REP.

5. During the scanning interval, the serving BS may buffer incoming data addressed to the MS and then transmit that data during any interleaving interval after the MS has exited the scanning mode.

Depending on the value of the scanning type field indicated in the MOB_SCN-REQ, the MS may request either scanning only or scanning with association. The association

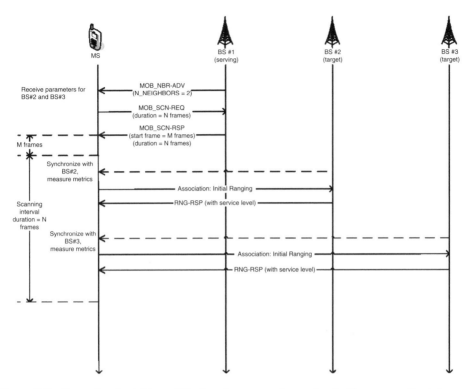

Figure 7.5 Example of neighbour BS advertisement and scanning (with non-coordinated association) by MS request [2].

procedure is an optional ranging phase that may be performed during the scanning interval. It enables the MS to acquire and record ranging parameters – by adjusting the time offset, the frequency and the power level – to be used to choose a potential target BS. The standard IEEE 802.16e [2] defines three levels of association:

- Association Level 0 – scan/association without coordination: the target BS has no knowledge of the scanning MS and only provides contention-based ranging allocations (c.f. Figure 7.5).
- Association Level 1 – association with coordination: the serving BS coordinates the association between the MS and the requested neighbouring BSs. Each neighbor (NBR) BS provides a ranging region for association at a predefined "rendezvous time" (corresponding to a relative frame number). It also reserves a unique initial ranging code and a ranging slot within the allocated region. The NBR BS may assign the same code or ranging slot to more than one BS but not both, so that no potential collision may occur between transmissions of different MSs.
- Association Level 2 – network-assisted association reporting: the procedure is similar to level 1 except that the MS does not need to wait for RNG-RSP from the NBR BS. The ranging response is sent by the NBR BS to the serving BS over the backbone, which then forwards it to the MS.

7.2.2   Handover Process

The handover is defined as the process in which a MS migrates from the air-interface provided by one BS (the serving BS) to the air-interface of another BS (target BS) [2]. It consists of the following phases:

7.2.2.1   Cell Reselection

Cell reselection refers to the process of an MS Scanning and/or Association with one or more BSs (as described in section 7.2.1) in order to determine their suitability, along with other performance considerations as a handover target [2]. The information acquired from the MOB_NBR-ADV message might be used by the MS to give insight into available neighbouring BSs for cell reselection considerations.

7.2.2.2   HO Decision and Initiation

The handover process begins with a decision that originates either from the MS, or the BS (The BS can force the MS to conduct handover), or on the network. A handover could be decided for many reasons; for example when the MS performance at a potential target BS is expected to be higher than at the serving BS. Note that the handover decision algorithm is beyond the scope of 802.16e standard.

Once a handover is decided, it is notified through a MOB_MSHO-REQ or a MOB_BSHO-REQ indicating one or more possible target BSs. If the handover request is formulated by the MS, it shall be acknowledged with a MOB_BSHO-RSP. When the handover is initiated by the BS, it could be either recommended or mandatory. If it is a mandatory handover, the MS will send MOB_HO-IND to the serving BS. The MOB_HO-IND may indicate a HO reject when the MS is unable to handoff to any of the recommended target BSs listed in the MOB_BSHO-REQ.

7.2.2.3   Synchronization to Target BS Downlink

MS will synchronize to downlink transmissions of target BS and obtain DL and UL transmission parameters. This process may be shortened in two cases: (i) if the MS had previously received a MOB_NBR-ADV message including target BSID, physical frequency, DCD and UCD, or (ii) if the target BS had previously received HO notification from serving BS over the backbone in which case the target BS may allocate a non-contention-based initial ranging opportunity for the MS.

7.2.2.4   Ranging and Network Re-Entry

After adjusting all the PHY parameters, the network re-entry process is initiated between the MS and the target BS. The network re-entry procedure normally includes the following steps (i–iv).

(i) Negotiation of basic capabilities: the MS and the target BS exchange their supported parameters such as the current transmit power or the security parameters support. This step is performed by exchanging SBC-REQ and SBC-RSP management messages.

(ii) Privacy key management (PKM) authentication phase: during this phase, the MS exchanges secure keys with the target BS. The MS sends a PKM-REQ message and the BS responds with a PKM-RSP message.

(iii) Traffic encryption keys (TEK) establishment phase.

(iv) Registration: the registration is the process by which the SS is allowed entry into the network [2]. The registration is performed by exchanging REG-REQ and REG-RSP between the MS and the target BS.

The network re-entry process may be shortened since the target BS may decide to skip one or more of these steps (i–iv) if it disposes of the corresponding information obtained from the serving BS over the backbone.

7.2.2.5 Termination of MS Context

The termination of the MS context is the final stage of the handover procedure. In this step, the serving BS proceeds to the termination of all the connections belonging to the MS along with their associated context (information in the queues, timers, counters, etc.).

Note that the handover procedure might be cancelled by the MS at any time prior to the expiration of Resource_Retain_Time interval after transmission of MOB_HO-IND message.

7.2.3 Fast BS Switching (FBSS) and Macro Diversity Handover (MDHO)

As mentioned before, in addition to the hard handover procedure previously described, the IEEE 802.16e standard defines two optional handover modes: MDHO and FBSS. The MDHO or FBSS capability can be enabled or disabled in the REG-REQ/RSP message exchange. In both modes, a Diversity Set is maintained. The Diversity Set is a list of selected BSs that are involved in the MDHO or FBSS process. These BSs should be synchronized in both time and frequency and are required to share the MAC context associated to the MS. The MAC context includes the parameters that are normally exchanged during the network entry along with the service flows associated to the MS connections.

7.2.3.1 Macro Diversity Handover (MDHO)

A MDHO begins with a decision for an MS to transmit to and receive from multiple BSs at the same time. This decision is communicated through MOB_BSHO-REQ or MOB_MSHO-REQ messages. When operating in MDHO mode, the MS communicates with all the BSs belonging to the Diversity Set for DL and UL unicast messages and traffic. For DL MDHO, two or more BSs provide synchronized transmission of MS data so that the MS performs diversity combining. For UL MDHO, the MS data transmission is received by multiple BSs so that they can perform selection diversity of the received information.

To monitor DL control information and DL broadcast messages, the MS can use one of the following two methods. The first method is the MS monitors only the Anchor BS – a BS defined among the Diversity Set – for DL control information and DL broadcast messages. In this case, the DL-MAP and UL-MAP of the Anchor BS may contain burst allocation information for the non-Anchor Active BS. The second method is the MS monitors all the BSs in the Diversity Set for DL control information and DL broadcast messages. In this case, the DL-MAP and UL-MAP of any Active BS may contain burst allocation information for the other Active BSs. The method to be used by MS is defined during the REG-REQ and REG-RSP handshake.

7.2.3.2 Fast BS Switching (FBSS)

FBSS HO begins with a decision for an MS to receive/transmit data from/to the Anchor BS that may change within the Diversity Set. A FBSS can start with MOB_BSHO-REQ or MOB_MSHO-REQ messages. When operating in FBSS mode, the MS is required to continuously monitor the signal strength of the BSs belonging to the Diversity Set. The MS will select a BS from its current Diversity Set to be the Anchor BS and report the selected Anchor BS on MOB_MSHO-REQ message. BS switching that is transition from the Anchor BS to another BS is performed without invocation of the handover procedure described in section 7.2.2.

The BS supporting MDHO or FBSS shall broadcast the DCD message that includes the H_Add Threshold and H_Delete Threshold. These thresholds are used by the FBSS/MDHO capable MS to determine if MOB_MSHO-REQ should be sent. When long-term CINR of a BS is less than H_Delete Threshold, the MS shall send MOB_MSHO-REQ to require dropping this BS from the Diversity Set; when long-term CINR of a neighbor BS is higher than H_Add Threshold, the MS shall send MOB_MSHO-REQ to require adding this neighbour BS to the diversity set. Figure 7.6 illustrates an example of a Diversity Set update – add of a new BS – during a MDHO procedure.

Discussion

From the description of the three handover modes, the hard handoff procedure consists of more steps and might cause intolerable delays for real-time traffic. Nevertheless, the two soft handover modes FBSS and MDHO cannot be a reliable alternative to the mandatory hard HO scheme for many reasons. On the one hand, as we have mentioned before, there are several restrictions on BSs working in MDHO/FBSS modes since they need to synchronize on time (same time source) and frequency and have synchronized frame structures which entails extra costs. On the other hand, in both FBSS and MDHO modes, the BSs in the same Diversity Set are likely to belong to the same subnet while a handover may occur between BSs in different subnets. Therefore, in the remainder of the chapter, more insight will be given into the hard handover scheme. More specifically, we will present some works aiming at optimizing the hard handover procedure in IEEE 802.16e networks.

7.3 Optimized 802.16e Handover Schemes

Improving the handoff process in mobile WiMAX networks is a topic that have received a lot of attention in the last few years. Indeed, in order to enable always-on connectivity,

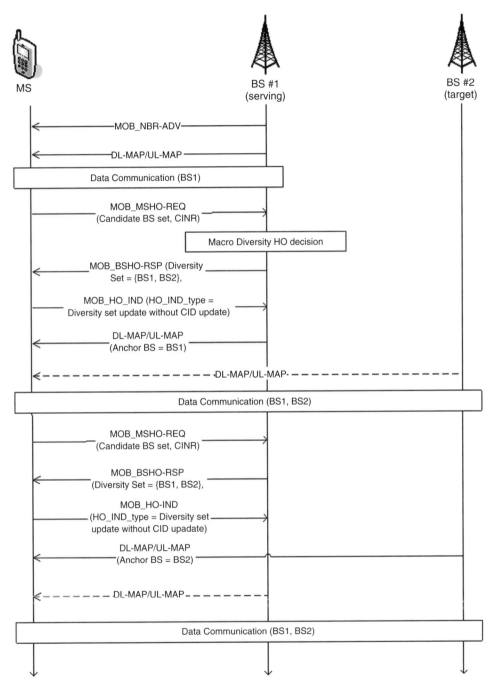

Figure 7.6 Example of macro diversity HO (Diversity Set Update: Add) [2]. Copyright © 2006 Institute of Electrical and Electronics Engineers, Inc. All rights reserved.

it is necessary to achieve a fast and smooth handoff over the network. To reach that goal, the research works addressing this issue have adopted mainly two approaches: improving the handover at Layer 2 or considering a cross-layer mechanism in which L2 and L3 collaborate to have better results.

7.3.1 L2 Handover Schemes

In order to reduce the handover delay, Lee *et al.* [5] have focused on eliminating the redundant processes existing in the handover procedure defined in the IEEE 802.16e standard [2]. The approach consists in using a target BS estimation algorithm to select a HO target BS instead of scanning, one by one all the neighbouring BSs. The target BS estimation algorithm assumes that the NBR BS with bigger mean CINR and and smaller arrival time difference is more likely to be the target BS. The MS does not need then to associate to the neighbouring BSs. However, by eliminating both the scanning and the association phases, the handover decision loses its accuracy since the MS does not dispose of information precise enough to make a handoff decision.

Instead of predicting the potential target BS, Chen *et al.* propose in [6] a pre-coordinated handover mechanism in which the handover time is predicted. In the proposed mechanism, the distance between the MS and the serving BS is calculated to estimate the needed handover time, then a pre-coordination is performed with the target BS.

In order to locate the position of the MS, the serving BS measures the signal-to-noise ratio (SNR) of the mobile station every 10 s. Based on that, the distance, the direction, and the velocity of the MS are derived. If the MS approaches the boundary h of the serving base station macrocell, the serving BS pre-coordinates a handover with the "only" target BS in that direction. The pre-coordination phase consists in sending a MOB_BSHO-REQ to the target BS which would respond with a MOB_BSHO-RSP in which it allocates – if it has enough resources – a fast ranging opportunity for the MS and specifies its PHY parameters. The target BS will have then to hold this request service for 10 s. When the MS requests to handoff (estimated to 10 s before the predicted handover time), the BS responds by MOB_NBR-ADV message in which it includes the information transmitted by the target BS. This would facilitate the migration of the MS to the new channel and thus reduce the disruption time. Nevertheless, the performance estimation algorithm needs further investigation to be reliable.

7.3.2 L2-L3 Cross-Layer Handover Schemes

In [7], Chen *et al.* have proposed a cross-layer handover scheme in which they use layer 3 to transmit MAC messages between the MS and the BSs (the serving BS and the NBR BSs) during the handoff process. In the proposed cross-layer scheme, two tunnels have been created to redirect and relay these messages: an L2 tunnel between the MS and the serving BS and an L3 tunnel between the serving and the neighbouring (target) BSs. The idea behind the creation of these tunnels is to minimize the delay due to direct messages transportation between the MS and NBR BSs which constitutes a source of latency in the handover process. When the handover is requested, the serving BS negotiates for the MS a fast ranging opportunity from the neighbouring BSs. The MS then switches to the channels to be scanned and tries to synchronize with each associated NBR BSs.

Once the synchronization is performed, the MS sends a MOB_RNG-REQ on each channel. However, unlike the regular handover procedure described in section 7.2, the MS does not need to wait for RNG-RSP from each scanned NBR BS. Instead, the MS informs the BS that the ranging request phase has finished by sending a RNG_RSP-REQ message (a new management message proposed by Chen *et al.* [7]) and restores the uplink transmission. Upon reception of the RNG_RSP-REQ, the serving BS understands that the MS is ready to receive the RNG_RSP messages. These messages have been encapsulated by the NBR BSs and sent to the serving BS which decapsulates and stores them before forwarding them to the MS. This way, the uplink transmission is restored faster.

Moreover, a fast re-entry procedure is proposed. Instead of disconnecting and connecting with the target BS as described in section 7.2, the MS sends all the messages to the serving BS which relays them to the target BS through the IP backbone.

The idea of combining L2 and L3 mechanisms to shorten the handover time and to allow handover between different subnets has been also investigated by Chang *et al.* in [8]. The authors have focused mainly on interleaving the authentication process with a fast handover mechanism to speedup the handover process while securing the whole mechanism. Chang *et al.* have based their proposal on a draft version of an RFC [9] – recently finalized by IETF – proposing Mobile IP fast handover mechanism over IEEE 802.16e networks.

7.3.3   *Mobile IPv6 Fast Handovers Over IEEE 802.16e Networks*

This section is dedicated to the description of the interleaving between 802.16e and fast mobile IPv6 (FMIPv6) handover mechanisms proposed by IETF in [9]. The handoff procedure is explained through two examples corresponding to the predictive (Figure 7.8) and reactive mode (Figure 7.9), respectively.

7.3.3.1   **Predictive Mode**

The different steps commented in this section are illustrated in Figure 7.8.

Access Router Discovery

1-3 When a new BS (Point of Attachment PoA) is detected through the reception of MOB_NBR-ADV or through scanning, the link layer of the MS triggers a NEW_LINK_DTECTED primitive to the IP layer.

4. When receiving the NEW_LINK_DTECTED from the link layer, the IP layer sends a router solicitation message RtSolPr (Router Solicitation for Proxy Advertisement) to the previous access router (PAR) to acquire the L3 parameters of the access router associated to the new PoA (the new BS). The PAR responds by sending a Proxy Router Advertisement (PrRtAdv) that provides information such as the router address and additional parameters about neighbouring links.

The objective of this step is to enable the quick discovery – in IP layer – of the access router associated to the new BS.

Handover Preparation

5. When the MN decides to change the PoA (because of a degradation in signal strength, or for better QoS, etc.) it initiates a handover procedure by sending a MOB_MSHO-REQ to the serving BS which will respond by a MOB_MSHO-RSP. As we have seen in section 7.2.2, the handover might also be initiated by the serving base station (MOB_BSHO-REQ).

6. Once a MOB_MSHO-RSP/MOB_BSHO-REQ is received, the link layer triggers a LINK_HANDOVER_IMPEND primitive, enclosing the decided target BS, to inform the IP layer that a link layer handover decision has been made and that its execution is imminent.

   Based on the information collected during the access router discovery phase, the IP layer checks whether the target BS belongs to a different subnet (cf. Figure 7.7). If the target network proves to be in the same subnet, the MN can continue to use the same IP address and thus, there is no need to perform FMIPv6.

7. Otherwise, based on the information provided by the PrRtAdv, the IP layer formulates a prospective NCoA (New Care of Address) and sends a Fast Binding Update (FBU) message to the PAR. When received successfully, the FBU is processed by the PAR and the NAR according to RFC 5268 (FMIPv6 [10]).

   The PAR sets up a tunnel between the PCoA (Previous Care of Address) and the NCoA by exchanging a HI (Handover Initiation) and HAck (Handover Acknowledgment) messages with the NAR. In the HAck message, the NCoA is either confirmed or re-assigned by the NAR. Finally, the NCoA is transmitted to the MN through the FBack (Fast Binding Acknowledgment) message in case of predictive mode

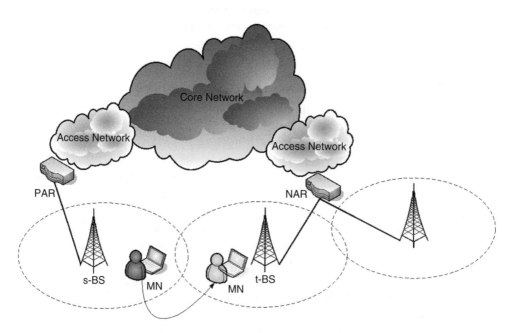

Figure 7.7   Example of a handover between two different subnets.

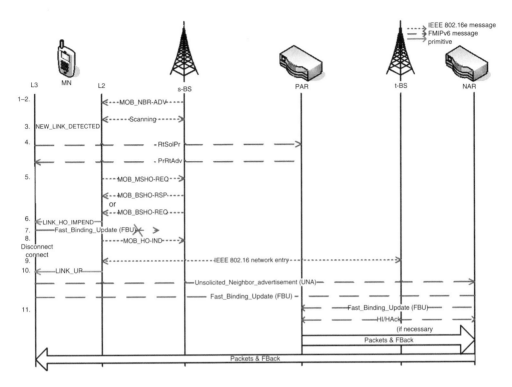

Figure 7.8 Predictive fast handover in 802.16e [9]. Copyright © The IETF Trust 2008.

(See Figure 7.8) and the packets destined to the MN are forwarded to the NCoA. The difference with the reactive mode will be explained at the end of this section.

Handover Execution

8. If the MN receives a FBack on the previous link, it sends a MOB_HO-IND message as a final indication of handover. Optionally, the LINK_SWITCH command could be issued by the IP layer upon the reception of FBack to force the MN to switch from an old BS to a new BS. This command forces the use of predictive mode even after switching to the new link.
9. Once the links are switched, the MN synchronizes with the new PoA (target BS) and performs the 802.16e network entry procedure. As we have mentioned before in section 7.2.2, this phase (or some of its steps) might be omitted if the serving BS had transferred the MN context to the target BS over the backbone.
10. Once the network entry is completed, the link layer triggers a LINK_UP primitive to inform the IP layer that it is ready for data transmission.

Handover Completion

11. When the MN IP layer receives the LINK_UP primitive, it checks whether the target network is the one predicted by the FMIPv6 operation. If it is the case, it sends an

Unsolicited Neighbour Advertisement (UNA) message to the NAR (predictive mode) using the NCoA as source IP address and starts performing the DAD (Duplicate Address Detection) for the NCoA.

12. As soon as the UNA message is received, the NAR transfers the buffered packets to the MN.

7.3.3.2   Reactive Mode

The different steps commented in this section are illustrated in Figure 7.9.

Access Router Discovery

1-4 The same procedure as in predictive mode.

Handover Preparation

5-7. The same procedure as in predictive mode. Nevertheless, note that the FBU has not reached the PAR, and so no FBack has been received by the MN either.

  8. Unlike in predictive mode, the MN issues a MOB_HO-IND without waiting for an FBack message. When receiving this final indication of handover (MOB_HO-IND),

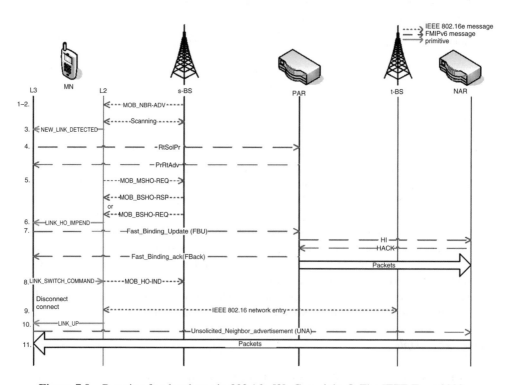

Figure 7.9   Reactive fast handover in 802.16e [9]. Copyright © The IETF Trust 2008.

the serving BS releases all the MN context which means that data packet transfer is no longer allowed between the MN and the BS (as we can see from Figure 7.9).

Handover Execution

9. The MN conducts handover to the target BS and performs the 802.16e network entry procedure.
10. The MN link layer triggers a LINK_UP primitive to inform the IP layer that it is ready for data transmission.

Handover Completion

11. Note that, in reactive mode, the MN has moved to the target network without receiving an FBack message in the previous link. Therefore, upon reception of the LINK_UP primitive, the IP layer sends (i) an UNA message to the NAR using the NCoA as source IP address to announce a link layer address change, and (ii) a FBU message to instruct the PAR to redirect its traffic towards the NAR.
12. When the NAR receives the UNA and the FBU from the MN, it exchanges a HI/HAck with the PAR. The FBack and Packets are then forwarded from the PAR and delivered to the MN through the NAR using the NCoA as destination IP address.

Discussion

Mobile IPv6 fast handovers, like all cross-layer handover management mechanisms in general, are based on the collaboration of different layers in order to enhance the mobility management. This idea of integrating information from different network layers helps to improve the HO management performances. Nevertheless, because these solutions usually require significant modifications in the network stack, their deployment becomes prohibitive [11].

7.4 Vertical Handover

Next generation networks will more likely consist of heterogeneous networks such as integrated WiFi/WiMAX networks, WiMAX/CDMA2000 or networks combining WiMAX and 3G/4G technology. In this section, we describe the deployment of such hybrid networks and discuss the main challenging issues that arise when inter-networking WiMAX and other technologies. We focus on the vertical handover mechanisms proposed in the literature to guarantee the service continuity without QoS degradation for users switching from one network to another. The second part of this section is dedicated to the media-independent handover (MIH) mechanism proposed by the IEEE 802.21 task group. The recently published IEEE 802.21-2008 Standard [12] enables handover and interoperability between heterogeneous network types including both 802 and cellular networks.

7.4.1   Vertical Handover Mechanisms Involving 802.16e Networks

For both horizontal and vertical handover, the main objective is to provide a fast and seamless handover. However, because of the heterogeneity of the networks involved in the vertical handoff process, ensuring a continuous connectivity is even more challenging.

To make a horizontal handover decision, considering only the radio signal strength was enough while in a hybrid network environment this metric is not sufficient. Indeed, more parameters need to be considered: available bandwidth, latency, packet error rate, monetary cost, power consumption, user preferences, etc. [13].

In this section, we present works that have investigated the vertical handoff mechanisms involving mobile WiMAX networks. Each of these works have focused on the enhancement of one or more of the three main phases of a vertical handover procedure which are:

1. Finding candidate networks: also referred to as system discovery phase during which the MS needs to know which networks can be used.
2. Deciding a handoff: during this phase, the MS needs to evaluate the reachable wireless networks and to decide whether to keep using the same network or to switch to another network. This decision could involve several criteria: the type of applications running, their QoS requirements, the access cost, etc. [14].
3. Executing a handoff: a critical phase during which the connections need to be rerouted in a seamless manner with transfer of the user's context.

According to another classification [12], the two first steps could be merged into a single phase called 'handover initiation' which encloses network discovery, network selection, and handover negotiation. Based on the same classification [12], the handover execution would correspond to two steps: handover preparation (L2 and L3 connectivity) and handover execution (connection transfer).

Whatever is the adopted classification, we notice that the phase on which most of the works have focused is the handover decision phase. In [15] for example, Dai et al. have proposed the use of two triggers: (i) connectivity trigger and (ii) performance trigger based on which the handoff between WiFi and WiMAX is decided. The first trigger is based on SINR indication to evaluate the risk of connection loss and would decide a handover if the SINR is below a certain SINR target and if other networks are detected. The performance trigger however, combines data rate and channel occupancy to derive an estimation of the current throughput and decide a potential handoff when needed (i.e. when the throughput is below a certain threshold).

In [16], the handover decision might be initiated either (i) by the user when it is moving and needs to gain in performance or (ii) by the WiMAX network to release resources and accommodate new calls (WiMAX calls) or VHO calls (from an UMTS network). The vertical handoff decision algorithm (VHDA) proposed in [16] depends of the improvement that could be gained from the handoff and the suitability of the target network. This gain is estimated based on two factors: the cost of the handoff C_h (function of the MS velocity V, the available bandwidth B, the service cost C, the power consumption P, the security level S, and the network performance F) and the QoS performance P_{QoS} (function of the handoff delay D, the packet loss ratio PLR, and the data rate R).

$$C_h = \frac{w_v . V_2}{max(V_1, V_2)} + \frac{w_b . \frac{1}{B_2}}{max\left(\frac{1}{B_1}, \frac{1}{B_2}\right)} + \frac{w_c . C2}{max(C_1, C_2)}$$

$$+ \frac{w_p . P_2}{max(P_1, P_2)} + \frac{w_s . \frac{1}{S_2}}{max\left(\frac{1}{S_1}, \frac{1}{S_2}\right)} + \frac{w_f . \frac{1}{F_2}}{max\left(\frac{1}{F_1}, \frac{1}{F_2}\right)}$$

$$P_{QoS} = w_{N_u} . \frac{1}{N_u} + w_{PLR} . \frac{1}{PLR} + w_D . \frac{1}{D} + w_R . R$$

where '1' represents WiMAX network, and '2' or 'u' represents UMTS network. The different weights (w_v, w_p, etc.) are chosen based on the significance of the associated network parameter in C_h and P_{QoS}. For transferring confidential data for example, w_s would have more importance than in regular cases.

Through these examples, we can see that, unlike in the optimized horizontal handover schemes presented in section 7.3, the vertical handover decision is more challenging and has to take several parameters into account before deciding a handoff.

7.4.2 IEEE 802.21, Media-Independent Handover Services

The IEEE 802.21 standard proposes a set of mechanisms that enhance the handovers between heterogeneous IEEE 802 networks and may facilitate handovers even between 802 (e.g. 802.11, 802.15, and 802.16) and non 802 systems (e.g. 3GPP and 3GPP2) [12]. In this section, we first define the core components of the general architecture proposed by the IEEE Std 802.21, then we present the main services provided by this media-independent handover (MIH) framework.

7.4.2.1 General Architecture

Figure 7.10 illustrates the IEEE 802.21 reference model within the protocol stack along with the different proposed MIH services. Note that the standard supposes that the MN is able to support several link-layer technologies.

MIH Function (MIHF)

The main role of the MIHF is to assist the network selector entity in making an effective network selection by providing all the necessary inputs for such a decision: QoS requirements, battery life constraints, monetary cost, user preferences, operators' policies, etc. This information are meant to facilitate the handover decision and to maximize its efficiency. To achieve this role, the MIHF communicates with lower layers through technology-specific interfaces and provides services to the upper layers (MIH users) in a unified and abstracted way. More details about the services provided by the MIHF are given in section 7.4.2.2.

MIH User (MIHU)

MIH users (MIHUs) are the entities responsible for mobility management and handover decision making. They reside at Layer 3 or above in the network stack. As examples

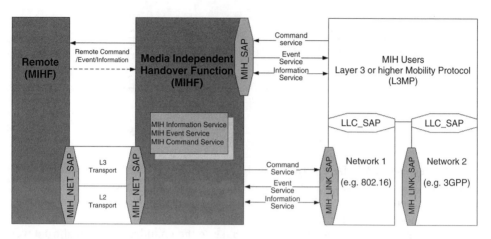

Figure 7.10 MIH Reference Model and Services [12]. Copyright © 2009 Institute of Electrical and Electronics Engineers, Inc. All rights reserved.

of MIH users, we can cite MIP at network layer, mobile Stream Control Transmission Protocol (mSCTP) at transport layer, and Session Initiation Protocol (SIP) at application layer [17]. The MIHU base their handover decisions on their own internal policy but also on the information provided by the MIHF.

SAPs

In order to make possible the communication between the different architectural components of the MIH framework, the IEEE 802.21 standard defines a set of SAPs with their associated primitives. Figure 7.10 shows the different SAPs interfacing the MIHF with other layers:

1. The media-independent SAP MIH_SAP allows the MIH users to access the MIHF services.
2. The link-layer SAPs MIH_LINK_SAP are media-dependent SAPs that allow the MIHF to gather link information and control link behaviour during handovers. Each link-layer technology (e.g 802.3, 802.16, 3GPP, etc.) specifies its own technology-dependent SAPs and the MIH_LINK_SAP maps to these technology-specific SAPs. As example of media-specific SAPs, we can cite the C_SAP, M_SAP, and CS_SAP that are defined in IEEE Std 802.16 to provide interfaces between the MIHF and different components of the 802.16 network stack; namely with the control plane (C_SAP), the management plane functions (M_SAP), and the service-specific Convergence Sublayer (CS_SAP).
3. The MIH_NET_SAP is another media-dependent SAP that provides transport services over the data plane and allows the MIHF to communicate with remote MIHFs.

7.4.2.2 MIHF Services

In order to facilitate the handover procedure across heterogeneous networks, the MIHF entity provides the three following categories of services to the MIH users: (i) MIH information service (MIIS), MIH event service (MIES), and MIH command service (MICS).

MIIS: MIH Information Service

The media independent information service allows the MIH users to acquire a general view about the networks present in the vicinity of the MN in order to enable a more effective handover decision. These information include for example the list of available networks, their link-layer static information (e.g. whether QoS and security are supported in a particular network), and other geographical positioning information that could be used further to optimize the handover decision.

MIES: MIH Event Service

Unlike the MIIS which provides a static (or rarely changing) information about the surrounding networks, the MIH event service (MIES) triggers dynamic changes in link conditions. Indeed, it provides event reporting about MAC and PHY state changes through triggers that indicate for instance that the L2 connection is broken (LINK_DOWN) or that the link conditions are degrading and the loss of connectivity is imminent (LINK_GOING_DOWN). Other triggers might report the failure/success of PDUs transmission (e.g. Link_PDU_Transmit_Status), or the handover status (e.g. Link_Handover_Complete).

MICS: MIH Command Service

The MIH command service (MICS) refers to the set of commands that originate (i) either from the MIH users: MIH commands, (ii) or from the MIHF: link commands and are directed to the lower layers. MIH_MN_HO_Candidate_Query is an example of a remote MIH command used by the MN to query and obtain handover related information about possible candidate networks. The link commands are local commands that are used to control and configure the link layers (e.g. Link_Configure_Thresholds through which is used to set link parameter thresholds) or to retrieve link-specific information (e.g. Link_Get_Parameters commands provides information about the SNR, the bit-error-rate BER, etc.)

Service Management

In order to benefit from the services provided by the MIHF, the MIH entities need to be configured properly using the following service management functions:

- MIH capability discovery

This step is necessary to the MN to discover local and/or remote MIHF capabilities in terms of MIH supported services. This could be performed either through the MIH protocol or via media-specific mechanisms (e.g. beacon frames for 802.11). For 802.16 networks for example, the MN can use the management messages such as downlink channel descriptor (DCD), or uplink channel descriptor (UCD) to retrieve such information.

- MIH registration

MIH registration is defined to query access to certain MIH services. This phase is either mandatory or optional depending on the required level of service support. Indeed, the registration allows the peer MIHF entities to communicate in a trusted manner and gives them access to extensive information [18]. Nevertheless, for security issues, this registration is valid only for a certain period of time and has to be re-established when needed.

- MIH event subscription

refers to the fact of subscribing to a particular set of events that are provided by the MIES of a local or remote MIHF. By subscribing to a set of events and commands, the MIHU expresses for example its interest in triggering specific link behavior. Each subscription request needs to be individually validated by a confirmation from the event source (e.g. the peer MIHF) [18].

Discussion

Because next generation networks will more likely consist of heterogeneous networks, the convergence towards a unified handover mechanism has become a must. From that perspective, the MIH mechanism offers an interesting alternative since it provides a generalized and standardized solution for handover across different access technologies. Nevertheless, its success highly relies on vendors support and willingness to integrate it in their future products [18].

7.5   Roaming

Roaming is the process through which a mobile user automatically gains access to the services of a different provider, when outside the coverage area of its home network provider. Roaming service is made possible through Network Service Providers (NSPs) that have cooperative agreements to grant each others' customers local access to their resources. The WiMAX roaming relationship between NSPs consists of a technical and a business relation. This section focus on the technical aspects of the roaming process as it is defined by the WiMAX forum.

Roaming provides significant advantages to customers, Home Network Service Providers (HNSP) and Visited Network Service Provider (VNSP) network operators. First, for users, they are able to use the network services even when traveling outside the coverage area of their HNSP. All the connectivity problems are transparent to them. From the HNSP point of view, roaming represents an increasing in the coverage footprint without incurring additional network capital costs. For the VNSP, roaming may provide additional revenue opportunities.

The roaming process may be considered outbound or inbound. For the HNSP a roaming is an outbound roaming, since the node is using the services of another operator. For the VNSP, it is an inbound roaming, since it is a user from another operator that is requesting to use the VNSP network.

Roaming can also be classified into national and international. National roaming occurs when the visited network is in the same country as the home network. International roaming occurs when the visited network is in a different country than the home network. Roaming can also occur between networks using different technologies, inter-standard roaming (which is referred to in this chapter as vertical handover), for example WiMAX and WiFi or WiMAX and GSM/CDMA [19].

To allow a more generic and flexible business model for the WiMAX technology, WiMAX forum identified and defined a series of business entities for the components of the WiMAX architecture that may, or may not, be implemented by the same real company. The defined business entities involved in the roaming process are [19]:

- Network Service Providers (NSPs) are business entities that provide IP connectivity and WiMAX services to WiMAX subscribers.
- Network Access Providers (NAPs) are business entities that provide WiMAX radio access infrastructure to one or more NSPs. NSPs may also have contractual agreements with other providers such as Internet Service Providers (ISPs).
- Home Network Service Provider (HNSP) is the service provider that has its users accessing the services of other operator's network through a roaming agreement.
- Visited Network Service Provider (VNSP) is the service provider that is hosting a node from another operator's network and with whom the VNSP has a roaming agreement.
- WiMAX Roaming Exchange (WRX) is an intermediary entity that can interconnect two or more NSPs to provide roaming service. NSPs may use the services of a WRX to handle specific functions while maintaining a bilateral roaming relationship with other NSPs, Hub Providers or Aggregators.

To enable a more broad and independent roaming process among operators the WiMAX forum defined WiMAX Roaming Interface (WRI). The definition of such interface does not prevent operators to exchange roaming information through proprietary interfaces, but it is a way to guarantee interconnection among different pairs that implements the interface.

Roaming agreements, when established between two NSPs that handle their respective networks and provide access services, are called bilateral roaming. This implies that users from one NSP can use the services of the other NSP and vice versa. The NSPs may be directly connected through a proprietary interface or using WRI. Alternatively one, or both NSPs may delegate some, or all, of the roaming functions to a third party WRX. In this case, the NSP will have a roaming agreement with the second NSP and a WRX agreement with a WRX provider. These two scenarios are described in Figure 7.11.

In the case of NSPs exchanging a high volume of roaming users, it is more likely that a direct connection should be established. However, direct connections can require a considerable amount of resources and management efforts. For NSPs relationships with smaller amount of roaming exchange, WRXs providers would probably be preferable.

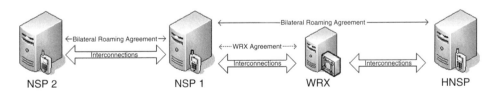

Figure 7.11 Direct and through WRX bilateral roaming agreement. Copyright © 2009 WiMAX Forum. All rights reserved.

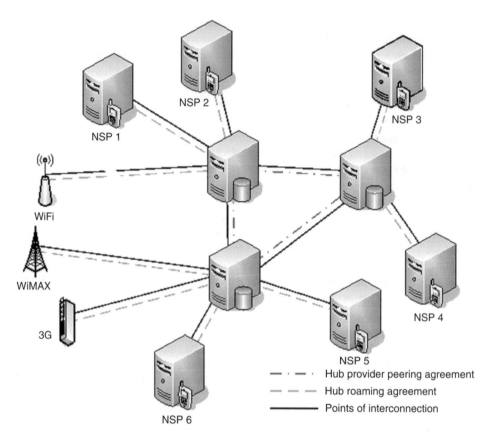

NSP 2

NSP 3

NSP 1

WiFi

WiMAX

3G

NSP 4

NSP 5

— · — · Hub provider peering agreement

— — — Hub roaming agreement

———— Points of interconnection

NSP 6

Figure 7.12   Possible components in a hub model architecture. Copyright © 2009 WiMAX Forum.

Unilateral roaming agreements occur when subscribers of one NSP roam onto the
network of a second NSP, without reciprocity. This can occur when the HNSP provides
wireless services to subscribers but does not operate a network itself [19].

One interesting business and technical model, defined by the WiMAX forum, to increase
the coverage of NSPs is the Hub Model. In this model, NSPs exchange roaming information through Hub Providers (Figure 7.12). A Hub provider may have relationships with
others Hub Providers, to increase their own services. However, the NSPs have contractual
agreements, and are financially liable, only to the Hub Provider they have agreements with.
The NSPs do not have a direct contractual relationship with each other. Hub Providers
act as intermediaries collecting payments from an HNSP and making payments to the
VNSP. The main attractive aspect of this model is the potential fast coverage growth it
can provide. Each new interconnection between a Hub Provider and a NSP may result in
new roaming between the new NSP users and the entire community of NSP members,
and vice versa.

A NSP may establish hundreds of roaming relationships to provide better services to
their clients and remain competitive in the market. However a small percentage of these

roaming relations may create the majority of the traffic. For these relationships, a direct bilateral roaming agreement may be more suitable to both HNSP and VNSP. However, roaming agreements normally present an incremental revenue source to an operator. For this reason, the Hub model may provide a more interesting, and less resource intensive, way to increase the NSP footprint and incomes.

7.5.1 WiMAX Roaming Interface

The WiMAX Roaming Interface (WRI) has the main objective of standardize the format and means for information exchange among the entities involved on the roaming process. The output for each phase of the roaming process is a file, in a specific format, which may be transmitted through File Transfer Protocol (FTP), or secure FTP, to the peer entity. Figure 7.13 shows the main components and files created and transferred along the roaming process.

7.5.2 The Roaming Process

The first phase of the WRI, Proxy Services, involve proxy of RADIUS messages, correlation and aggregation of session records, validating roaming agreements between the roaming partners and the transfer of aggregated session records to the Wholesale Rating and Fraud Management functions. The Proxy service will validate attributes which are present in the RADIUS records and will create files (aggregated sessions) to be sent to the Wholesale Rating function via the X2 interface [20]. The X2 file contains aggregated and correlated user sessions. Remote Access Dial-In User Service (RADIUS), is an IETF standard [21], that defines the functions of the authentication server and the protocols to access those functions. WRI uses RADIUS to communicate with the AAA server.

The wholesale rating function applies wholesale charging principle including taxes to the sessions of the visiting user. This phase receives X2 interface files, rates WiMAX subscriber sessions and generates X3 interface files. The Wholesale Rating function processes the X2 file sessions by applying the wholesale charging principle, in accordance to the specific Roaming agreement, and calculates the applicable charges and taxes. These processed sessions are then summarized in an X3 file, which is then forwarded to the Clearing function [22].

Clearing determines the accounts payable and receivable between WiMAX operators. Clearing also facilitates financial processes in support of revenue assurance, invoice balancing, and reconciliation.

The Clearing Module receives multiple X3 files from the Wholesale Rating module and processes these files to create one X5 file per roaming partner on a daily basis. The VNSP then forwards the X5 file the proper HNSP. After receiving the X5 file, the HNSP validates the charges and session based on its own X2 and X3 files. If the validation fails then the HNSP creates a Reject file and sends it to the VNSP. The reject sessions are negotiated between the NSPs until they can reach an agreement [23].

Financial settlement involves validating the correctness of the financial values assigned during data clearing, calculating an overall financial position for each NSP, performing foreign exchange for NSPs with different currencies, making the corresponding payments, and tracking any outstanding financial obligations. The Financial Settlement module

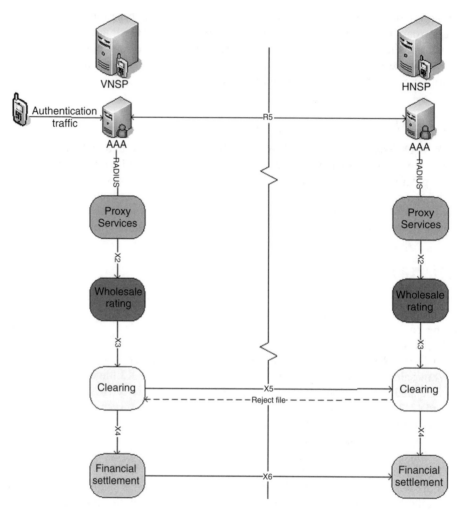

Figure 7.13 Components and files involved in the roaming process. Copyright © 2008 WiMAX Forum. All rights reserved.

receives the X4 files from the Clearing module and calculates the Net Financial position in generating an X6 – an Electronic Invoice File – which is sent to the HNSP [24].

7.6   Mobility Management in WiMESH Networks

In opposition to the PMP networks handoff, the WiMAX mesh networks handoff normally involve multi-hop communications, that is, the handoff process may be carried for many hops. For this reason, the handoff process should have a small signaling traffic to save bandwidth and decrease the handoff delay.

The existing solutions normally do not consider all the particular aspects of mesh networks, for example, Mobile IP can provide a solution to the inter-domain movement in WiMESH, but it is not suitable for the intra-domain movement, which is much more frequent than the inter-domain movement [25]. The reason for that is the cost of the communication, if the whole mobile IP protocol was to be implemented in each mesh node, the latency and signaling cost would be prohibitive for a mesh network environment.

Akyildiz *et al.* [26] classify the mobility management strategies for wireless networks as Full Connection Rerouting, Route Augmentation, Partial Connection Rerouting, and Multicast Connection Rerouting. Full Connection Rerouting maintains the connection by establishing a completely new route for each handoff just as initiating a new call. Route Augmentation relays on the route from current station and just increases the path adding a hop to the mobile node's next destination. Partial Connection Rerouting reestablishes segments of the original connection, while preserving the remainder. Multicast Connection Rerouting combines the former three protocols but includes the maintenance of potential handoff connection routes to support the original connection, to reduce the latency of finding a new route for handoff. All these rerouting protocols aim to solve different handoff problems and part of them may incur packet loss during handoff process [27].

More specifically for WiMAX mesh mode networks, several schemes were proposed to handle mobility management. Here we will present four of these proposals with the intention of illustrating the handoff process on WiMAX mesh networks.

Zhou *et al.* [28] propose a dynamic hierarchical architecture, based on a mesh structure, for the next generation wireless metropolitan area networks. The work uses the most interesting characteristics of both WiMAX and WiFi mesh networks to provide high quality access to mobile users. The WiFi mesh network offers mobility to the users, while WiMAX structure offers long distance backhaul and last mile solution. The main components defined by the architecture are: Mesh Points (MPs), Mesh Access Points (MAPs), Mobile Stations (MSs), Subscriber Stations (SSs) and Mesh Portal Points (MPPs). MPPs and MAPs are WiFi stations responsible, respectively, for the network maintenance and for providing access to end user nodes, the Mobile Stations.

The SSs and MPPs are WiMAX nodes, the first acts as a backhaul and keeps track of every node attached to the mesh network. MPPs stations have two interfaces and act as gateways between the WiFi and WiMAX networks. The Dynamical Hierarchical Mobility Management (DHMM) mechanism is composed of two phases: Registration and Fast Handoff procedures. In the registration procedure a new node, when arriving to a mesh access point, uses the standard IEEE 802.11 procedure. The mesh access point then warns the local subscriber station, a WiMAX node, about the presence of the new node. The subscriber station then contacts the mobile node home agent to inform it of the node new location. In the fast handoff procedure, when a MS is moving from one MAP to another, it sends a move notification to the first one that starts to buffer the MS messages. When the MS is attached properly to the new MAP this one performs a registration procedure with the previous MAP that starts than to forward the stored packets. The path is not rebuilt, the MS keeps using the ancient MAP to forward its message, unless one of the two things happens, either the chain length threshold is reached or the MS moves to a MAP that announces to be two hops closer to a MPP than the MPP in the forwarding chain. This fast handoff procedure primes for local communications which decreases signalling traffic in the network and leads to lower loss rates of in-flight packets.

An interesting idea, proposed by Du *et al.* [27], is to perform interactive local rerouting, based on mobile nodes speed, to create better and more optimized routes among nodes.

When a handoff occurs in a multi-hop network, some handoff algorithms try to find the nearest common node on the route to reconstruct the path between source and destination nodes. This technique may not always lead to an optimal route, especially when the nodes are far apart. Du *et al.* [27] propose that the rerouting scheme should try to find out the best common node in the next k hop nodes. With this mechanism, the new route can achieve the lowest route cost in original link backtracking and rebuilds the communication route from this common node in an iterative approach. The faster a node is moving, the lowest will be the value of k. When a node receives a location update message, it forwards all the packets through the new optimal route to the destination node. This avoids losing packets, the so-called 'missed on the flight data'.

Huang *et al.* [25] propose the Mesh Mobility Management (M^3) scheme for wireless mesh networks. M^3 works with a three level hierarchy composed by: gateway, Superior Routers (SR) and Access Points (AP). Superior Routers are the APs linked directly to the gateway that have a specific role on the protocol. SRs act as delegates of the gateway and share the signalling traffic. For small networks, where the access to the gateway does not represent a bottleneck, the SRs may be omitted. Unlike other Mobility management models that organize the nodes in trees, M^3 allows communications along the paths which are not in the tree. The advantage of this is that geographically adjacent APs have shorter communication paths other than the one existent along the tree. On this method, each node receives an IP number when attaching to the network, this number is used for authentication, authorization and accounting (AAA) information and identifies the nodes QoS profiles. The gateway and the serving AP store each node's IP number. Downstream traffic is tunnelled from the source to the destination AP, that unpacks the message and delivers it to the final destination. For upstream packets no tunneling is created since each AP can use the default routes to forward packets to the gateway. When a handoff occurs the moving client informs the former AP's ID to the new one. The new AP then sends a handoff request to the old one that replies with the corresponding subscriber information. The former AP creates a temporary entrance on its routing table to express the new position of the mobile station. When the timer expires, the information is removed from the former AP routing table. If the mobile station changes AP again, the process repeats and a forward chain may be formed.

Hoang *et al.* [29] have focused their work on the specific problem of handover for maritime WiMAX Mesh networks. When the MS joins the maritime WiMAX mesh, it first finds all the mesh Base Stations it could connect to. After that, the MS chooses the best two BSs, given some criteria, and registers to both BSs. The MS uses the best of the two BSs as the default server, and keeps the second one as a backup. During all the connected time the MS continuously collects information about the near BSs. A node can change the serving BS if the backup BS starts to present a better quality than the previously chosen one. In this case the Backup BS becomes the serving BS and vice versa. If a new BS appears and is better than the present serving BS, the SS registers to the new BS, unregisters from the old serving BS, and switches all communication the new BS. Another scenario could consist in having a BS presenting better characteristics than the backup one. In this case, the MS just unregisters from the present backup BS and registers to the new one signalling it as backup.

7.7 Conclusion

The WiMAX forum estimates that more than 133 million of people will be using the WiMAX technology by the year 2012. From these users, more than 70 % are expected to be using the mobile implementation of the technology. Mobility management is a key aspect to provide access for these potential 70 % of WiMAX users.

This book chapter has addressed some of the most important aspects and challenges related to mobility management in WiMAX networks in both PMP and Mesh modes. The crucial concept for mobility management is the handoff which is the process of transferring an ongoing session from one base station to another. The handoff has been studied in this chapter in all its forms: intra-WiMAX technology (horizontal handoff), inter-technologies (vertical handoff) and inter-providers (roaming). First we have described the different handoff mechanisms proposed by the IEEE 802.16e standard. Then, we have presented some of the works aiming at optimizing these procedures. We have classified the proposed works into two categories: those improving the handoff at layer 2 and those adopting an L2–L3 cross-layer approach in which the two layers collaborate to enhance the handoff performances. Among these cross-layer mechanisms, we have described more in details the fast MIPV6 handover mechanisms over 802.16e, proposed by IETF in [9].

The vertical handoff in heterogeneous networks – including WiMAX systems – has been considered first through some works proposed in the literature, then through the MIH framework proposed by the IEEE 802.21 standard. Roaming, which is a key concept to increase the coverage of WiMAX network has also been addressed in this chapter. The last section of this chapter has been dedicated to handover mechanisms in WiMESH networks.

7.8 Summary

Through this Summary we point out the main conclusions derived through this book chapter:

- Considering the technical and business merits of the interoperability profiles proposed by the NWG could help to choose the most appropriate architecture when deploying a technological solution.
- The IEEE 802.16e standard proposes three handover modes, the hard handoff procedure consists of more steps and might cause intolerable delays for real-time traffic. Nevertheless, the two soft handover modes FBSS and MDHO cannot be a reliable alternative to the mandatory hard HO scheme for many reasons. On the one hand, there are several restrictions on BSs working in MDHO/FBSS modes since they need to synchronize on time and frequency and have synchronized frame structures which entails extra costs. On the other hand, in both FBSS and MDHO modes, the BSs in the same Diversity Set are likely to belong to the same subnet while a handover may occur between BSs in different subnets.
- Mobile IPv6 fast handovers, like all cross-layer handover management mechanisms in general, are based on the collaboration of different layers in order to enhance the mobility management. This idea of integrating information from different network layers

helps to improve the HO management performances. Neverthless, because these solutions usually require significant modifications in the network stack, their deployment might be too complex.

- Because next generation networks will more likely consist of heterogeneous networks, the convergence towards a unified handover mechanism has become a must. From that perspective, the MIH mechanism offers an interesting alternative since it provides a generalized and standardized solution for handover across different access technologies. Nevertheless, its success highly relies on vendors support and willingness to integrate it in their future products.

- A key concept to increase the coverage of network providers is roaming. Through roaming a mobile user may automatically access the services of a different provider, when outside the coverage area of its home network provider. This allows a more generic, flexible and extensible business model for the WiMAX technology. To enable a more broad and independent roaming process among operators, the WiMAX forum defined the WiMAX Roaming Interface. The objective of this interface is to standardize the format and means for information exchange among the entities involved in the roaming process in order to provide a broader and more independent information exchange among different entities.

- The handoff process for the WiMAX mesh mode is fairly different from the one defined for the PMP mode. In opposition to the PMP networks handoff, the WiMAX mesh networks handoff normally involve multi-hop communication, as the node may not be directly connected to the provider's access point. Akyildiz *et al.*'s classification divides the existent mesh mobility management techniques into Full Connection Rerouting, Route Augmentation, Partial Connection Rerouting, and Multicast Connection Rerouting. However, the existent methods for mesh networks are not sufficient to overcome all the needs of WiMAX Mesh mode. The research on new techniques leaded to the development of concepts such as dynamic hierarchical architectures, interactive local rerouting, etc.

References

[1] IEEE Std 802.16-2004, IEEE Standard for Local and metropolitan area networks–Part 16: Air Interface for Fixed Broadband Wireless Access Systems. 2004.

[2] IEEE Std 802.16e 2005, IEEE Standard for Local and metropolitan area networks Part 16: Air Interface for Fixed and Mobile BWA Systems-Amendment 2: Physical and Medium Access Control Layers for Combined Fixed and Mobile Operation in Licensed Bands and Corrigendum 1. 2005.

[3] WiMAX Forum, WiMAX Forum Network Architecture (Stage 2: Architecture Tenets, Reference Model and Reference Points) [Part 1]. volume Release 1, Version 1.2, Jan. 2007.

[4] R.Q. Hu, D. Paranchych, Mo-Han Fong and Geng Wu, On the evolution of handoff management and network architecture in WiMAX. In *Proceedings of the Mobile WiMAX Symposium, 2007*, pp. 144–149, Mar. 2007.

[5] D.H. Lee, K. Kyamakya and J.P. Umondi, Fast handover algorithm for IEEE 802.16e broadband wireless access system. In *Proceedings of the 1st International Symposium on Wireless Pervasive Computing, 2006*, p. 6, Jan. 2006.

[6] J. Chen, C.-C. Wang and J.-D. Lee, Pre-Coordination Mechanism for Fast Handover in WiMAX Networks. In *Proceedings of the The 2nd International Conference on Wireless Broadband and Ultra Wideband Communications, 2007. AusWireless 2007*, Aug. 2007.

[7] L. Chen, X. Cai, R. Sofia and Z. Huang, A Cross-Layer Fast Handover Scheme For Mobile WiMAX. In *Proceedings of the IEEE 66th Vehicular Technology Conference, 2007. VTC-2007 Fall*, pp. 1578–82, Oct. 2007.

[8] C.-K. Chang and C.-T. Huang, Fast and Secure Mobility for IEEE 802.16e Broadband Wireless Networks. In *Proceedings of the International Conference on Parallel Processing Workshops, 2007. ICPPW 2007*, pp. 46–52, Sept. 2007.

[9] H. Jang, J. Jee, Y. Han, S. Park and J. Cha, Mobile IPv6 Fast Handovers over IEEE 802.16e Networks. RFC 5270 (Informational), 2008.

[10] E.R. Koodli, Mobile IPv6 Fast Handovers. RFC 5268 (Standards Track), 2008.

[11] M. Kassab, *Layer-2 Handover Optimization for Intra-technologies and Inter-technologies Mobility*. PhD thesis, Institut Télécom-Télécom Bretagne, UR1 - Université de Rennes 1, 2008.

[12] IEEE 802.21-2008, IEEE Standard for Local and metropolitan area networks Part 21: Media Independent Handover. Jan. 2008.

[13] J.-Y. Baek, D.-J. Kim, Y.-J. Suh, Eui-Seok Hwang and Young-Don Chung, Network-Initiated Handover Based on IEEE 802.21 Framework for QoS Service Continuity in UMTS/802.16e Networks. In *Proceedings of the IEEE Vehicular Technology Conference, 2008. VTC Spring 2008.*, pp. 2157–61, May 2008.

[14] J. McNair and F. Zhu, Vertical handoffs in fourth-generation multinetwork environments. *IEEE Wireless Communications* **11**(3): 8–15, June 2004.

[15] Z. Daia, R. Fracchiaa, J. Gosteaub, P. Pellatia and G. Vivier, Vertical handover criteria and algorithm in IEEE 802.11 and 802.16 hybrid networks. In *Proceedings of the IEEE International Conference on Communications, 2008. ICC'08*, pp. 2480–84, May 2008.

[16] Y. Liu and C. Zhou, A Vertical Handoff Decision Algorithm (VHDA) and a Call Admission Control (CAC) policy in integrated network between WiMAX and UMTS. In *Proceedings of the Second International Conference on Communications and Networking in China, 2007. CHINACOM'07*, pages 1063–68, Aug. 2007.

[17] L. Sarakis, G. Kormentzas and F. Moya Guirao, Seamless Service Provision For Multi Heterogeneous Access. *IEEE Wireless Communications*, Oct. 2009.

[18] E. Piri and K. Pentikousis, IEEE 802.21: Media Independent Handover Services. *The Internet Protocol Journal* **12**(7): 27, June 2009.

[19] WiMAX Forum, WiMAX Forum® Roaming Models White Paper. In *WMF-T48-001-v01 Approved Version 1*, volume Version 1, Apr. 2009.

[20] WiMAX Forum, WiMAX Forum® Interface based on WiMAX Forum Certified™ Products, Stage 2: Part 1–AAA Proxy. In *Release 1.0 Approved Specification*, volume Version 1.0.0, Nov. 2008.

[21] C. Rigney, S. Willens, W. Simpson and A. Rubens, IEEE Remote Authentication Dial In User Service (RADIUS), RFC 2865, IETF Network Working Group. RFC 2865 (Standards Track), 2000.

[22] WiMAX Forum, WiMAX Forum® Interface based on WiMAX Forum Certified™ Products, Stage 2: Part 2–Wholesale Rating. In *Release 1.0 Approved Specification*, volume Version 1.0.0, Nov. 2008.

[23] WiMAX Forum, WiMAX Forum® Interface based on WiMAX Forum Certified™ Products, Stage 2: Part 3–Clearing. In *Release 1.0 Approved Specification*, volume Version 1.0.0, Nov. 2008.

[24] WiMAX Forum, WiMAX Forum® Interface based on WiMAX Forum Certified™ Products, Stage 2: Part 4–Financial Settlement. In *Release 1.0 Approved Specification*, Nov.

[25] R. Huang, C. Zhang and Yuguang, A mobility management scheme for wireless mesh networks. In *Proceedings of the IEEE Global Telecommunications Conference, GLOBECOM'07*, Nov. 2007.

[26] I. Akyildiz, J. McNair, J. Ho, H. Uznalioglu and Weyne Wang, Mobility management in next generation wireless systems. In *Proceedings of the IEEE*, pp. 1347–84, 1999.

[27] W. Du, W. Jia and W. Lu, Backtracking Based Handoff Rerouting Algorithm for WiMAX Mesh Mode. *Springer, Lecture Notes in Computer Science*, 4159, 2006.

[28] H. Zhou, C. Yeh and H.T. Mouftah, A Dynamic Hierarchical Mobility Management Protocol for Next Generation Wireless Metropolitan Area Networks. In *Proceedings of the 4th IEEE Consumer Communications and Networking Conference, CCNC 2007*, Jan. 2007.

[29] V.D. Hoang, M. Ma, R. Miura and M. Fujise, A Novel way for Handover in Maritime WiMAX Mesh Network. In *Proceedings of the 7th International Conference on ITS Telecommunications, ITST '07*, June 2007.

Part D

Advanced Topics

8

QoS Challenges in the Handover Process

Marina Aguado and Eduardo Jacob
ETSI, Departamento de Electronica y Telecommunicaciones, University of the Basque Country, Spain

Marion Berbineau
Institut National de Recherche sur les Transports et leur Securite (INRETS), France

Ivan Lledo Samper
Bournemouth University, UK

8.1 Introduction

Currently, research is being carried out to develop a new generation of wireless mobile networks that provide broadband data communication in the high speed vehicular scenario. International Telecommunication Union, Radiocommunication Section (ITU-R) has proposed International Mobile Telecommunications (IMT-Advanced) technical requirements for supporting such usage scenarios. IMT-Advanced identifies those mobile communication systems with capabilities which go further than those of IMT-2000. The IEEE 802.16 standard, supported under the WiMAX network, has evolved from a fixed scenario, in IEEE 802.16d, towards a mobile typical vehicular scenario (up to 120 km/h) with IEEE 802.16e. In the near future, the IEEE 802.16m specification will cover mobility classes and scenarios supported by IMT-Advanced, including the high-speed vehicular scenario (up to 350 km or even up to 500 km/h).

IEEE802.16 initial standards adopted Data Over Cable Service Interface Specification (DOCSIS) Quality of Service (QoS) mechanisms. Similarly, the IEEE802.16 Medium Access Control (MAC) Security Sublayer, responsible for providing security mechanisms such as privacy, authentication and encryption over the air link, was also based on DOCSIS

WiMAX Security and Quality of Service: An End-to-End Perspective Edited by Seok-Yee Tang, Peter Müller and Hamid Sharif
© 2010 John Wiley & Sons, Ltd

standard. However, DOCSIS is a wired based technology and QoS mechanisms in fixed wireless technologies, while sharing many of the features of QoS mechanisms for wired technologies, face some extra limitations including bandwidth limitations, longer end-to-end delays and higher packet losses owing to channel-induced bit errors.

Additionally, QoS mechanisms in mobile broadband wireless technologies represent a step further in complexity. Time variability and the unpredictability of the channel become more acute and the main challenge arises from the need to hand over sessions from one cell to another as the user moves across their coverage boundaries. During this handover process, it is still necessary to provide session continuity and to offer the previously negotiated end-to-end QoS and security levels.

In this context, and from the end-to-end QoS point of view, the packet loss and additional latency introduced by the handover process is an issue which needs to be tackled. And from the security point of view, a new goal comes to light: to minimize the impact of security procedures on the performance of the handover process. Stronger security mechanisms, that is using stronger encryption methods and multiple layers of security or changing encryption keys more frequently, come at the price of compromising QoS performance (i.e. increasing processing time and therefore higher end-to-end delays).

This trade-off, security versus QoS performance, is highlighted during the handover process. This chapter focuses on enhancement techniques in the handover process that represent an improvement of global end-to-end QoS indicators.

This chapter is structured as follows. Section 8.1 describes the challenge that the handover process represents from the point of view of QoS performance indicators in the full mobility scenario. It describes the application of QoS requirements for the full mobility scenario. These requirements are related to end-to-end performance but will also apply to sessions involving handovers.

Section 8.2 is a necessary overview of the handover process in the IEEE802.16 standard; timing and performance considerations illustrate each stage in the handover process. Section 8.3 describes the Media Independent Handover (MIH) Initiative or IEEE802.21. Section 8.4 presents a survey on the different handover enhancement strategies found in the literature. These strategies are grouped, classified and discussed. Section 8.5 covers the efficient scheduling of the handover process and its influence on handover performance and end-to-end quality of service. To conclude, in section 8.6, a handover performance analysis is carried out.

8.2   Handover in WiMAX

When mobile broadband wireless technologies migrate from a nomadic scenario to a typical vehicular or a high speed vehicular usage scenario, the supported mobile speed increases, the dwelling time within a cell decreases, and the time variability as well as the channel unpredictability become more acute.

The time during which the mobile node is involved in handover processes compared to normal operation increases. Consequently, end-to-end QoS indicators, such as delay time or data loss, are significantly more affected by QoS handover performance indicators. Therefore, as WiMAX technology reaches higher mobility scenarios, a heavy burden is placed upon the performance of the handover process and mobility management solutions.

This leads one to consider the handover process as a fundamental research topic and the critical issue to be considered as WiMAX technology reaches higher mobility scenarios.

WiMAX architecture is expected to support six different usage scenarios: fixed, nomadic, portable, simple mobility, full mobility and the high speed vehicular scenario. In the simple mobility scenario at least one of the mobile nodes involved in the communication flow performs its trajectory at a speed of up to 50 km/h. In the full mobility usage scenario at least one of the mobile nodes involved in the communication flow performs its trajectory at a speed of up to 150 km/h. The characterization of this full mobility usage scenario from a telecommunication point of view is that one in which:

- the mobile node stays connected to the network and experiences no performance degradation at mobility speeds up to 150 km/h;
- there is session continuity;
- real and non real time applications are supported while moving, including during the handover process;
- any pre-negotiated QoS levels are supported across multiple Base Stations (BSs) at all times;
- the total handover latency is below 50 ms;
- there is bounded packet loss (e.g., <1 %) during handovers.

These features may be viewed as the requirements that Mobile WiMAX networks are expected to meet when supporting real time applications in the full mobility scenario.

8.3   The IEEE802.16 Handover Process

In order to propose a strategy for enhancing the handover process it is necessary to study the IEEE 802.16 handover mechanisms and internal features, along with timing considerations. This overview allows one to identify the main points where enhancement strategies may be implemented. This section details some specific IEEE802.16e standard concepts and procedures that are involved in the handover process. First, subsection 8.2.1 details the specific network entry procedure. Second, in subsection 8.2.2, some of the details of the IEEE802.16e specific network topology advertising and network topology acquisition strategy are outlined. The next subsection covers the association procedure. We conclude this section detailing the different stages in the handover process.

8.3.1   The Network Entry Procedure

The first process to be considered when studying the handover process is the network entry procedure. It must be taken into account that a handover process worst case scenario represents a complete re-entry procedure. Figure 8.1 presents the network entry procedure. The implementation of phases represented in the light grey blocks is optional.

8.3.1.1   Scan for DL Channel and Establish Synchronization with BS

The Mobile Station (MS) begins to scan the possible channels of the downlink (DL) frequency band of operation until it finds a valid DL signal. Once the physical (PHY)

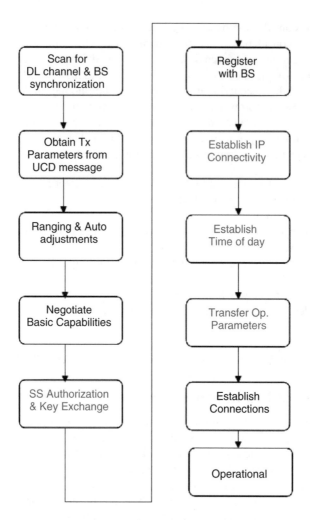

Figure 8.1 Initialization of an MS (Mobile Station), neither error paths nor timeout values being considered (IEEE802.16e).

layer has obtained synchronization, as given by a PHY trigger indication, the MAC layer will attempt to acquire the channel control parameters for the DL and then the uplink (UL). The MS searches for the DL frame preambles. The preamble initiates the downlink subframe. When one is detected, the MS can synchronize itself with respect to the DL transmission of the BS. The MS then obtains the PHY DL synchronization.

The MS then listens to the various MAC management messages, such as Frame Control Header (FCH), Downlink Channel Descriptor (DCD), Uplink Channel Descriptor (UCD), Downlink MAP (DL-MAP), and Uplink MAP (UL-MAP), that follow the preamble. The MS achieves MAC synchronization once it has received at least one DL-MAP message and is able to decode the DL burst profiles contained in the message. Then, the Link Detected (LD) Trigger is fired. An MS remains in MAC synchronization as long as it continues to receive the DL-MAP and DCD messages for its channel.

8.3.1.2 Obtain UL Parameters

After synchronization, the MS waits for a UCD message from the BS in order to retrieve a set of transmission parameters for a possible UL channel. These messages are transmitted periodically from the BS for all available UL channels and are addressed to the MAC broadcast address. If no UL channel can be found after a suitable timeout period, then the MS continues scanning to find another DL channel.

If the channel is suitable, the MS extracts the parameters for this UL from the UCD. Then, the SS will wait for a bandwidth allocation map for the selected channel. It may begin transmitting UL in accordance with the MAC operation and the bandwidth allocation mechanism.

The uplink synchronization latency, or time involved in this second phase, has the greatest significance when compared to the other latencies. The reason why is that the DL_MAP and UL_MAP messages which contain the burst allocation decided by the BS are sent normally in each frame. The standard defines a timeout with a maximum value of 600 ms. However, DCD and UCD messages are not sent normally in each frame.

8.3.1.3 Ranging

As defined in 802.16e2005 : 'Ranging is the process of acquiring the correct timing offset and power adjustments so that the MS's transmissions are aligned with the BS receive frame and received within the appropriate reception thresholds'.

Uplink ranging consists of two procedures: initial ranging and periodic ranging. Initial ranging allows an MS to join the network to acquire correct transmission parameters, such as time offset and Tx power level, so that the MS can communicate with the BS. Following initial ranging, periodic ranging allows the MS to adjust transmission parameters so that the MS can maintain UL communications with the BS (802.16e2005).

The **ranging latency** or latency corresponding to the initial ranging activity depends on the backoff window size and on the number of initial ranging opportunities per frame.

8.3.1.4 Negotiate Basic Capabilities

Immediately after completion of ranging, the MS sends an SBC-REQ message informing the BS of its basic capabilities (PHY and bandwidth allocation parameters). The BS responds with an SBC-RSP message. The PHY and bandwidth-allocation parameters can be the same as the informed MS's capabilities or a subset of them.

8.3.1.5 Registration

Registration is the process by which the MS is allowed to enter the network and receive secondary Channel Identifiers (CIDs). To register with a BS, the MS will send a REG-REQ message to the BS. The BS responds with a REG-RSP message.

8.3.1.6 Establishing Provisioned Connections

After registration the BS will send DSA-REQ messages to the MS to set up connections for pre-provisioned service flows belonging to the MS. The MS responds with DSA-RSP messages.

All the phases previously described can be observed from Figure 8.2.

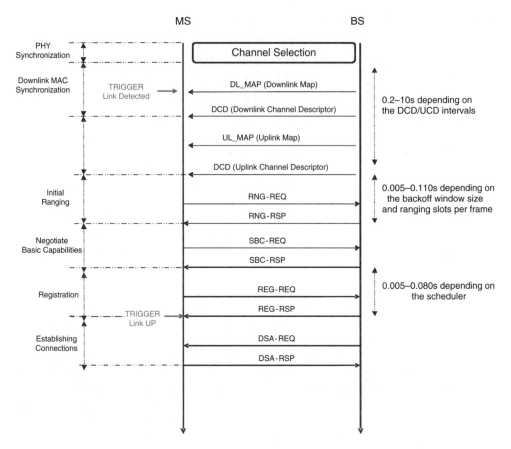

Figure 8.2   The network entry process. Exchange message based on IEEE802.16e standard and values based on [8].

8.3.2   Network Topology Advertising and Acquisition

Previous to performing a handover process, the mobile node has to be aware of the network topology so as to be able to identify the possible base stations that are available and capable of providing/able to provide the access service. The IEEE802.16 standard provides two complimentary mechanisms designed to achieve this.

8.3.2.1   Neighbouring Advertising

Base stations in the IEEE802.16 standard are configured to belong to one or various neighbourhoods within the network topology. Each base station may receive channel information related to its neighbouring BSs over the backbone network. Each base station, similar to the access point beacon in wireless LAN, broadcasts the network topology relative to its neighbourhood using the MOB_NBR-ADV message. This message is sent through the broadcast connection and it is sent at periodic intervals. It has a variable size that depends on the number of neighbouring BSs in the same neighbourhood. It contains

DCD and UCD settings for each base station in the neighbourhood. A base station may belong to more than one neighbourhood.

The MS receives this MOB_NBR-ADV message and creates and updates a list of neighbouring BSs. The mobile node will then be aware of its target BSs for handover and scanning purposes. Every time that the MS connects to a new serving BS, it will receive a new MOB_NBR-ADV message from its current serving base station and the list of neighbouring base stations will be updated. The new list may differ from the previous list of neighbouring BSs.

The information contained in MOB_NBR-ADV message and relative to DCD and UCD settings for each neighbouring base station *facilitates* MS synchronization with neighbouring BSs by removing the need to monitor transmission from the neighbouring BS for DCD/UCD broadcast messages; *the most critical and time consuming step*. These messages describe the physical characteristics for each burst indicated in the DL-MAP and UL-MAP messages. The modulation is not included since it varies adaptively with link quality.

Receiving on-time neighbouring information properly enhances handover performance.

8.3.2.2   Cell Reselection or Scanning Process

The main purpose of this procedure is the MS node monitoring target BS nodes in the neighbourhood and finding out if they are suitable for handover. The time during which the MS scans for available BSs will be referred to as the *scanning interval*.

This procedure may be initiated by either the BS or the MS nodes. A BS may allocate time intervals to the MS for scanning purposes. In this case a MOB_SCN-RSP message is sent by the serving BS. This message includes information relative to the start frame, the scanning process duration (N frames), the interleaving interval (P frames) and the iteration (T times).

As it can be observed from Figure 8.3, the scanning interval and interleaving interval will be repeated with the number of scan iterations (T).

An MS may also initiate the scanning procedure by sending a MOB_SCN-REQ message to the serving BS. Upon reception of the MOB_SCN_RSP message, the BS responds with a MOB_SCN-RSP message with the previously identified data. Following the reception of a MOB_SCN-RSP message granting the request sent previously by the MS, an MS may scan for one or more target BS during the interval allocated in the message. The MS may attempt to synchronize with its DL transmissions and estimate the quality of the PHY channel. The serving BS buffers incoming data addressed to the MS during the scanning interval. The BS then transmits that data during any interleaving interval or once the scanning mode is over.

It is important to note that the event that initiates the whole scanning procedure (sending the MOB_SCN-REQ message for MS, or sending the MOB_SCN-RSP message for BS) is not covered by the standard. It is implementation dependent. In most of the implementations, initiation of the scanning procedure is triggered by power related measurements such as Received Signal Strength (RSS), Received Signal Strength Interference (RSSI), Carrier to Interference-plus-Noise Ratio (CINR), Signal-to-Noise-Ratio (SNR), performed either at the MS or the BS. However, different scanning initiation policies may also be implemented.

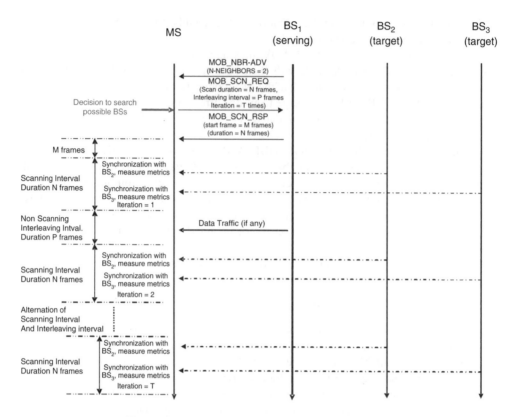

Figure 8.3 The scanning procedure (IEEE802.16e).

Some implementations define what is called as a *scanning threshold*. This pre-defined scanning threshold value will be compared with MS or BS measurements to determine when the scanning procedure can start. Each node, MS or BS may have its own scanning thresholds pre-defined. And a node may have multiple values for the scanning threshold. Each of them will be associated with a specific set of scanning parameters.

It may be easily understood that the longer the scanning interval duration, the bigger the queues in the BS and consequently the end-to-end delay measurements will increase. However, the scan procedure is necessary. Measurements taken during scanning mode may trigger a handover procedure. In the scan procedure proper timing is also important. It should start with enough time in advance so that the measurements can be taken and contribute accurately to the handover decision-making process. To summarize, the scanning interval and related parameters have to be adjusted properly and carefully to optimize global performance.

8.3.3 The Association Procedure

During the scanning process there may or may not be an association procedure. This association procedure is an optional initial ranging process between the MS and one of the neighbouring BSs. The main goal of the association process is to enable the MS to acquire and record ranging parameters and service availability information for the purpose of expediting a potential future handover to this target BS. There are three levels of association:

- *Association Level 0: Scan/Association without coordination.* This is the most basic level of association. The MS performs ranging without coordination from the network. The target BS has no knowledge of the MS and thus it will provide only contention-based ranging allocations. This association level is described in more detail in Figure 8.4.
- *Association Level 1: Association with coordination.* In this case, the serving BS will coordinate the association procedure between the MS and neighbouring BSs. Each neighbouring BS assigns a unique code number and a transmission opportunity within the allocated region to each MS. The serving BS will provide the pre-assigned ranging information via the MOB_SCN-RSP message.
- *Association Level 2: Network assisted association reporting.* When adopting the network assisted association reporting, the MS will include a list of neighbouring BSs with which it wishes to perform the association. The association will then be coordinated in a similar manner to Association Level 1. However, in this case the MS is only required to send the Code Division Multiple Access (CDMA) ranging code to the neighbouring BS, and not to wait for the RNG-RSP. Instead, the RGN-RSP information on PHY offsets will be sent by each neighbouring BS (over the backbone) to the serving BS.

8.3.4   Handover Stages in the IEEE 802.16 Standard

This subsection describes the different stages in the handover process. Special attention is drawn to timing considerations. These details and understanding are necessary so as to identify the main variables and factors that influence handover performance. The IEEE802.16 handover process, in which a MS migrates from the air interface provided by one BS to the air interface provided by another BS, consists of the following stages from a general point of view: normal or regular operation, cell reselection, handover decision, handover initiation and handover execution.

8.3.4.1   Stage1: Normal or Regular Operation

The MS is connected to the serving BS in the packet scheduling process. Periodic ranging takes places during all of the time the MS is connected to the serving BS.

8.3.4.2   Stage 2: Cell Reselection or Scanning

The main purpose of this stage is to collect information relative to the neighbouring BSs. As it is illustrated in Figure 8.5, this information forms part of the Handover (HO) decision process. In this stage, the MS uses neighbouring BS information acquired from a decoded MOB_NBR-ADV message. The MS may also make a request to the serving BS to schedule scanning intervals for the purpose of evaluating the MS's interest in handover to a potential target BS. This can be observed from Figure 8.4.

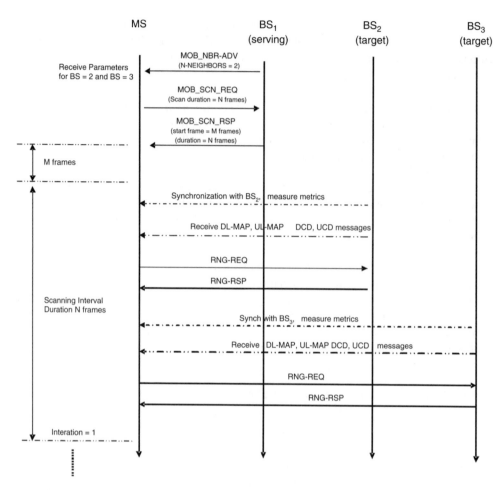

Figure 8.4 Neighbouring BS advertising and scanning with non coordinated association by MS request (IEEE802.16e).

The relationship between the scanning process and the handover process is represented in the block diagram in Figure 8.5.

8.3.4.3 Stage 3: Handover Decision-Making Process

There are a set of different concepts involved in the decision-making process. On one hand, there are the handover policies – a set of rules that contribute to shape the handover decision for a mobile node – on the other hand, there are the triggers or events from the different layers that will be used by the handover policies. The handover policy itself, although it plays a pivotal role in handover performance, is not within the scope of the IEEE802.16 standard and remains implementation-dependent.

Over recent years, a huge amount of research has addressed handover policies for wireless communications systems in general. Most of the currently implemented handover

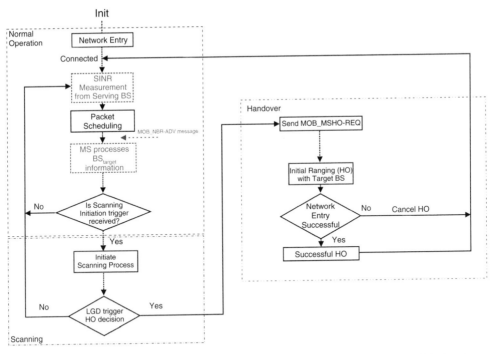

Figure 8.5 Block diagram model detailing relationship between scanning process and the handover process.

decision strategies or policies are based on CINR measurements. However, the MS may use any of the following available *information* to support its decision-making process:

- Channel quality measurements through instantaneous or weighted average values of link layer performance parameters from the serving BS and neighbouring BSs, such as RSS, RSSI, SNR, BER (Bit Error Rate), etc.
- Accurate MS position obtained through various radio location techniques such as hyperbolic position using either the time difference of signals arriving from neighbouring BSs, global positioning system equipment or balise (beacon) readers.
- The serving BSs may trigger a handover in accordance with RRM (Radio Resources Management) information provided by the RRM network entity. For a successful handover, it is also necessary to take into account whether radio resources are available in the target BS to handle the session. To minimize the possibility of dropping sessions owing to a lack of resources at the target BS, some system designs may reserve a fraction of network resources solely for accepting handover sessions. This may lead to the consumption of additional radio resources and consequently to decreased spectral efficiency. A better approach is to incorporate radio resource information in the handover decision.

Some of the most common **handover policies** in general mobile communication systems are:

- *HO decision based on CINR* – the decision is made when:

$$CINR_{BStarget} > CINR_{BSserving}$$

This method provokes too many unnecessary handovers when the current base station signal is still adequate. The fluctuations of signal strength associated with shadow fading cause a communication to be handed over back and forth repeatedly between neighbouring BSs. This is called the ping-pong effect.

- *HO decision based on RSS with a threshold* – the decision is made when:

$$(CINR_{BSserving} < Threshold) \wedge (CINR_{BStarget} > CINR_{BSserving})$$

The performance of this method is dependent on the threshold value. If the threshold value is too small, numerous unnecessary handovers may be processed. If the threshold value is too large, the handover initiation delay increases and consequently degrades the QoS performance indicators. The effectiveness of this method depends on prior knowledge of crossover signal strength.

- *HO decision based on RSS with hysteresis* – the decision is made when:

$$CINR_{BStarget} > (CINR_{BSserving} + Hysteresis)$$

This method prevents the ping-pong effect.

- *HO decision based on RSS with hysteresis and threshold* – the decision is made when:

$$(CINR_{Bstarget} < Threshold) \wedge (CINR_{Bstarget} > (CINR_{Bsserving} + Hysteresis))$$

Apart from these strategies some others may be found in the literature which are also related to RSS measurements:

- *HO decision based on expected future value of RSS, predictive technique.*
- *HO decision based on candidate's maximum RSS.*

These are a variant of the relative signal strength method that includes multiple *handover candidates.*

- *HO decision based on candidate's maximum RSS plus timer.*

Similar to the previous scheme but in order to prevent the ping-pong effect, the handover is allowed only after a timer expires.

- *HO decision based on RSS with hysteresis plus dwell timer.*

This time the timer is related to the mobile node trajectory performed.

There are some other approaches based on the application of non-standard control techniques that include neural networks, fuzzy logic, hypothesis testing and dynamic programming in order to identify the optimum set of parameters for each scenario [1], [2]. They decrease handover latency and the number of unnecessary handovers by changing the RSSI average window according to the MS's speed. It is worth mentioning that these algorithms are complex and are not easy to implement in practical systems.

8.3.4.4 Stage 4: Handover Initiation

In most IEEE802.16e implementations, the decision to initiate a handover process is taken typically by mobile stations as a result of their own measurement of the quality of the signal from the neighbouring base stations; but it can also be initiated by the network (serving base station) under special circumstances. In accordance with the entity that makes the decision, this handover decision process is consummated by a notification of MS intent to handover through a MOB_MSHO-REQ message or MOB_BSHO-REQ message.

When a MOB_MSHO-REQ message is sent by a MS, the MS may indicate one or more possible target BSs. When a MOB_BSHO-REQ message is sent by a BS, the BS may indicate one or more possible target BSs. The MS may evaluate possible target BSs through scanning and association activity performed previously.

The serving BS may negotiate the location of common time intervals where *dedicated initial ranging* transmission opportunities for the MS are provided by all potential target BSs. This information may be included in the MOB_BSHO-RSP message. *Dedicated allocation* for transmission of RNG-REQ means that channel parameters collected by the MS autonomously during association with that BS are considered to be valid during a sufficient time and can be reused for actual network re-entry without being preceded by CDMA ranging.

The complete set of messages exchanged during the HO process can be observed from Figure 8.6.

8.3.4.5 Stage 5: HO Execution

At this stage, once the decision to handover an on-going session to a target BS has been made, the procedures related to the network entry process, including synchronization and ranging with the new target BS, take place.

8.3.5 Handover Execution Methods

This section provides an overview of the different handover execution techniques in the IEEE 802.16 standard.

Regarding the handover procedure or execution, the handover methods supported within the IEEE 802.16 standard can be classified into hard and soft handover, as represented in Figure 8.7. The four supported methods are Hard Handover (HHO), Optimized Hard Handover (OHHO), Fast Base Station Switching (FBSS) and Macro Diversity Handover (MDHO). Out of these, only the HHO or hard handover *(break-before-make)* is mandatory.

The OHHO takes place when the network re-entry is shortened by a target BS's previous knowledge of MS information obtained from a serving BS over the backbone network.

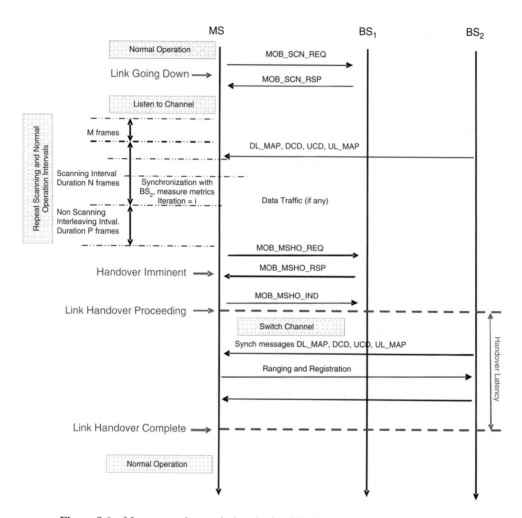

Figure 8.6   Message exchange during the hard handover process (IEEE802.16e).

In this way, OHHO assumes inter-BSs' backbone communication. Depending on the information provided, a target BS might decide whether to skip one or several stages in network entry process such as negotiating basic capabilities, the Privacy Key Management (PKM) authentication phase, Traffic Encryption Key (TEK) establishment phase or REG-REQ message phase.

The last two optional methods (FBSS and MDHO) are understood as soft handovers or *make-before-break* handovers because the MS maintains a valid connection simultaneously with more than one BS. In the FBSS case, the MS monitors a set of candidate BSs continuously, performs ranging and maintains a valid connection ID with each of them. The MS, however, communicates with only one BS, called the anchor BS. When a change of anchor BS is required, the connection is switched from one base station to another without performing handover signalling [3].

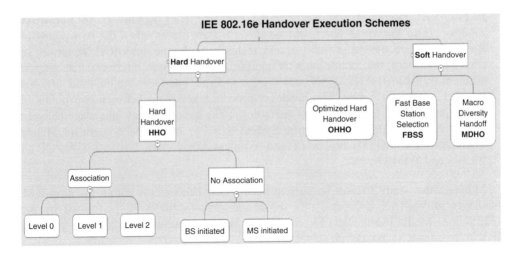

Figure 8.7 IEEE 802.16e Handover execution schemes (IEEE802.16e).

Macro Diversity Handover is similar to FBSS, except that the MS communicates simultaneously on the downlink and the uplink with all of the base stations in a diversity set. The well-known diversity-combining (for the downlink) and diversity selection (for the uplink) techniques are employed. Both FBSS and MDHO offer superior performance to HHO. However, they require that the base stations be synchronized and use the same carrier frequency.

8.4 The Media Independent Handover Initiative – IEEE 802.21

Before introducing the Media Independent Handover Initiative, it is necessary to present a commonly accepted handover taxonomy. A handover process in mobile wireless networks consists basically of the process that a mobile node or station carry out when moving from one Point of Attachment (PoA) or base station to another PoA while the mobile node moves across the BSs' cell boundaries. This transfer process may be motivated by signal fading, interference levels, etc. Consequently, the MS needs to connect to another BS with a higher signal quality or to another BS where the MS can be serviced with higher QoS, as in 802.16-2009.

The main features of these two points of attachment along the mobile node's route define the handover process taxonomy [4]:

- *Intradomain or Interdomain handover*: The two points may or may not belong to the same administrative domain; this classification is also commonly known as *micromobility or macromobility* handover.
- *Intrasubnet or Intersubnet handover*: The two points of attachment may or may not belong to the same or different subnet; The Inter subnet handover requires that the mobile node acquire a new Layer 3 identification IP address in the new subnet during the handover process and possibly undergo a new authentication process.
- *Intratechnology or Intertechnology handover*: The two points of attachment may or may not support the same access technology. This last classification is also known

as Intra-RAT (Radio Access Technology) and Inter-RAT handover. An Intra-RAT handover (also known as *horizontal handover*) takes places when the two points of attachment share the same radio access technology. In the Inter-RAT handover or *vertical handover,* the mobile node is equipped with multiple interfaces that support different technologies. Nevertheless, although multiple interfaces are required, just one interface is used at a time. The handover process takes place between two points of attachment that use different access technology. The handover may take place between Universal Mobile Telecommunication System (UMTS) and Global System for Mobile Communications (GSM) or UMTS and Wireless Local Area Network (WLAN) or WLAN and WiMAX, etc.

Although this study focuses on current handover enhancements within the intra-RAT handover in WiMAX networks, this subsection reviews the work performed by the IEEE802.21 Media Independent Handover working group and its approach to supporting link layer Inter-RAT Handover.

Session continuity during the inter-RAT handover demands a huge amount of research effort. One of the most difficult issues is the high number of standardization groups that are involved. The MIH initiative, IEEE 802.21, is leading this effort. The IEEE802.21 working group is a regulatory group that started its work in March 2004, which was standardized in 30 January 2009.

This standard's main goal is to provide independent mechanisms that enable the optimization of intra-RAT handovers between media types specified by Third Generation Partnership Project (3GPP), 3G Partnership Project 2 (3GPP2), Long Term Evolution (LTE) initiative and both wired and wireless media in the IEEE802 family of standards, including IEEE802.16 specification.

In order to maintain uninterrupted user connections during handover across different networks, IEEE802.21 defines a common media independent handover function between Layer 2 and Layer 3 of the OSI protocol stack. As it can be observed from Figure 8.8, along with the MIH Function, there are three services that allow for messages to be passed across the protocol stack.

- The Media Independent Event Service provides event classification, event filtering and event reporting corresponding to dynamic changes in link characteristics, link status and link quality.
- The Media Independent Command Service refers to the commands sent by the upper layers to the lower layers in the reference model. These commands mainly carry the upper layer decisions to the lower layer, and control the behaviour of lower layer entities.
- The Media Independent Information Service provides the capability for obtaining the necessary information for the handover process including neighbour maps, link layer information and availability of services.

It is important to emphasize that all the different media types covered in the MIH initiative, the current standards and future versions, have extended their architecture to include MIH services directives. However, the mapping of the common link layer triggers, as defined by the MIH initiative, to the specific link layer characteristics in the different media types is technology dependent, and in some cases is even implementation specific.

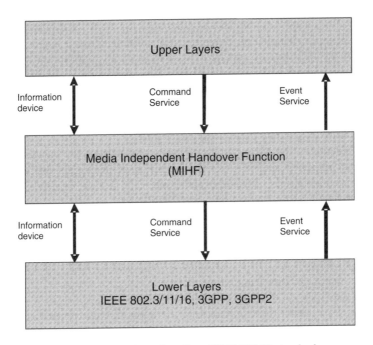

Figure 8.8   MIH services from IEEE 802.21 standard.

As an example, an IEEE 802.16 entity may send or receive a MOB_MIH-MSG message to or from the peer IEEE802.16 entity in order to convey MIH frames carrying the IEEE802.21 MIH protocol messages. The IEEE802.16 standard then provides support for IEEE 802.21 specific features and functions.

The IEEE802.16m new project will support IEEE802.21 MIH services and the mobility procedures will be fully compatible with the IEEE802.16 Network Control and Management Services (NCMS) defined in IEEE802.16g.

8.4.1 MIH Interactions with Layer 2 and Layer 3 Protocols

Within the inter-RAT handover context, the mobility management problem has to be solved in the link layer and in the network layer. It may be understood that the MIH standard occupies a L2.5 layer in the OSI protocol stack.

8.4.2 MIH Scope and Limitations

It is necessary to note that the following items **are not** within the scope of the MIH standard:

- Intra-technology handover (except for handovers across extended service sets (ESSs) in the case of IEEE 802.11). Homogeneous (Horizontal) handovers within a single Network (Localized Mobility) are handled by the specific standard: (IEEE802.11r, IEEE802.16e, 3GPP, 3GPP2).

- Handover control, handover policies and other algorithms involved in handover decision making are generally handled by communication system elements that do not fall within the scope of the IEEE802.21 standard. The IEEE802.21 contribution to the handover decision process is that MIH services provide the information about different networks and their services, thus enabling a more effective handover decision to be made across heterogeneous networks. The IEEE802.21 based Media Independent Handover (MIH) mechanism presents different types of triggers on Layer 2. However, IEEE802.21 does not specify how to generate these handover triggers.
- Handover execution. The MIH standard focuses primarily on the handover preparation and handover initiation stages; while the definition of the trigger mechanism and the handover execution itself is outside of its scope.
- Security mechanisms involved in inter-RAT handover.
- The MIH framework defines a set of triggers which may be used between layers to communicate specific events and that can be used to facilitate both vertical and horizontal handover. For example:
  - Link Going Down (LGD) trigger, after a MOB_BSHO-RSP message. This trigger indicates that a Handover is imminent;
  - Link Down (LD), on a MOB_HO-IND message;
  - Link Up (LUP), after the completion of the network entry process.

The Layer 2 triggers firing the MIH operation are identified in Figure 8.9.

8.5   Enhancing the Handover Process

This section presents the existing work on enhancement techniques in the handover process.

Currently the handover process is one of the most active research topics. The new enhancement handover mechanisms that come out frequently in each new draft release contribute to emphasizing this fact.

This section first introduces the two in-built mechanisms in the IEEE802.16 latest releases aimed at enhancing handover performance. Second, the different initiatives found in the literature are grouped in accordance with the corresponding handover stage that they propose to enhance.

8.5.1   Fast Ranging Mechanism

The fast ranging mechanism consists of the serving BS's capability, for the sake of expediting network re-entry process of the MS with the target BS, to negotiate with the target BS the allocation of a non-contention-based ranging opportunity for the MS, that is an unsolicited UL allocation for transmission of the RNG-REQ message. The MS, in this case, may ascertain the required ranging parameters from the target BS at the time of the handover.

The serving BS should indicate the time of the fast (i.e., non-contention-based) ranging opportunity, negotiated with the potential target BSs in the MOB_BSHO-REQ/RSP message. The target BS indicates the fast ranging allocation in the UL-MAP via Fast_Ranging_IE to the MS. Fast_Ranging_IE and zero BRH transmission are optional mechanisms

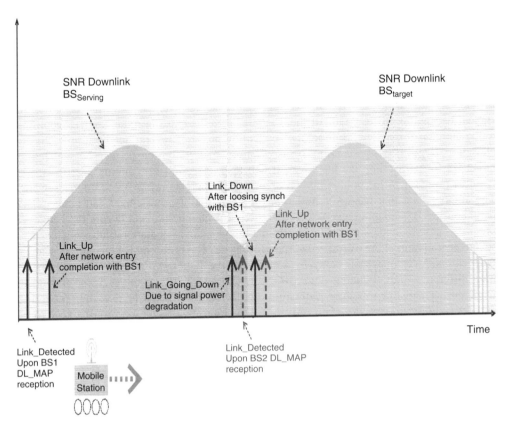

Figure 8.9 MIH pre-defined L2 triggers.

in the IEEE 802.16e-2005. However, they are mandatory in the WiMAX forum's mobile WiMAX system profile release 1 document.

8.5.2 Seamless Handover Mechanism

This new execution strategy was introduced in latest release of the IEEE802.16 standard, IEEE802.162009. In addition to optimized handover, the MS and the BS may perform what is called *seamless handover* to reduce handover latency and message overhead.

In order to perform a seamless handover, all the different entities (MS, serving BS and target BS) involved must support the seamless handover feature. This capability has to be included in the connection identifier descriptor Type, Length and Value (TLV) in the DCD message. The capability of performing seamless handover is negotiated by the MOB_MSHO_REG_REQ and the MOB_MSHO_REG_RSP messages.

In *seamless handover*, a target BS calculates primary management Connection Identifier (CID), secondary management CID, and Transport CIDs for an MS by using the descriptor. During the seamless handover, a serving BS includes the pre-allocated basic CID in MOB_BSHO-REQ/RSP for an MS. When a BS pre-allocates a basic CID to an

MS during seamless handover, the primary management CID is allocated autonomously without explicit assignment in the message. By doing so, the ranging step is shortened.

8.5.3 Initiatives in the Cell Reselection Stage

In 2006, Lee *et al.* [5] and Wang [6], highlighted some deficiencies in the IEEE802.16 standard scanning process. One of them is related to the existing redundancy. According to the authors, at the network topology acquisition phase, if several neighbouring BSs are chosen as target BSs for scanning or association, since only one BS can be selected as target BS for performing the handover, it will lead to redundant scanning processes.

Lee's proposal in [5] is also known as the single neighbouring BS scanning scheme. The authors state that handover initiation timing is not clearly defined and unnecessary neighbouring BS scanning and association are performed before and during the handover process. These redundant processes result in a long handover operation time, which causes severe degradation of the overall system performance. The proposed Fast handover scheme reduces handover operation delay by identifying the target BS using CINR and arrival time difference, thus reducing unnecessary neighbouring BS scanning. Some redundant work during network topology acquisition and scanning process is also shortened. The authors propose to associate with only one neighbouring BS, the one most likely to be the target BS.

Boone, in [7] introduces a strategy to enhance scanning process latency by reducing the number of frequencies checked during each scanning operation. It incorporates the history of successful scanning frequencies in order to guide the MS in choosing frequencies for future scanning operations. An MS builds a history of handovers between BSs and uses this to determine the most likely neighbouring target BS for a handover. Since the MOB_NBR-ADV messages provide the MS with the list of neighbours and their parameters, knowing which BS is the most likely handover target improves the scanning operation. The strategy requires no additional network support and only limited memory and computational resources from the MS.

In February 2007, Rouil from the National Institute of Standards and Technology (NIST) proposed the Adaptive Channel Scanning algorithm, to enhance the handover mechanism [8]. This minimizes the disruptive effects of scanning on the application traffic by using information regarding the QoS traffic requirements, the available bandwidth and the number of concurrent scanning stations to define the set of parameters of the scanning configuration. It is assumed that neighbouring BSs exchange information over the backbone and that accurate measurements of available bandwidth are available. The main objective is to configure the scanning parameters correctly so the necessary measurements are taken without losing a significant amount of throughput. Simulation results showed that by using the proposed algorithm, it is possible to minimize the impact of channel scanning on the data traffic.

8.5.4 Initiatives in the Execution Stage

One of the conclusions reached by Rouil *et al.* in [9], from their quantifying study on IEEE802.16 network entry performance, is that the delay contributed by the synchronization component is the most significant (0.2 s to 10 s depending on the DCD/UCD intervals for a 5 ms frame). The authors also conclude that when both UCD and DCD messages are synchronized the delay between the downlink and the uplink synchronization is minimized.

Synchronization also occurs during the handover execution stage and according to Rouil, in [9] any prior knowledge for synchronization (channel descriptor messages) is critical in speeding up the handover execution. If the information contained in the DCD and UCD messages acquired during scanning can be used to decode the DL_MAP and UL_MAP during the handover, this would represent a major enhancement. Again, a timely schedule for the scanning process is noted.

However, reducing the synchronization time by increasing the frequency of the channel descriptor messages comes generally at the cost of a higher bandwidth overhead. Additional improvements can be achieved if the BSs synchronize over the backbone [9].

Choi et al., in [10] introduce a fast handover scheme for real-time downlink services in IEEE802.16 networks. Their focus is on reducing the service disruption during hard handover for real time services by allowing the MS to receive downlink data just after synchronization with a target BS and before the establishment of the MS registration and authorization. This feature is called the Fast DL_MAP_IE HO scheme for real-time downlink services. However, it only promotes downlink services. Moreover, this proposal was not adopted by later releases of the standard, whereas some other strategies such as FAST-RNG_IE, suggested in order to reduce the ranging time, have been adopted.

8.6   Handover Scheduling

As introduced in section 8.3, under the MIH initiative the scheduling of the layer 2 handover process is provided by the link layer triggers that are fired at the PHY and MAC layer and that may communicate either to the MIHF function or the handover function in the WiMAX NRM (open WiMAX Network Reference Model).

One of the most important triggers is the predictive link-going-down (LGD) trigger that implies that a broken link is imminent; see Figure 8.9 The second is the link-down (LD) trigger which represents that no information is decidable further, and that therefore MAC synchronization is over.

According to Rouil *et al.* [8], major improvements in handover performance are obtained when LGD triggers **are involved in the handover process,** as compared to a single LD trigger strategy.

Why is an efficient scheduling handover process important? And what does an efficient or timely handover schedule mean? If the LGD is prompted too late, it may happen that the LD is triggered before establishing a new link. In such a case, a full network entry procedure would necessarily take place, increasing enormously the service disruption time. It may also happen that the LD is not triggered; however the link deteriorates to the point that crucial handover signalling messages are lost. Then, IEEE802.16 pre-defined timers start counting down; increasing disruption time. Probably before the timer expires, a LD is triggered. Closely related to this problem is the fact that one of the performance criteria commonly used when designing handover schemes is the average signal strength during handover. In summary, the major risk involved if the LGD is initiated too late is related to the fact that critical handover signalling messages can be missed and a new full re-entry process may take place, leading to a significant degree of handover delay [11].

If the LGD is generated too early, a loss of a 'working connection' takes place. Changing too soon to an improper interface may represent reduced bandwidth and QoS. Moreover, when there is a large time gap between the LGD and the LD, frequent roll-back events or handover cancellations may occur [12].

Link trigger generation is not covered by any standard. It is implementation dependent. The handover policies involved in the handover decision process play a decisive role in scheduling the handover process efficiently. Handover policies take charge of trigger generation.

As a basic rule, predictive events such as LGD need to occur in a timely fashion in order to prepare for a handover. The LGD trigger should be invoked prior to an actual LD event by at least the time required to prepare and execute a handover. This is known as *anticipation factor* or *optimum threshold value*. Thus, one of the most relevant attributes for timely link triggering is previous knowledge of the required time for handover execution.

However, the timely generation of a LGD trigger by determining the value for this anticipation factor or optimum threshold value accurately is a difficult task. It depends on several parameters that change over time, such as:

- the mobile station (MS) speed (dwell timer);
- the time required for performing a handover;
- the neighbouring network conditions;
- the wireless channel conditions (which are dynamic in time, owing to factors such as the MS's movement and shadowing).

8.7   Handover Performance Analysis

The most common metrics used to assess handover algorithm performance are handover latency and data loss. This section introduces a performance analysis of the mandatory handover strategy in IEEE802.16 networks: the hard handover.

The set of variables involved in the HHO strategy are listed on Table 8.1. Some of them are handover design parameters to be defined by the handover algorithm designer. Others are dynamic information obtained from RSSI measurements or location information sources.

Other variables are related to MS data profile and the data application being supported. Last but not least, it is necessary to take into account the service class and scheduler that are involved and the classifier table.

The next step in our analysis is to calculate the handover interruption time as a function of this set of variables.

The handover interruption time, also known as handover latency or handover delay, represents here the time duration during which a MS is not receiving service from any BS during a handover. It is defined as the time interval between when the MS disconnects from the serving BS until the start of transmission of the first data packet from the target BS. Handover delay is a key metric for evaluating and comparing various handover schemes as it has a direct impact on perceived application performance.

It is worth pointing out that sometimes the handover delay concept found in the literature means handover latency and at other times handover delay refers to the handover initiation delay, which is closely related to the deviation of handover location and the cell-dragging effect and consequent interference.

Table 8.1 Variables involved in a HHO handover strategy. Range values for the variables involved based on IEEE802.16e

Configuration parameters	Range
Frame size	2 ms to 20 ms
Neighbour Advertisement Interval	
Ranging parameters	
Backoff Window Size	
Ranging Backoff Start	0–15
Ranging Backoff End	0–15
T3 = Timeout value for receiving a valid Ranging code	0 – 200 ms
Ncs = Contention Area *	> 2 * 6
Number of retries to send contention ranging requests (T33)	> 16
Scanning parameters	
Scan Duration (N)	0–255 frames
Interleaving Interval (P)	0–255 frames
Scan Iterations (T)	0–255 frames
Start Frame (M)	0–15 frames
T44 Scan request retransmission timer	0–100 ms

Dynamic information	QoS configuration in the MS and BS
MS speed	Service Class & Schedulers being used
Distance between BS serving and MS	Downlink and Uplink Service flows (admitted and active)
Distance between BS target and MS	Classifier table been used
CINR BStarget	Queuing/buffering per connection
CINR BStarget2	Buffer size associated to service class and Service Flow
MS Data traffic Profile	Initial Modulation chosen for each service flow
Minimum jitter and latency from application profile	
Traffic Load	

*(Number of Symbol times > 2 per number of Subchannels > 6) or number of slots per frame in Single Carrier PHY model

Total handover latency is broken down into two latency elements:

- The Radio Layer Latency or elapsed time between MS disconnection from the serving BS and when the MS achieves PHY layer synchronization at the target BS.
- The Network Entry and Connection Setup Time or elapsed time between MS synchronization and transmission of the first data packet from the target BS.

In order to estimate mathematically the handover delay and for the sake of simplicity, some reasonable assumptions are made. They are listed below:

- The radio propagation delay is much smaller than the frame duration, so we omit it from our analysis.

- WiMAX frame size may vary from 2 ms to 20 ms. However, at least initially, all WiMAX equipment supports only 5 ms frames. So, the calculations introduced here for considerations of latency will be based on this value.

T_{frame} : IEEE802.16e OFDMA (Orthogonal Frequency-Division Multiple Access)

Frame duration

- The message processing time in each node is much lower than the frame duration. Besides, the OFDMA frame is split into the downlink and uplink sub frame. Thus, a MS is able to receive a message in frame N, process it and send the answer back in frame N+1 when the basic connection is used, as in most management MAC messages. This argument is based on timing considerations observed in P802.16j, 2008, p. 147.

Based on these assumptions, the total HHO handover execution delay, with no security considerations, $T_{handover}$ is equal to:

$$T_{handover} = T_{ranging} + T_{SBC} + T_{REG} + T_{DSA}$$

Where,

$T_{ranging}$ = time required for MS to carry out the initial ranging process,

T_{SBC} = time required for MS to inform on basic capabilities, SBC-REQ and SBC-RSP message exchange,

T_{REG} = time required for MS registration with target BS, REG-REQ and REG-RSP message exchange,

T_{DSA} = time required for the DSA-REQ and DSA-RSP messages exchange for provisioning service flows.

Observing Figure 8.10 and considering that these messages are exchanged sequentially between MS and BS, and each message delay is T_{frame}

$$T_{handover} = T_{ranging} + 6T_{frame}$$

The maximum initial ranging latency, before any retry, responds to the following expression:

$$T_{ranging} = \left\lceil \frac{2^{B_{exp}}}{N_{cs}} \right\rceil \times T_{frame} + T_3$$

Where:

B_{exp} = backoff exponent,

T_{frame} = frame duration,

T_3 = Timeout value for receiving a ranging response (50ms to 200ms),

N_{cs} = number of slots per frame in Single Carrier PHY model or contention area in OFDMA PHY profile.

The backoff exponent is been characterized by the Ranging backoff Start and End values. These attributes determine the maximum range over which the ranging backoff

MS   BS

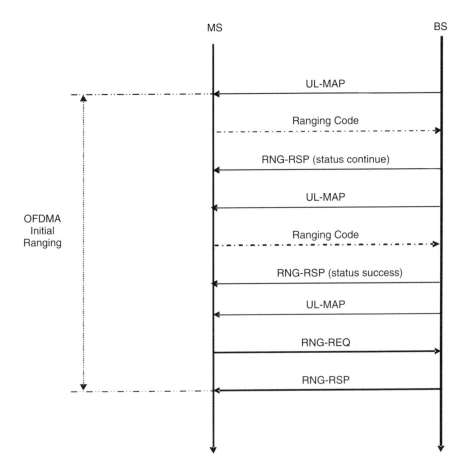

Figure 8.10 OFDMA ranging detail.

window is picked randomly. At the first transmission, the range extends over Ranging Backoff Start slots. With perceived failure, the backoff range is doubled up to the maximum range specified by the Ranging Backoff End attribute. The attribute value is expressed as a power of 2. Only values between 0 and 15 are allowed. For instance, if this attribute is set to 3, the backoff window is chosen from a range no larger than $2^3 = 8$ slots.

The Ncs attribute defines the extent (in time and frequency) of the contention area reserved for Initial Ranging within the frame. For the purpose of Initial Ranging, the minimum recommended value is 6 subchannels by 2 symbol times. There is also a timer (T33) that represents the contention ranging number of retries. In accordance with Table 342 of IEEE 802.16-2004 standard, the number of retries is no less than 16. A retry to send a ranging requests (message or CDMA code) happens when T3 seconds elapse since the last try, without a response arriving to the BS in this T3 seconds interval.

The following dependencies are identified:

- Regarding backoff window size: a higher value of this parameter in a high density scenario will result in higher values for handover delay.

- The time out value for receiving a ranging response (default 50 ms) outlines the maximum value for the ranging activity delay.
- The higher the number of initial ranging opportunities per frame, the better.

8.8 Summary

As WiMAX technology reaches higher mobility usage scenarios, a heavier burden is placed upon the performance of the handover process. This chapter describes different enhancement techniques in the handover process that contribute to improve QoS end-to-end performance indicators, such as delay time or data loss.

One of the most critical stages in the handover process is the decision-making process. The decision-making process itself, and the handover policies that rule the scheduling of the handover process, are not within the scope of IEEE802.16. They are implementation dependent.

Timely handover scheduling involves a handover policy and a set of link layer triggers fired at the right time. The predictive LGD trigger needs to occur in a timely enough manner so as to prepare for the handover. Different techniques may be implemented that help to solve the difficult task of generating a LGD trigger in this high mobility usage scenario.

In conclusion, the QoS performance indicator *handover delay* is calculated as a function of a set of variables involved in the handover process.

References

[1] A. Majlesi and B.H. Khalaj, 'An Adaptive Fuzzy Logic Based Handoff Algorithm for Interworking between WLANs and Mobile Networks'. IEEE Proceedings of the 13th International Symposium on Personal, Indoor and Mobile Radio Communications, Vol. 5, pp. 2446–51, 2002.

[2] A. Majlesi, B.H. Khalaj, A. Mehbodniya and J. Chitizadeh, 'An Intelligent Vertical Handoff Algorithm for Next Generation Wireless Networks', IEEE Proceedings of the 2nd International IFIP Conference on Wireless and Optical Communications Networks (WOCN'05), p. 244–249, 2005..

[3] J.G. Andrews, A. Ghosh and R. Muhamed, *Fundamentals of WiMAX: Understanding Broadband Wireless Networking*, 2007.

[4] A. Dutta, D. Famolari, S. Das, Y. Ohba, V. Fajardo, K. Taniuchi, R. Lopez and H. Schulzrinne, 'Media-Independent Pre-Authentication Supporting Secure Interdomain Handover Optimization' [architectures and protocols for mobility management in all-IP mobile networks], *Wireless Communications*, IEEE [see also *IEEE Personal Communications*] 15[2]: 55–64, 2008. Notes: U–HO.

[5] D.H. Lee, K. Kyamakya and U.P. Umondi, 'Fast Handover Algorithm for IEEE 802.16 e Broadband Wireless Access System', IEEE Proceedings of the first International Symposium on Wireless Pervasive Computing, 2006.

[6] Y. Wang, P.H.J. Chong, L. Qiu, L. Chen, E. Lee, L.C. Seck and D. Cheung, 'Research and Software Development of TETRA & TETRAPOL Networks Models for IP-Based Data Services Using OPNET', 2006.

[7] P. Boone, M. Barbeau and E. Kranakis, 'Strategies for Fast Scanning and Handovers in WiMAX/802.16' IEEE Proceedings of the 2nd International Conference on Access Networks '07, pp. 1–7, 2007.,

[8] R. Rouil and N. Golmie, 'Adaptive Channel Scanning for IEEE 802.16 e', Proceedings of 25th Annual Military Communications Conference (MILCOM 2006), Washington, DC, October 23–25, pp. 1–6, 2006.

[9] R. Rouil and N. Golmie, 'Effects of IEEE 802.16 Link Parameters and Handover Performance for Select Scenarios'. IEEE 802, pp. 21–6.

[10] S. Choi, G.H. Hwang, T. Kwon, A.R. Lim and D.H. Cho, 'Fast Handover Scheme for Real-Time Downlink Services in IEEE 802.16 e BWA System', Vehicular Technology Conference. VTC – Spring, IEEE 61st 3, 2005.

[11] N. Golmie, 'Seamless and Secure Mobility', 9th DOD Annual Information Assurance Workshop, 2005.

[12] S.J. Yoo, D. Cypher and N. Golmie, 'Predictive Link Trigger Mechanism for Seamless Handovers in Heterogeneous Wireless Networks', published online at www3.interscience.wiley.com, Wireless Communications and Moblie Computing, 2008.

9

Resource Allocation in Mobile WiMAX Networks

Tara Ali Yahiya
Computer Science Laboratory, Paris-Sud 11 University, France

9.1   Introduction

Mobile WiMAX based on the IEEE 802.16e standard (the mobile version of the IEEE 802.16-2004 standard) is expected to shift from fixed subscribers to mobile subscribers with various form factors: Personal Digital Assistant (PDA), phone, or laptop [1]. Thus, it is expected that Mobile WiMAX will not only compete with the broadband wireless market share in urban areas with DSL, cable and optical fibers, but also threaten the hotspot based WiFi and even the voice-oriented cellular wireless market. This is due to the variety of fundamentally different design options. For example, there are multiple physical layer (PHY) choices: a Single-Carrier-based physical layer called Wireless-MAN-SCa, an Orthogonal Frequency Division Multiplexing (OFDM) based physical layer called Wireless MAN-OFDM and an Orthogonal Frequency Division Multiple Access (OFDMA) based physical layer called Wireless-OFDMA. Similarly, there are multiple choices for medium access control (MAC) architecture, duplexing, frequency band of operation, etc. [2].

However, for practical reasons of interoperability, the scope of the standard needs to be reduced, and a smaller set of design choices for implementation needs to be defined. The WiMAX Forum does this by defining a limited number of system and certification profiles. Accordingly, the WiMAX Forum has defined mobility system profiles for IEEE 802.16e, which we use in this chapter for the design of the wireless communication system and the performance evaluation of Mobile WiMAX in the link and the system levels.

In this chapter, we present a concise technical overview of the emerging Mobile WiMAX solution for broadband wireless. The purpose here is to provide a summary of the most important issues related to Quality of Service (QoS) in Mobile WiMAX.

WiMAX Security and Quality of Service: An End-to-End Perspective   Edited by Seok-Yee Tang,
Peter Müller and Hamid Sharif
© 2010 John Wiley & Sons, Ltd

This is an important step towards understanding the problem of resource allocation in mobile WiMAX, and thus understanding the proposed solutions in the literature that deal with this issue. We explain such aspects before offering more detail about the proposed contributions for resource allocation in Mobile WiMAX in the subsequent sections of this chapter.

9.2   Background on IEEE 802.16e

Mobile WiMAX based on IEEE 802.16e is a wireless broadband solution that offers a rich set of features with a great deal of flexibility in terms of deployment options and potential service offering. These features provided by both MAC and PHY layers are as follows.

9.2.1   The Medium Access Control Layer – MAC

The MAC layer of Mobile WiMAX provides a medium-independent interface with the PHY layer and is designed to support the wireless PHY layer by focusing on efficient radio resource management. The MAC layer supports both Point-to-Multipoint (PMP) and Mesh network modes and is divided into three sublayers: the *service-specific convergence* sublayer, *common part* sublayer and *security* sublayer. The primary task of the *service-specific convergence* sublayer is to classify external Service Data Units (SDU) and associate each of them with a proper MAC Service Flow (SF) identifier and connection identifier. The function of the *common part* sublayer is to (i) segment or concatenate the SDUs received from higher layers into the MAC Protocol Data Units (PDU), (ii) retransmit MAC PDUs that were received erroneously by the receiver when Automated Repeat Request (ARQ) is used, (iii) provide QoS control and priority handling of MAC PDUs belonging to different data and signaling bearers and (iv) schedule MAC PDUs over the PHY resources. The *security* sublayer handles authentication, secure key exchange and encryption [2]. Among all of these functions, we will emphasise on QoS related functions and mechanisms that are associated with our field of interest and our proposed solutions, as follows.

9.2.1.1   Channel Access Mechanism

In mobile WiMAX, the MAC layer at the Base Station (BS) is fully responsible for allocating bandwidth to all Mobile Stations (MSs), in both uplink *and* downlink. It supports several mechanisms by which an MS can request and obtain uplink bandwidth. Depending on the particular QoS and traffic parameters associated with a service, one or more of these mechanisms may be used by the MS. The BS allocates dedicated or shared resources periodically to each MS, which it can use to request bandwidth. This process is called *polling*. Mobile WiMAX defines a contention access and resolution mechanism for the case when more than one MS attempts to use the shared resource. If it already has an allocation for sending traffic, the MS is not polled. Instead, it is allowed to request more bandwidth by (i) transmitting a stand-alone bandwidth request or (ii) piggybacking a bandwidth request on generic MAC packets.

9.2.1.2 Quality of Service

Support for QoS is a fundamental part of the mobile WiMAX MAC layer design, strong QoS control is achieved by using a connection-oriented MAC architecture, where all downlink and uplink connections are controlled by the serving BS. Before any data transmission happens, the BS and the MS establish a unidirectional logical link, called a *connection*, between the two MAC-layer peers. Each connection is identified by a Connection Identifier (CID), which serves as a temporary address for data transmission over the particular link.

Mobile WiMAX also defines a concept of a *service flow*. A SF is a unidirectional flow of packets with a particular set of QoS parameters and is defined by a Service Flow Identifier (SFID). To support a variety of application, mobile WiMAX defines five SFs:

1. **Unsolicited grant services (UGS)**: This is designed to support fixed-size data packets at a Constant Bit Rate (CBR). Examples of applications that may use this service are T1/E1 emulation and VoIP without silence suppression. The SF parameters that define this service are maximum sustained traffic rate, maximum latency, tolerated jitter and request/transmission policy.
2. **Real-time polling services (rtPS)**: This service is designed to support real-time SFs such as MPEG video, that generate variable-size data packets on a periodic basis. The mandatory SF parameters that define this service are minimum reserved traffic rate, maximum sustained traffic rate, maximum latency and request/transmission policy.
3. **Extended real-time variable rate (ertPS) service**: This service is designed to support real-time applications, such as VoIP with silence suppression, that have variable data rates but require guaranteed data rate and delay. The mandatory SF parameters that define this service are minimum reserved traffic rate, maximum sustained traffic rate, maximum latency and request/transmission policy.
4. **Non real-time polling service (nrtPS)**: This service is designed to support delay-tolerant data streams, such as an FTP, that require variable-size data grants at a minimum guaranteed rate. The mandatory SF parameters to define this service are minimum reserved traffic rate, maximum sustained traffic rate, traffic priority and request/transmission policy.
5. **Best-effort (BE) service**: This service is designed to support data streams, such as Web browsing, that do not require a minimum service-level guarantee. The mandatory SF parameters to define this service are maximum sustained traffic rate, traffic priority and request/transmission policy.

9.2.1.3 Mobility Support

In addition to fixed broadband access, mobile WiMAX envisions four mobility related usage scenarios: Nomadic, Portable, simple mobility up to 60 kmph and full mobility up to 120 kmph [1].

9.2.2 *The Physical Layer – PHY*

The mobile WiMAX physical layer is based on OFDMA. OFDMA is the transmission scheme of choice to enable high-speed data, video and multimedia communications and is used by a variety of commercial broadband systems, including DSL, WiFi, etc. In this section, we cover the basics of OFDMA and provide an overview of the Mobile WiMAX physical layer.

9.2.2.1   Subchannalization in Mobile WiMAX: OFDMA

OFDM is a very powerful transmission technique. It is based on the idea of dividing a given high-bit-rate data stream into several parallel lower bit-rate streams and modulating each stream on separate carriers often called subcarriers. OFDM is a spectrally efficient version of multicarrier modulation, where the subcarriers are selected so that they are all orthogonal to one another over the symbol duration, thereby avoiding the need to have non overlapping subcarrier channels to eliminate intercarrier interference [3].

In order to have multiple user transmissions, a multiple access scheme such as Time Division Multiple Access (TDMA) or Frequency Division Multiple Access (FDMA) has to be associated with OFDM. An OFDM signal can be made from many user signals, giving the OFDMA multiple access [4]. Multiple access has a new dimension with OFDMA. A downlink or an uplink user will have a time and a subcarrier allocation for each of its communications. However, the available subcarriers may be divided into several groups of subcarriers called *subchannels*. Subchannels may be constituted using either contiguous subcarriers or subcarriers pseudorandomly distributed across the frequency spectrum. Subchannels formed using distributed subcarriers provide more frequency diversity. This permutation can be represented by Partial Usage of Subcarriers (PUSC) and Full Usage of Subcarriers (FUSC) modes.

The subchannelization scheme based on contiguous subcarriers in mobile WiMAX is called Band Adaptive Modulation and Coding (AMC). Although frequency diversity is lost, band AMC allows system designers to exploit multiuser diversity, allocating subchannels to users based on their frequency response [5]. In this chapter we are interested in the band AMC, since multiuser diversity can provide significant gains in overall system capacity, if the system strives to provide each user with a subchannel that maximizes its received Signal-to-Interference-plus-Noise Ratio (SINR). Therefore, all explanations related to resource allocation will be based on AMC mode.

9.2.2.2   Slot and Frame Structure in OFDMA based Mobile WiMAX

Before providing detail of mobile WiMAX frame structure, it is worth mentioning that the downlink and uplink transmissions co-exist according to one of two duplexing modes: Time Division Duplex (TDD) or Frequency Division Duplex (FDD). They are sent through the downlink and uplink subframes. However TDD is favoured by a majority of implementations of the WiMAX forum [6]. The frame is divided into two subframes: a downlink subframe followed by an uplink subframe after a small guard interval. Figure 9.1 shows an OFDMA frame when operating in TDD mode. The downlink subframe begins with a downlink preamble that is used for physical layer procedures, such as time and frequency synchronization and initial channel estimation. The download preamble is followed by a Frame Control Header (FCH), which provides frame configuration information, such as the MAP message length, the modulation and coding scheme and the usable subcarriers.

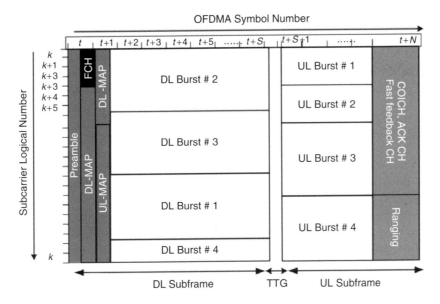

Figure 9.1 A simple mobile WiMAX OFDMA frame structure for the TDD mode.

Multiple users are allocated data regions within the frame, and these allocations are specified in the uplink and downlink MAP messages (DL-MAP and UL-MAP) that are broadcast following FCH in the downlink subframe. MAP messages include the burst profile for each MS, which defines the modulation and coding scheme used in that link.

The uplink subframe is made up of several uplink bursts from different users. A portion of the uplink subframe is set aside for contention-based access that is used for a variety of purposes. This subframe is used mainly as a ranging channel to perform closed-loop frequency, time and power adjustments during network entry as well as periodically afterwards. The ranging channel may also be used by MS to make uplink bandwidth requests. Besides the ranging channel and traffic bursts, the uplink subframe has a Channel Quality Indicator Channel (CQICH) for the MS to feed back channel-quality information that can be used by the BS scheduler and an Acknowledgment (ACK) channel for the MS to feed back downlink acknowledgements.

According to that which we described in the previous section, once higher layer data have been classified into SFs and scheduled by the MAC layer, they are assigned to OFDMA slots by a slot allocator. A slot is the basic resource unit in the OFDMA frame structure as it expresses a unit of (one subchannel at one symbol). One may consider that the data region (OFDMA frame) is a two-dimensional allocation which can be visualized as a rectangle. Allocating OFDMA slots to data in the downlink is done by segmenting the data after the modulation process into blocks that fit into one OFDMA slot. It is useful to note that the definition of an OFDMA slot depends mainly on the mode of permutation of subcarriers in an OFDMA subchannel, that is, FUSC, PUSC and AMC. Given that we focus on AMC mode for resource allocation in this chapter, we will detail only the structure of slots in an AMC OFDMA frame in the following section.

9.2.2.3   OFDMA Slot Structure in AMC Permutation Mode

Mobile WiMAX supports a variety of AMC schemes and allows for the scheme to change on a burst-by-burst basis per link, depending on channel conditions. In the downlink, Quadrature Phase Shift Keying (QPSK), 16 Quadrature Amplitude Modulation (QAM) and 64 QAM are mandatory for mobile WiMAX; 64 QAM is optional in the uplink. Forward Error Correction (FEC) using convolutional codes is mandatory. Convolutional codes are combined with an outer Reed-Solomon code in the downlink for OFDM-PHY. The standard also supports optionally turbo codes and Low Density Parity Check (LDPC) codes at a variety of code rates. A total of 52 combinations of modulation and coding schemes is defined in mobile WiMAX as burst profiles. In order to constitute an OFDMA frame in the AMC, it is important to understand the slot structure that is called band AMC mode.

Unique to the band AMC permutation mode, all subcarriers constituting a subchannel are adjacent to each other. Therefore, taking a microscopic view of an OFDMA frame in AMC mode, nine adjacent subcarriers with eight data subcarriers and one pilot subcarrier are used to form a bin, as shown in Figure 9.2. Four adjacent bins in the frequency domain constitute a band. An AMC slot consists of six contiguous bins within the same band. Thus, an AMC slot can consist of one bin over six consecutive symbols, two consecutive bins over three consecutive symbols, or three consecutive bins over two consecutive symbols. Therefore, in each frame, MSs are allocated a successive set of slots, forming bursts. Each allocation is represented in the DL-MAP message by the slot offset and the number of slots in the allocation frame.

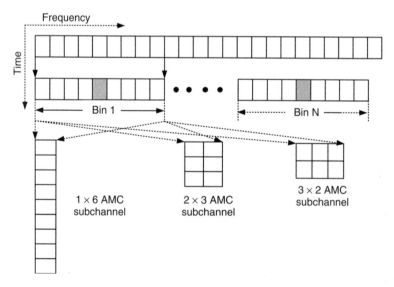

Figure 9.2   Mapping OFDMA slots to subchannels and symbols in IEEE 802.16e downlink based on Band AMC subcarrier permutation. Reproduced with permission from *Computer Communications*, An adaptive cross-layer design for multiservice scheduling in OFDMA based mobile WiMAX systems, © Elsevier 2009.

Generally, the BS receives periodically the Channel Quality Indicator Feedback (CQI) from the MSs indicating information such as channel-quality measurements, for example Received Signal Strength Indicator (RSSI) and SINR. Then, the BS supporting AMC allows the scheduler to exploit multiuser diversity by allocating each MS to its corresponding strongest subchannel; thus increasing the overall throughput of the system.

Based on the detailed description of subchannalization in OFDMA system, we can formulate resource allocation in OFDMA as a constrained optimization problem that can be classified into either (1) minimizing the total transmit power with a constraint on the user data rate [7], [8] or (2) maximizing the total data rate with a constraint on total transmit power [9], [10], [11], [12]. The first objective is appropriate for fixed-rate applications, such as voice, whereas the second is more appropriate for bursty applications, such as data and other IP applications. Therefore, in this section, we focus on the rate-adaptive algorithms (category 2), which are more relevant to Mobile WiMAX systems. However, achieving high transmission rates depends on the ability of the system to provide efficient and flexible resource allocation. Recent studies [8]–[13] on resource allocation demonstrate that significant performance gains can be obtained if frequency hopping and adaptive modulation are used in subchannel allocation, assuming knowledge of the channel gain in the transmitter, that is, the Base Station (BS).

In a multiuser environment, a good resource allocation scheme leverages multiuser diversity and channel fading [14]. It was shown in [15] that the optimal solution is to schedule the user with the best channel at each time – this is the so-called multiuser diversity. However, in this case the entire bandwidth is used by the scheduled user; this idea can also be applied to the OFDMA system, where the channel is shared by the users, each owing a mutually disjoint set of subchannels, by scheduling the subchannel to a user with the best channel. Of course, the procedure is not simple since the best subchannel of a user may also be the best subchannel of another user who may not have any other good subchannels. The overall strategy is to use the peaks of the channel resulting from channel fading. Unlike in the traditional view where channel fading is considered to be an impairment, here it acts as a channel randomizer and increases multiuser diversity [14].

Recent studies consider further QoS application requirements in the allocation of subchannels [16]. QoS requirements are defined here as achieving a specified data transmission rate and BER of each user's application in each transmission. In [8] a Lagrange-based algorithm to achieve a dramatic gain is proposed. However, the prohibitively high computational complexity renders this impractical. To reduce the complexity in [8], a heuristic subcarrier allocation algorithm is proposed in [17], [18]. The two schemes both assume fixed modulation modes.

However, none of the aforementioned adaptive algorithms have taken into account the impact of a radio resource allocation scheme on different classes of services. For example, there is no doubt that voice service and data service co-exist in both current systems and future mobile communication systems. Voice and data users have quite different traffic characteristics and QoS requirements. Voice traffic requires a real time transmission but can tolerate a moderate bit error rate. While data traffic can accept the varied transmission delay but it requires a lower BER. In this chapter, we propose a radio resource allocation scheme supporting a multi-traffic class, whose objective is to guarantee the QoS requirements for the different classes of services along with improving the performance of the system in terms of spectral efficiency.

9.3   System Model

The architecture of a downlink data scheduler with multiple shared channels for multiple MSs is shown in Figure 9.3. OFDMA provides a physical basis for the multiple shared channels, where the total frame is divided into slots. Thus, an OFDMA frame is divided into K subchannels in the frequency domain and T symbols in the time domain. Let $M = \{1, 2, \ldots, m\}$ denote the MSs index set. In each OFDMA frame there are $T \times K$ slots and each MS may be allocated one or more such slots according to its application requirements. One of the advantage of this model is that a wide range of data rates can be

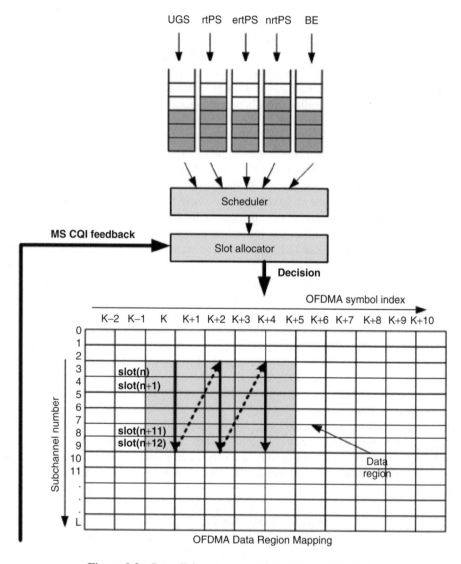

Figure 9.3   Downlink system model for Mobile WiMAX.

supported and it is thus very suitable for the the Mobile WiMAX system. For simplicity we denote the slot on the kth subchannel at the tth symbol as (kth,tth) slot.

We suppose that the CQI of the whole frame is known perfectly at the BS through the Channel Quality Indicator Channel (CQICH) feedback message. Thus, the BS serves simultaneously M MSs, where each of these MSs has queues to receive its incoming packets for their different SFs. The scheduler, with the help of slot allocator at the BS, can schedule and assign effectively slots and allocate power on the downlink OFDMA slots exploiting the knowledge of the wireless channel conditions and the characteristics of the SFs.

9.4 OFDMA Key Principles–Analysis and Performance Characterizations

Since the system model is based on the OFDMA technique, it is necessary to provide a discussion of the key principles that enable high performance in OFDMA: AMC and multiuser diversity. We then analyze the performance characterization of OFDMA frame capacity and protocols.

9.4.1 Multiuser Diversity

In an environment, when many users fade independently, there is at any time a high probability that one of the users will have a strong channel. By allowing only that user to transmit, the shared channel resource is used in the most efficient way and the total system throughput is maximized. This phenomenon is called multiuser diversity. Thus, the larger the number of users, the stronger tends to be the strongest channel, and the more the multiuser diversity gain [19].

To illustrate multiuser diversity, we consider a two-user case in Figure 9.4, where the user with the best channel condition is scheduled to transmit signals. Therefore, the equivalent Signal-to-Noise Ratio (SNR) for transmission is max{$\text{SNR}_1(t)$, $\text{SNR}_2(t)$}. When there

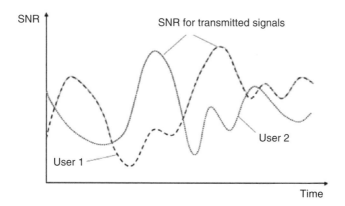

Figure 9.4   Multiuser diversity – Scheduling for two users case.

are many users served in the system, the packets are with a high probability transmitted at high data rates since different users experience independent fading fluctuations.

9.4.2   Adaptive Modulation and Coding – Burst Profiles

Mobile WiMAX systems use AMC in order to take advantage of fluctuations in the channel. The basic idea is quite simple: Transmit at as high data rate as possible when the channel is good and transmit at a lower rate when the channel is poor, in order to avoid excessive dropped packets. Lower data rates are achieved by using a small constellation, such as Quadrature Phase Shift Keying (QPSK) and low-rate error-correcting codes, such as rate convolutional or turbo codes. The higher data rates are achieved with large constellations, such as 64 Quadrature Amplitude Modulation (QAM) and less robust error correcting codes; for example, rate convolutional, turbo, or Low Density Parity Check (LDPC) codes. In all, 52 configurations of modulation order and coding types and rates are possible, although most implementations of Mobile WiMAX offer only a fraction of these. These configurations are referred to as *burst profiles*.

Both Table 9.1 and Figure 9.5 show that by using six of the common Mobile WiMAX burst profiles, it is possible to achieve a large range of spectral efficiencies. This allows the throughput to increase as the SNR increases following the trend promised by Shannon's formula $C = \log_2(1 + SNR)$ [20]. In this case, the lowest offered data rate is QPSK and rate 1/2 turbo codes; the highest data-rate burst profile is with 64 QAM and rate 3/4 turbo codes. The achieved throughput normalized by the bandwidth is defined as in (9.1):

$$Se = (1 - BLER)\ \delta\ \log_2(N)\ \text{bps/Hz} \qquad (9.1)$$

where BLER is the block error rate, $\delta \leq 1$ is the coding rate, and N is the number of points in the constellation.

9.4.3   Capacity Analysis – Time and Frequency Domain

Given that an OFDMA frame is partitioned in frequency and time domain (subchannel and symbol), that is, slot, each connection is converted to slots according to the instantaneous SNR value that is derived from the channel model. In order to analyze the capacity of the two-dimensional frequency-time domain, we use the Additive White Gaussian Noise

Table 9.1   Transmission modes in IEEE 802.16e

Modulation	Coding Rate	bit/symbol	Received SNR (dB)
QPSK	1/2	1.0	9.4
	3/4	1.5	11.2
16 QAM	1/2	2.0	16.4
	3/4	3.0	18.2
64 QAM	2/3	4.0	22.7
	3/4	4.5	24.4

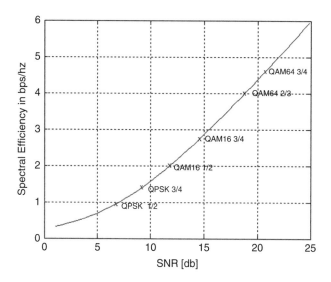

Figure 9.5 Throughput versus SNR, assuming that the best available constellation and coding configuration are chosen for each SNR. Reproduced with permission from *Computer Communications*, An adaptive cross-layer design for multiservice scheduling in OFDMA based mobile WiMAX systems, © Elsevier 2009.

(AWGN) capacity, or Shannon capacity,

$$C_{awgn} = \log_2(1 + SNR) \tag{9.2}$$

where $SNR = P_0/(N_0 B)$ is the instantaneous SNR over the whole frequency band B. P_0 and N_0 denote the total transmission power and the noise power spectral density, respectively. Radio resources are allocated in both the frequency and time domain with equal power allocation, which fully exploits the channel time variant characteristic, that is, time diversity as well as frequency diversity. In this case, the achievable data rate for one frame is written as

$$R = \frac{1}{T} \sum_t \sum_k B_k \log_2(1 + \alpha * SNR) \tag{9.3}$$

$$= \frac{1}{T} \sum_t \sum_k B_k \log_2 \left(1 + \alpha * \frac{g_{k,t} P_{av}}{N_0 B_k}\right)$$

$$= \frac{1}{T} \sum_t \sum_k B_k \log_2 \left(1 + \alpha * g_{k,t} \frac{P_0}{(K B_k) N_0}\right)$$

$$= \frac{1}{T} \sum_t \sum_k B_k \log_2(1 + \alpha * g_{k,t} * SNR)$$

where $g_{k,t}$ and B_k determine the channel gain and bandwidth of the kth subchannel respectively. While $P_{av} = P_0/N$ is the equal power allocated over all subchannels in one slot. The α is the constant BER specified as $\alpha = 1.5/ln\,P_{ber}$, and P_{ber} is the target BER. Then, the capacity is written as

$$C = \frac{R}{B} = \frac{R}{K * B_k} = \frac{1}{T * K} \sum_t \sum_k \log_2(1 + \alpha * g_{k,t} * SNR) \qquad (9.4)$$

As shown in Figure 9.1, the OFDMA frame is partitioned in both frequency and time domains, therefore, for the slot (k, t), according to [21], the achievable bits of the m'th MS can be written as

$$r_m[k, t] = \Delta B\ \Delta T \log_2(1 + \alpha_m \gamma_m[k, t]) = \Delta B\ \Delta T \log_2(1 + \alpha_m \frac{g_m[k, t] P_m[k, t]}{N_0 \Delta B})$$
$$(9.5)$$

where ΔB and ΔT are the frequency bandwidth and the symbol length of one slot, respectively and $\gamma_m[k, t]$ is the instantaneous SNR at symbol t for subchannel k corresponding to MS m, which can be calculated as

$$\gamma_m[k, t] = \frac{g_{m,k,t} P_m[k, t]}{N_0 \Delta B} \qquad (9.6)$$

Assume that L is the time duration of an OFDMA frame, then the mth connection achievable data rate (bps) for one frame is

$$u_m = \frac{1}{L} \sum_{t=1}^{T} \sum_{k=1}^{K} r_m[k, t] \rho_m[k, t] \qquad (9.7)$$

where $\rho_m[k, t]$ is the slot assignment indicator for the mth MS, $\rho_m[k, t] = 1$ indicates that slot (k, t) is allocated to the mth MS otherwise $\rho_m[k, t] = 0$ when the slot is not allocated. Then equation (9.7) yields an overall throughput of one frame as

$$Thr = \frac{1}{L} \sum_{m=1}^{M} \sum_{t=1}^{T} \sum_{k=1}^{K} u_m[k, t] \rho_m[k, t] \qquad (9.8)$$

9.4.4 Mapping Messages

In order for each MS to know which slots are intended for it, the BS must broadcast this information in DL–MAP messages. Similarly, the BS tells each MS which slots to transmit in a UL–MAP message. In addition to communicating the DL and UL slot allocations to the MS, the MS must also be informed of the burst profile used in the DL and the UL. The burst profile is based on the measured SINR and BLER in both links and identifies the appropriate level of modulation and coding.

9.5 Cross-Layer Resource Allocation in Mobile WiMAX

Cross-layer resource allocation shows promise for future wireless networks. The mechanism of exploiting channel variations across users should be used in scheduling and

Medium Access Control (MAC) designs to improve system capacity, fairness and QoS guarantees. Owing to variable data rates and stochastic transmission inherent in channel-aware networks, the issue of cross-layer is becoming very challenging and interesting.

Since Mobile WiMAX is based on OFDMA, decisions to which timeslot, subchannel and power level for communication are determined by the intelligent MAC layer which seeks as maximize the Signal-to-Interference-Ratio (SINR) for every Mobile Station (MS). This allows MSs to operate at the maximum modulation rates obtainable given the radio frequency conditions at the MS location. Accordingly, this allows service providers to maximize the number of active users whether they are fixed, portable, or mobile [22].

The intelligent MAC layer mentioned above requires adaptability with the PHY layer in response to different application services. The MAC layer has to distinguish the type of Service Flow (SF) and its associated QoS parameters, and then allocates the SF to the appropriate physical layer configurations, that is, Adaptive Modulation and Coding (AMC) mode permutation. Therefore, in this chapter, we propose a cross-layer scheme with guaranteed QoS for the downlink multiuser OFDMA based mobile WiMAX. The scheme defines an adaptive scheduling for each type of connection scheduled on OFDMA slots that integrates higher layer QoS requirements, SF's types and PHY layer Channel Quality Indication (CQI). Based on the adaptive scheduling mechanism (in MAC layer) combined with slot allocation scheme (in PHY layer), a fair and efficient QoS guarantees in terms of maximum delay requirement for real-time SFs and minimum reserved data rate for non real-time SFs are achieved.

9.6 Channel Aware Class Based Queue (CACBQ) – The Proposed Solution

The solutions described in the previous section can be used either for real-time or non real-time classes of services. No combination is possible for both types of SF. Besides, users with bad channels are heavily penalized as compared with users with good channels. Therefore, in this section we describe our solution which considers these two main problems, by introducing two algorithms in both the MAC and PHY layers. Both algorithms interact adaptively to constitute a cross-layer framework that tries to find a solution for a cost function in order to make a tradeoff among channel quality, application rate and QoS requirements for each type of SF.

9.6.1 System Model

We consider a Point-to-Multipoint (PMP) MAC layer mode. At the BS each MS can be backlogged with packets of different QoS requirements concurrently. Based on QoS requirements all packets transiting the network are classified into c SF and indexed by i. Let w_i be the weight assigned to SF_i with $w_i > w_j$ if $i > j$ and $\sum_{i=1}^{c} w_i \leq 1$, that is, SF_i requires better QoS than SF_j. We refer to the tuple (i, m), that is, MS_m to exchange the HOL packet in queue SF_i as a connection. The input parameters to the scheduler for SF_i are: (a) delay constraint W_i, (b) weight w_i, (c) feedback F_i to monitor fairness, and (d) predicted instantaneous transmission $r_m[k, t]$ of MS_m's link with the serving BS. The basic design principles of the scheduler are

- packets belonging to the same SF but to be scheduled to different MSs are queued in the different logical queue. Packets in each queue are arranged in the order of arrival to the queue. Packet (re)ordering in a queue can also be based on (earliest) delay deadlines specially for real-time SFs;
- only *HOL* packet P_{HOL} in each queue is considered in each scheduling decision;
- w_i and W_i of each $P_{HOL,i}$ and $r_m[k, t]$ of the MS to receive $P_{HOL,i}$ are jointly used in the scheduling policy.

We expect the IP layer to communicate to the MAC layer the traffic QoS-related parameters w_i and W_i in the IP packet header field. Our goal is to achieve fairness among the heterogenous SFs while assuring their QoS requirements. Since UGS SF has a fixed size grant on a real-time basis, its maximum sustained traffic rate is equal to its minimum reserved traffic rate, while the data rate for rtPS, ertPS and nrtPS is bounded by the maximum sustained traffic rate and the minimum reserved traffic rate [2]. This is due to their tolerance of some degradation in their QoS requirements. Hence, the problem to be solved is to find a policy by which a connection is scheduled, such that

$$(i, m) = \arg \max_{i,m} Z_{i,m}[k, t] \tag{9.9}$$

where $Z_{i,m}[k, t] \triangleq \text{function}(r_m[k, t], F_i, w_i, W_i)$ is the cost function, that is, priority value for connection (i, m). Note the coupling between queue state and channel state through information obtained from higher and lower layers. However, using cost function to select the connection is not convenient since all the parameters involved to select the connection have the same importance; therefore, we cannot assign the same weight to all of them. The problem become more complicated when we know that each parameter has a constraint associated with it, as shown in the following equations:

$$r_m[k, t] \geq c_{max} \forall SF \in \{UGS\} \tag{9.10}$$

$$W_i \leq D_i \forall SF \in \{UGS, ertPS, rtPS\} \tag{9.11}$$

$$c_{min} \leq r_m[k, t] \leq c_{max} \forall SF \in \{ertPS, rtPS \text{ and } nrtPS\} \tag{9.12}$$

where c_{min} and c_{max} denote minimum reserved traffic rate and maximum sustained traffic rate for these SFs. While D_i is the maximum latency for real-time SFs. Note that the search for a feasible policy that takes into consideration (9.10), (9.11) and (9.12) is hard to obtain since a trade-off among these parameters is required. Thus, the decision to schedule which type of SF under which condition cannot be made by a simple cost function. The constraint associated with each involved parameter of QoS such as delay, minimum sustained traffic rate and maximum sustained traffic rate is related to the allocation of slots in an OFDMA frame. Thus, we need mechanisms for slot allocation in a way that they satisfy these restraints on QoS parameters. Consequently, SF's scheduler in MAC layer and slot allocator in PHY layer need to interact with each other. Therefore, we propose some functional entities in both MAC and PHY layer that are linked to each other by information measurement and feedback exchanging. This is the reason behind the proposition of our cross-layer scheme called Channel Aware Class Based Queue (CACBQ) [23].

9.6.2   Channel Aware Class Based Queue (CACBQ) Framework

The proposed CACBQ solution is based on a cross-layer scheme which is composed of two main entities: the general scheduler at the MAC layer and the Slot Allocator at the PHY layer. The conceptual framework for CACBQ is depicted in Figure 9.6. The general scheduler includes two principal cooperative modules: *Estimator* and *Coordinator*. The *Estimator* is based on a priority function that estimates the number of slots for each connection (i, m) according to its channel quality which is provided by the PHY layer through CQI feedback message. While the *Coordinator* monitors the decision of the *Estimator* for slot allocation and control the level of satisfaction for each type of SF. Thus, it ensures that the real-time SFs or the non real-time SFs do not monopolize the slots on the OFDMA frame. Generally, the three functions distinguished by CACBQ can be stated as follows:

 (i) An estimation of slot numbers for the SF through the *Estimator*.
 (ii) Decision making is done to verify whether a SF is satisfied or not. Satisfaction should distinguish between real-time SF and non real-time SF in terms of delay and throughput. Whenever dissatisfaction occurs, the *Coordinator* either performs priority changing of the dissatisfied SF to the highest one or decreases the number of slots estimated for the SF with the lower priority.
(iii) Finally, after determining the number of slots for each user, the *Slot Allocator* will determine which slot is to be allocated for each SF through a specified allocation policy.

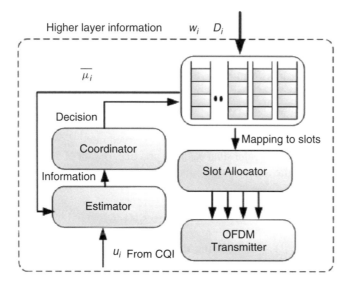

Figure 9.6   CACBQ cross-layer scheduler. Reproduced with permission from *Computer Communications*, An adaptive cross-layer design for multiservice scheduling in OFDMA based mobile WiMAX systems, © Elsevier 2009.

Thus the main functional elements of the framework are described as follows:

1) *Estimator*: The estimator estimates the number of slots used by each SF over an appropriate time interval, to determine whether or not each SF has been receiving its slot sharing bandwidth. In each turn, the scheduler selects a SF knowing not only its packet rate but also its physical data rate, that is, $u_m[k, t]$ (See equation 9.7). By knowing this information, the estimator estimates how many slots can be allocated for each packet in each turn. Once the number of slots are estimated for each SF, the estimator send this information to coordinator.

2) *Coordinator*: The coordinator uses the information received by the estimator to adjust dynamically the priority for SFs. The work of the coordinator can be divided into two parts. In the first part, a coordinator should realize whether the allocated slots are enough for each SF. If a SF does not obtain enough slots, then the coordinator starts the second part of the work; coordinating the priorities of all SFs to fulfil the QoS requirements of those that are dissatisfied. In doing so, the coordinator should distinguish between real-time and non real-time SFs satisfaction methods. Since the QoS requirements for each SF are different, the coordinator calculates the level of satisfaction in terms of delay for real-time SF and the minimum reserved data rate for non real-time SF. The delay satisfaction indicator for real-time SFs can be calculated as in [24]:

$$F_i = \frac{D_i - W_i}{T_g} \qquad (9.13)$$

where T_g is the guard time. Thus, the delay satisfaction indicator is defined as the ratio of waiting time packet i to the guard time. If $F_i(t) < 1$, that is, the time that a packet i can continue to wait is smaller than the guard time T_g. Thus, the packets of SF_i should be sent immediately to avoid packet drop due to delay outage; therefore, the priority of this queue is changed to the highest one. Then, the scheduler will verify if there are unallocated remaining slots from the whole number of slots S in order to assign them to the given dissatisfied SF. Otherwise, packet i will exceed the maximal delay and will be considered invalid and then will be discarded. However, if the queues have the same priorities, then the tie is broken and one of them will be selected randomly.

For nrtPS connection guaranteeing the minimum reserved rate c_{min} means that the average transmission rate should be greater than c_{min}. In practice, if data of connection i are always available in the queue, the average transmission rate at time t is usually estimated over a windows size t_c:

$$\eta_i(t)(1 - 1/t_c) + r_i(t)/t_c \qquad (9.14)$$

We aim to guarantee $\eta_i(t) \geq c_{min}$ during the entire service period. Then, the throughput indication will be

$$F_i = c_{min}/\eta_i(t) \qquad (9.15)$$

If $F_i(t) < 1$, then packets of connection i should be sent as soon as possible to meet the rate requirement; in this case, the priority of this queue will be changed to the highest one and will be served directly.

3) **Slot Allocator**: Once packets are scheduled by the general scheduler, the second phase includes an algorithm by which slots are allocated to these packets in AMC mode permutation. The algorithm iterates all SFs' packets, sorted by their current priority. In each iteration step, the considered SF is assigned the best slots available in term of channel gain value g. Afterwards, these slots are removed from the list of available slots. To achieve fairness among the lowest and highest priority SFs in terms of slot allocation, we introduce additional information – the weight – about the slot used. When considering a particular SF for slot assignment, the weight of a slot expresses how well this slot might be used by all other SFs with a lower priority than the currently considered one. A weight $\omega_{k,t,i}$ of a slot (k, t) for a SF i is given by the sum of all channel gain values of this slot regarding all SFs with lower priority than SF i has

$$\omega_{i,k,t} = \sum_{\forall j \text{ SF with lower priority than } i} g_{j,k,t} \qquad (9.16)$$

The algorithm selects always the highest possible weight between gain value and weight. The weight ratio of slot (k,t) with respect to SF i is defined as

$$\frac{g_{i,k,t}}{\omega_{i,k,t}} \qquad (9.17)$$

A SF i is assigned those slots with largest weight ratio. After an assignment of slots to a SF, weights for all unassigned slots are recomputed and sorted with respect to the next SF to be assigned. An algorithmic example is given below:

Algorithm 1

1: Let $S = \{1, 2, \ldots, s\}$ denote the set of unallocated slots and $Ga = \{1, 2, \ldots, g\}$ denote the set of all channel gains
2: Sort the connections according to their orders of scheduling specified by the satisfaction function F
3: **for** every $SF \in \{$rtPS, nrtPS and ertPS$\}$ **do**
4:     Calculate the weight as specified in (16)
5:     Calculate the weight ratio as in (17)
6:     Sort the weight ratio according to each SF
7:     Assign the slot of the highest weight ratio to the SF with the highest priority
8:     Remove this slot from the list of available slots
9: **end for**
10: Iterate 3: until $U = \phi$

9.7 Summary and Conclusion

This chapter presented an overview of the principal issues in Mobile WiMAX that are important to consider for resource allocation. These issues combine scheduling and method of channel access for different SFs in the MAC layer and burst profiles based on AMC slot structure in the OFDMA frame. Multiuser resource allocation which involves OFDMA,

AMC and multiuser diversity is proposed for the downlink mobile WiMAX networks. Furthermore, CACBQ which is an adaptive cross-layer for scheduling and slot allocation is introduced. The proposed cross-layer consists of two basic functional entities: estimator and coordinator. These entities provide an adaptive interaction with the change of quality of channel by sending feedback to the higher layers to offer fairness and QoS guarantees. Thus, approaches such as cross-layer are needed in mobile WiMAX systems since such type of approach includes different parameters that influence the performance of the network.

References

[1] J. G. Andrews, A. Ghosh and R. Muhamed, *Fundamentals of WiMAX Understanding Broadband Wireless Networking*, Pearson Education, Inc., 2007.

[2] IEEE Std 802.16e, 'IEEE Standard for Local and Metropolitan Area Networks – Part 16: Air Interface for Fixed and Mobile Broadband Wireless Access Systems,' Feb. 2006.

[3] G. Song and Y. Li, 'Cross-layer Optimization for OFDM Wireless Networks–Part I: Theoretical Framework' *IEEE Transactions on Wireless Communications* 4(2): 614–24, Mar. 2005.

[4] D. Tse and P. Viswanath, *Fundamentals of Wireless Communication*, Cambridge University Press, 2005.

[5] T. Ali-Yahiya, A. L. Beylot, and G. Pujolle, 'Cross-Layer Multiservice Scheduling for Mobile WiMAX Systems', *Proceedings of IEEE Wireless Communications and Networking conference*, pp. 1531–35, Mar. 2008.

[6] WiMAX Forum, 'Recommendations and Requirements for Networks Based on WiMAX forum Certified^{TM} Products', *Release 1.0*, 2006.

[7] D. Kivanc, G. Li and H. Liu, 'Computationally Efficient Bandwidth Allocation and Power Control for OFDMA', *IEEE Transactions on Communications*, 6(2): 1150–58, Nov. 2003.

[8] C. Wong, R. Cheng, K. Letaief and R. Murch, 'Multiuser OFDM with Adaptive Subchannel, Bit, and Power Allocation', *IEEE Journal on Selected Areas in Communications* 17(10): 1747–58, Oct. 1999.

[9] J. Jang and K. Lee, 'Transmit Power Adaptation for Multiuser OFDM Systems,' *IEEE Journal on Selected Areas in Communications* 21(2): 171–8, Feb. 2003.

[10] Y. J. Zhang and K. B. Letaief, 'Multiuser Adaptive Subchannel-and-Bit Allocation with Adaptive Cell Selection for OFDM Systems', *IEEE Transactions on Communications* 3(4): 1566–75, Sept. 2004.

[11] C. Mohanram and S. Bhashyam, 'A Sub-Optimal Joint Subchannel and Power Allocation Algorithm for Multiuser OFDM,' *IEEE Communications Letters* 9(8): 685–87, Aug. 2005.

[12] A. Gyasi-Agyei and S. Kim, 'Cross-Layer Multiservice Opportunistic Scheduling for Wireless Networks', *IEEE Communications Magazine* 44(6): 50–7, June 2006.

[13] W. Rhee and J. M. Cioffi, 'Increase in Capacity of Multiuser OFDM System Using Dynamic Subchannel Allocation', *Proceedings of IEEE VTC-spring*, Vol. 2, pp. 1085–9, May 2000.

[14] P. Viswanath, D. Tse, and R. Laroia, 'Opportunistic Beamforming Using Dumb Antennas', *IEEE Transactions on Information Theory* 48(6): 1277–94, June 2002.

[15] R. Knopp and P. Humblet, 'Information Capacity and Power Control in Single Cell Multiuser Communications', *Proceedings of IEEE International Conference on Communications(ICC)*, Vol. 1, pp. 331–5, June 1995.

[16] G. Song and Y. G. Li, 'Adaptive Subcarrier and Power Allocation in OFDM Based on Maximizing Utility,' *Proceedings of IEEE Vehicular Technology Conference*, Vol. 2, pp. 905–9, Apr. 2003.

[17] C. Y. Wong, C. Y. Tsui, R. S. Cheng and K. B. Letaief, 'A Real-Time Subchannel Allocation Scheme for Multiple Access Downlink OFDM Transmission,' *Proceedings of IEEE Vehicular Technology Conference*, Vol. 2, pp. 1124–8, Sept. 1999.

[18] S. Pietrzyk and G. J. M. Janssen, 'Multiuser Subchannel Allocation for QoS Provision in the OFDMA Systems', *Proceedings of IEEE Vehicular Technology Conference*, Vol. 5, Oct. 2002.

[19] T. Taiwen and R. W. Heath, 'Opportunistic Feedback for Downlink Multiuser Diversity', *IEEE Communications Letters* 9(10): pp. 948–50, Oct. 2005.

[20] C. Shannon, ' The Zero Error Capacity of a Noisy Channel,' *IEEE Transactions on Information Theory* 2(3): 8–19, Sept. 1956.

[21] X. Zhang and W. Wang, 'Multiuser Frequency-Time Domain Radio Resource Allocation In Downlink OFDM Systems: Capacity Analysis and Scheduling Methods,' *Computers & Electrical Engineering, Elsevier* **32**(1): 118–34, Jan. 2006.

[22] S. Brost, 'User-level Performance of Channel-Aware Scheduling Algorithms in Wireless Data Networks', *IEEE/ACM Transactions on Networking*, **13**(13): 636–47, June 2005.

[23] T. Ali-Yahiya, A. L. Beylot and G. Pujolle, 'Radio Resource Allocation in Mobile WiMAX Networks Using Service Flows', *Proceedings of IEEE 18th Symposium on Personal, Indoor and Mobile Radio Communications*, PIMRC'07, pp. 1–5, 2007.

[24] Q. Liu, X. Wang, and G. B. Giannakis, 'A Cross-Layer Scheduling Algorithm With QoS Support in Wireless Networks', *IEEE Transaction Vehicular Technology* **55**(3): 839–47, May 2006.

10

QoS Issues and Challenges in WiMAX and WiMAX MMR Networks

Kiran Kumari, Srinath Narasimha and Krishna M. Sivalingam
Indian Institute of Technology Madras, Chennai, INDIA
University of Maryland, Baltimore County, Baltimore, MD, USA

10.1   Introduction

Internet connectivity from "anywhere, anytime" is an important goal for the current generation of networking technologies. Existing wireless standards, such as IEEE 802.11 (WiFi) and cellular networks (GPRS, CDMA2000) provide a part of the access solution. However, they have the limitation of either low coverage areas (WiFi) or low data rates (cellular). To overcome the shortcomings in existing technologies, a new wireless standard, IEEE 802.16 was published in 2001. The IEEE 802.16e standard, which includes support for mobile users, was presented in 2005 [6, 27, 29]. Currently, the standards group is working on the development of IEEE 802.16m [18]. The term WiMAX ('Worldwide Interoperability for Microwave Access') refers to a subset of the IEEE 802.16 standards that will be developed as products. The WiMAX standards are set by the WiMAX Forum [36], a worldwide consortium of interested companies, which also develops the certification procedures for IEEE 802.16 based products.

WiMAX is thus a potential broadband wireless access (BWA) technology designed to provide high-speed, last-mile wireless Internet connectivity and network access over a wide area as compared to conventional wireless technologies, such as WiFi. It can be used as an alternative to cable access networks and digital subscriber lines (DSL) owing to its low infrastructure cost and ease of deployment. One of the applications of WiMAX is to

WiMAX Security and Quality of Service: An End-to-End Perspective   Edited by Seok-Yee Tang,
Peter Müller and Hamid Sharif
© 2010 John Wiley & Sons, Ltd

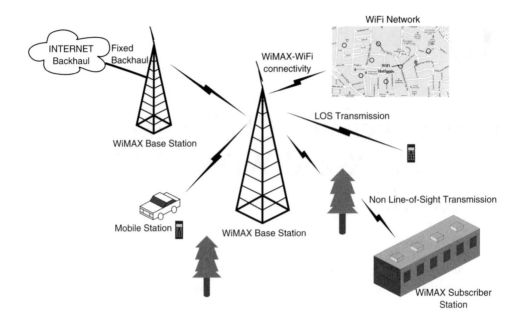

Figure 10.1 IEEE 802.16 standard architecture.

support fixed and mobile user's needs by establishing interoperability between existing wireless technologies. The WiMAX Forum plays a major role in ensuring conformance with the vertical and horizontal interoperability of wireless technologies and devices.

The WiMAX network architecture is shown in Figure 10.1. A typical WiMAX network consists of a base station (BS) that serves fixed and mobile users, called subscriber stations (SS) and mobile stations (MS), respectively. WiMAX provides two types of wireless service–line-of-sight (LOS) and non-line-of-sight (NLOS). In LOS, the receiver antenna (at SS) and the WiMAX BS are in line of sight; hence, the connection is more stable owing to less interference and better signal strength. LOS transmissions operate at a higher frequency band of 11–66 GHz covering a range of 50 Km. The NLOS service operates at 2–11 GHz because lower frequency transmissions are able to diffract or bend and hence are not disrupted easily by any physical obstacle. The operating range for a NLOS service is around 6–8 Km. In the infrastructure mode of operation WiMAX works based on downlink (BS to SS) and uplink (SS to BS) data exchanges. The link establishment process follows certain steps that include ranging and negotiation between BS and SS, authentication and registration and finally the establishment of Radio Link Control (RLC) [11].

10.1.1 *Motivation*

The current generation of wireless networks has been designed primarily to support voice and more recently data traffic. However, it is expected that the next generation of wireless networks (referred to as 4G networks) will be required to support additional high-bandwidth and interactive traffic such as streaming video, IPTV, games, video conferencing and other entertainment. Hence, support for different types of traffic, including ensuring service quality, is an important requirement.

Although WiMAX networks could prove to be an effective solution for high data rate connectivity with a large coverage area, they face challenges such as providing effective Quality of Service (QoS). QoS in WiMAX has been achieved primarily by developing a contention-free scheduling based medium access control (MAC) layer. This is in contrast to the IEEE 802.11 WiFi MAC protocol, which uses primarily the contention-based distribution control function (DCF) mode. In WiMAX, users in a given service area make reservation requests to the BS, which then schedules the requests according to QoS specifications. Thus, the scheduling algorithm is an important component of future wireless networks. With this in mind, we focus on MAC-level QoS scheduling algorithms in WiMAX networks which support multimedia traffic.

We will then discuss scheduling algorithms designed for WiMAX mobile multi-hop relay (MMR) networks [31]. WiMAX networks (and all wireless networks, in general) will have coverage holes (due to obstacles, fading, etc.) and cell-edge connectivity problems. One way to overcome this problem is to deploy additional less-complex *relay stations*. The relay stations will be used to extend the coverage of a base station. Scheduling in such MMR network introduces additional challenges. The second part of this chapter will focus on scheduling algorithms for MMR networks.

The rest of the chapter is organized as follows. Section 10.2 discusses the characteristics of real-time traffic and different codecs used for voice and video. Section 10.3 provides an introduction to the limitations of wireless LAN technologies in handling real-time data. This is followed by an introduction to the MAC layer of the WiMAX network stack. A short description of a few algorithms on uplink scheduling for real-time traffic in WiMAX networks is also provided in this section. Section 10.7 focuses on MMR-based WiMAX networks and downlink scheduling schemes for MMR-based WiMAX networks. Sections 10.8–10.10 present algorithms related to WiMAX MMR networks. Section 10.11 discusses material for further reading and Section 10.12 concludes the chapter.

10.2 Multimedia Traffic

The term 'multimedia' refers to multiple forms of media integrated in a single document or interaction: such as text, audio, graphics, animation, video, etc. An example of a multimedia document is a web page containing the audio, video and text of a certain song. Multimedia also stands for interactive types of media, such as video games and virtual reality. We will focus on support for voice, video and data.

We first describe the characteristics of commonly used voice and video generation coders-decoders (codec). We will then present typical Quality of Service (QoS) metrics for real-time applications including voice and video conversations. In addition, we address the issues involved in achieving real-time traffic communication in existing wireless technologies and how these shortcomings can be resolved using WiMAX technology.

Table 10.1 Voice Codecs

Codec	Algorithm	Bit Rate (Kbps)
ITU G.711	PCM (Pulse Code Modulation)	64
ITU G.722	SBADPCM (Sub-Band Adaptive Differential Pulse Code Modulation)	48, 56 and 64
ITU G.723	Multi-rate Coder	5.3 and 6.4
ITU G.726	Adaptive Differential Pulse Code Modulation	16, 24, 32, and 40
ITU G.727	Variable-Rate ADPCM	16–40
ITU G.729	Conjugate-Structure Algebraic-Code-Excited Linear Prediction (CS-ACELP)	8
ILBC	Internet Low Bitrate Codec	13.33 and 15.20
GSM - Full Rate	RPE-LTP (Regular Pulse Excitation Long-Term Prediction)	13
GSM - Enhanced Full Rate	ACELP (Algebraic Code Excited Linear Prediction)	12.2
GSM - Half Rate	CELP-VSELP (Code Excited Linear Prediction - Vector Sum Excited Linear Prediction)	11.4
DoD FS-1016	CELP (Code Excited Linear Prediction)	4.8

10.2.1   Voice Codecs

In a circuit-switched telephone network, voice conversations are digitized using a sampling rate of 8000 samples per second (twice the 4 KHz allotted bandwidth) with 8 bits per sample, leading to a constant 64 Kbps capacity requirement per voice conversation. In order to reduce the network bandwidth requirement for voice communication, coders/decoders (also called as codecs) are used at the sender and receiver ends. The objective of the voice codec is to represent the high-fidelity audio signal with a minimum number of bits without any degradation in quality for transmission across IP networks. Codecs for VoIP systems, also referred as vocoders (voice encoders), support silence suppression and compression to save network bandwidth utilization along with handling of small packet loss. Codecs also regulate the traffic by deciding the periodicity of the frames. Some of the common voice codecs used for VoIP are provided in Table 10.1. The reader is referred to [30] for detailed information.

10.2.1.1   G-Series

The G.711 codec uses pulse code modulation (PCM) of voice frequencies at a rate of 64 Kbps, as explained earlier. The voice quality achieved with this codec is very good since it uses no compression. This leads to less computation and hence negligible latency. The downside of this codec is that it requires more per-call bandwidth than other codecs, which leads to a limited number of active calls per network. Many extensions to G.711 have been designed, of which G.729 is one of the most recent. This is considerably more suitable for VoIP applications owing to its low bandwidth requirement. The G.729 codec samples voice data at 8 KHz with a 16 bit resolution and generates a stream of 8 Kbps using compression. Extensions to G.729 can provide rates of 6.4 Kbps and 11.8 Kbps for marginally worse and better speech quality respectively. Although the standard supports low bandwidth communication it suffers from other drawbacks such as high computation complexity.

10.2.1.2 GSM Full Rate (GSM FR)

The GSM FR codec was the first digital speech coding standard used in a GSM digital mobile phone system. The bit rate of the codec is 13 Kbps. The quality of the GSM FR coded speech is poor as compared to G.711 owing to the compromise between computational complexity and quality. Extensions to GSM FR are Enhanced Full Rate (EFR) and Adaptive Multi-Rate (AMR) standards, which provide much higher speech quality with a lower bit rate of around 12.2 Kbps.

10.2.1.3 DoD FS-1016

This is a 4.8 Kbps bit rate CELP (Code-Excited Linear Prediction) codec. It divides the speech stream into 30 ms frames, which is further divided into four 7.5 ms sub-frames. For each frame the encoder models the vocal track of the speaker by calculating a set of 10 filter coefficients for the short-term synthesis filter. The excitation for the synthesis filter is determined for each subframe and is given by the sum of scaled entries from two codebooks. The decoder decodes the scaled entries from the two codebooks by passing it through synthesis filter to give the reconstructed speech. Finally, a post-filter is used to improve the perceptual quality of the speech. The drawback of the standard is that the low bit rate coding can lead to noisy speech quality (lower than commercial cellular speech codecs).

10.2.2 Video Codecs

Consider a simple video encoder that is generating 15 frames per second (half of NTSC's 30 frames per second), with 800×600 8-bit pixels per frame. This will lead to a video stream encoded at 57.6 Mbps. A compression-based PAL video (Phase Alternating Line, colour-encoding system used in broadcast television systems) requires an estimated bandwidth of 216 Mbps, whereas high definition TV requires bandwidth of approximately 1 Gbps. Compared to voice's bitrate requirements, this is significantly higher. Hence, video compression algorithms need to be even more sophisticated to achieve reasonably low bitrates in order to transmit over wireless networks.

Codecs can be implemented either using advanced hardware technologies or in software, each having advantages and disadvantages over the other such as use of hardware versions gives faster response time with less noisy video quality whereas software versions are much more flexible which allows for updates of algorithm and codes used without the need of extra hardware devices.

Some commonly used video codec standards are described below.

10.2.2.1   MPEG-2 (Moving Picture Expert Group, Part 2)

This is an international standard accepted by the International Standards Organization (ISO), and is commonly used for digital video broadcasting and cable distribution systems. The MPEG-2 codec exploits the presence of redundant sections in video data such as the background image. It supports a low bit rate of less than 1.5 Mbps. MPEG-2 removes the temporal and spatial redundancies and hence reduces the overall bandwidth requirement. It supports progressive scanning and interlaced video for better quality. To comply with limited bandwidth requirements, a constant bit rate for different compression levels is maintained either by buffers or by dropping some packets with minimum degradation in quality.

10.2.2.2   MPEG-4 [21]

This is an extended version of MPEG-2 with the additional features of enhanced compression, object oriented coding and security with a bitrate lower than 1.5 Mbps. This standard enables the integration of the production, distribution and content access paradigms of the fields of interactive multimedia, mobile multimedia, interactive graphics and enhanced digital television. It is a suite of standards having many 'parts', where each part standardizes various entities related to multimedia, such as audio, video and file formats.

MPEG-7, an extension of MPEG 'parts', is a multimedia content description standard that provides fast and efficient searching of material owing to the association of description with its content. MPEG-7 uses XML to store meta-data and can be attached to timecode (Numeric codes generated at regular intervals by a timing system for synchronization) in order to tag particular events, or synchronize the lyrics of a song.

10.2.2.3   H.261

This is an ITU-T video coding standard defined by the ITU-T Video Coding Experts Group (VCEG) for video conferencing over Public Switched Telephone Network (PSTN) synchronous circuits. It is designed to run at multiples of 64 Kbps data rates from 1x to 30x. It supports two video frame sizes: 352 × 288 luma (brightness; achromatic image without any color) with 176 × 144 chroma (colour information) and 176 × 144 luma with 88 × 72 chroma using a 4:2:0 sampling scheme with support for backward-compatibility for sending still pictures with 704 × 576 luma resolution and 352 × 288 chroma resolution. This codec optimizes bandwidth for luminance over colour due to high human vision sensitivity to luminance ('black and white') as compared to colour.

10.2.2.4   H.264/MPEG-4 Part 10/AVC

This is a block oriented motion compensation based codec standard developed by ITU-T VCEG together with MPEG. It is designed to provide better video quality at substantially lower bit rates as compared to its predecessors (i.e. MPEG-2, H.261 and MPEG-4 Part 2) in addition to improved perceptual quality. It also provides DVD quality video at less than 1 Mbps and can be used for full motion video over wireless, satellite and ADSL Internet connections. The enhanced compression and perceptual quality of the standard is obtained by a motion estimation technique, which minimizes temporal redundancies; intra estimation, which minimizes spatial redundancies; transformation of motion estimation and intra estimation into the frequency domain; reduction of compression artifacts; and entropy coding, which assigns a smaller number of bits to frequently encountered symbols and a larger number of bits to infrequently encountered symbols.

Like the previous standards, MPEG-4 Part 10 or the Advanced Video Coding standard (AVC) does not define a specific codec. Instead, it defines the syntax of the encoded video stream and the method of decoding it. There are many algorithms which increase the efficiency of the encoder at various stages of encoding. They can be broadly classified as: (i) Encoder Parameter Selection Algorithms: The basic aim of these algorithms is to set the parameters such as number of reference frames, quantized transform co-efficients, resolution of motion vectors, etc. so that the distortion complexity points are close to optimal; (ii) Parallel algorithms for encoding: Parallel algorithms speed up considerably the encoding process since it is a computation intensive task; (iii) Mode selection algorithms: Each macroblock can be broken down into a number of smaller blocks using the many modes provided by H.264. Deciding the mode which has to be used to break down a given macroblock is done by a rate distortion optimization (RDO) algorithm; (iv) Rate control Algorithms: The main purpose of rate control algorithms is to vary the parameters so that the bit rate can be achieved and maintained.

10.2.3   QoS Specifications

Quality of Service (QoS) refers to a network's ability to provide a preferential delivery service for the real-time applications with assurance of sufficient bandwidth, latency and jitter control and reduced data loss. The network characteristics, such as available bandwidth and traffic load, have a different impact on different types of media including voice and video. For example, voice or audio requires the timely delivery of IP data packets and hence are more sensitive to delay. In order for the network to support voice and video efficiently, applications typically specify their QoS requirements in terms of bandwidth, latency, jitter and packet loss.

Bandwidth refers to the bitrate that a given voice or video conversation requires. The bitrate can be specified as continuous bit rate (CBR) – when compression is not used; or variable bit rate (VBR) – when compression (and silence/activity detection in case of voice) is used. The bitrate specifications depend on the codec used, as explained in the previous sections. In a packet-switched voice-over-IP (VoIP) network, voice traffic is characterized by small packet sizes (80 to 256 bytes) and varying bit rate traffic. After compression, streaming video codecs generate variable bit rate traffic with a packet size ranging from 65 bytes to 1500 bytes.

Latency refers to the time taken by the time packet generation at source and delivery at the destination node. One of the major contributing factors is the queuing delay at intermediate nodes in a packet-switched network. If the traffic load on a network is low, latency is likely to be low as compared to heavy traffic load scenario. Unlike data traffic, late packets are not useful for voice and video traffic, and are mostly discarded.

Jitter is variation in the arrival times of different packets within a data stream (i.e. variation in latency). There are different definitions of latency: the difference between the maximum and minimum packet delays over a time period; or inter-packet jitter (j) as given by:

$$j_i = l_i - l_{i-1} \tag{10.1}$$

where, l_i is the latency of packet i and l_{i-1}: Latency for packet $i-1$. Jitter is usually handled by buffering data at the receiver side; thus, it has an important effect on the buffering requirements of the receiver's playback application.

Packet loss leads to voice and video quality degradation. Depending upon the application, the tolerance can vary. Unlike data transmissions, lost packets are not re-transmitted since the packet's contents would arrive too late to be useful.

For voice communication, the maximum acceptable standard value for delay is between 0 ms and 200 ms [1], jitter is between 0 ms and 50 ms and acceptable maximum packet loss is 1.5 %. Typical video playback applications buffer about 4–5 seconds of video data to smooth out the jittery traffic. Also, higher packet loss of 5 % can be tolerated by video as compared to voice communication.

10.2.4 QoS Effectiveness Measures

As mentioned earlier, data traffic such as file transfer and web browsing are time insensitive and hence can tolerate delay and/or jitter. On the other hand, time sensitive data (such as voice or video) suffer degradation in performance due to the various network constraints such as low network bandwidth. The application can specify the QoS requirements described in the previous section. The network attempts to meet these requirements to the extent possible. However, it is also important to understand the user's perception of voice/video quality in the presence of non-ideal delay, jitter and packet losses. Two different methods, that is, subjective method and objective method, are used for estimating quality of the traffic.

Mean Opinion Score (MOS) is a subjective method in which average quality of perception is considered. For example, a number of listeners rate the quality of voice over the communications circuit. A listener then gives each sentence a rating between 1 and 5 as bad, poor, fair, good and excellent, respectively. The arithmetic mean of all the scores gives the MOS value. A MOS value of 4.0 typically denotes good voice quality.

In the objective method, the quality of video/voice is estimated based on actual and reconstructed image (or voice) information. Some of the objective methods used for calculating the quality of data are R-Score [35], Peak Signal to Noise Ratio (PSNR), Perceptual Evaluation of Audio Quality (PEAQ) and Perceptual Evaluation of Speech Quality (PESQ). Two of the objective methods, R-Score (audio) and PSNR (video) are described below.

R-Score [35] estimates the quality of the voice on the basis of signal-to-noise impairment (I_s), equipment impairment (I_e), impairment caused by mouth-to-ear delay (I_d), and a compensation factor A (Expectation factor; compensates for the various impairments under various user conditions). The R-score is given by:

$$R = 100 - I_s - I_e - I_d + A \tag{10.2}$$

The value of R-score ranges from 0 to 100. A value of more than 70 indicates that the voice is of acceptable quality.

PSNR estimates the quality of the video by computing the signal-to-noise ratio, in decibels, between the original (at source node) and a compressed video image (at destination node) as given in equation 10(3). The higher the value of PSNR, the better is the quality of the compressed, or reconstructed video image.

$$PSNR = 10log_{10}\left(\frac{R^2}{MSE}\right) \tag{10.3}$$

where R is the maximum fluctuation in the input image data type (e.g. if the input image has an 8-bit unsigned integer data type then R is 255) and MSE is the Mean Square Error, given by:

$$MSE = \frac{\sum_{M,N}[I_1(m,n) - I_2(m,n)]^2}{M*N} \tag{10.4}$$

where, M, N are the indexes of the image matrix, I_1 is the source image and I_2 is the reconstructed image.

In this section, we have described multimedia encoding, QoS specifications and effectiveness measures.

10.3 Multimedia: WiFi versus WiMAX

In this section we will cover the limitations of wireless LAN technologies in order to understand the need for WiMAX technology. An introduction to WiMAX MAC frame format and WiMAX QoS architecture will also be covered. We will discuss in detail the various conventional and extended scheduling mechanisms designed to support real-time traffic.

10.3.1 Limitations of Wireless LAN Technologies

Today, most of the wireless technologies have to support real-time communication owing to increasing use of multimedia based communications. Unfortunately, the current IEEE 802.11 standard (WiFi) does not guarantee real-time communications owing to limited support for QoS. The bandwidth required for supporting multimedia data is quite high, as seen earlier. Even though WiFi offers a maximum of 54 Mbps channel bandwidth, a single user system can get only 50 % of the capacity. As the number of users sharing a channel increases, the per-user bandwidth drops even further. In addition, owing to the equal priority being assigned to all traffic types, real-time communication is not guaranteed. For example, owing to the same priority allocation, a voice frame can be queued behind some large data frame and hence could be delayed until the delivery of the data frames.

To overcome these limitations, a new standard, IEEE 802.11e, has been designed. The IEEE 802.11e standard provides mechanisms designed to deal with QoS. It defines two parts for supporting QoS, namely Enhanced Distributed Channel Access (EDCA) and HCF Controlled Channel Access (HCCA). The HCF (Hybrid Coordination Function) mechanism schedules the station's access to the channel by allocating transmission opportunities (TXOP).

EDCA is an extension of DCF (Distributed Coordination Function) which uses Carrier Sense Multiple Access Collision Avoidance (CSMA/CA) to control medium access. It prioritizes different traffic classes using queues called Access Categories (AC). This algorithm gives higher priority to voice as compared to video and text data, thus voice gets faster access to the medium. Also, owing to traffic separation, congestion in one traffic type does not affect the other traffic types. Apart from prioritization, EDCA supports admission control and provides varying transmit opportunities to each class of traffic. For example, video, which has high data rate traffic requirements, gets the highest bandwidth share, followed by voice and data, respectively. With the additional mechanisms (i.e. admission control, prioritization) EDCA reduces the latency and bandwidth problem. But it does not prove to be a potential technology for real-time communication owing to more jitter and periodic bandwidth requirements.

HCCA, on the other hand, is an extension of PCF (Point Coordination Function), a centralized polling scheme. It offers a mechanism designed to guarantee periodic bandwidth with reduced delay and jitter. It is a centralized approach and as such does not suffer from delay caused by medium contention. Applications at the mobile stations ask for bandwidth and polling intervals from the central scheduler located at access point (AP). The AP then assigns the bandwidth to the stations using strict admission control protocols that allow communication with reduced delay and jitter.

Although the IEEE 802.11e standard has incorporated many changes to support QoS, it does not fully guarantee real-time secure communication owing to limitations like more packet drops, limited security and low channel capacity. Also, there are few vendor implementations of IEEE 802.11e. The focus at present in the 802.11 standards is more on IEEE 802.11n, based on MIMO technologies.

Because of these limitations of IEEE 802.11, WiMAX is being considered as a potential solution for real-time data communication in wireless networks. Sub-channelization, different coding scheme and flexible scheduling mechanisms make end-to-end QoS possible in WiMAX. Also, the wide coverage (e.g. 50 Kms for LOS) and throughput of up to 70 Mbps makes it a better competitor for multimedia and other applications. WiMAX (i.e. IEEE 802.11m) is one of the two competing technologies under consideration for the IMT-A 4G standard. We will now explain how the WiMAX MAC layer was designed to better support multimedia and Quality-of-Service.

10.3.2   WiMAX MAC Layer

The IEEE 802.16 MAC [34] was designed for point-to-multi-point BWA applications to support QoS for both up-link (SS to BS) and down-link traffic (BS to SSs), power management, mobility management and security. The primary task of the WiMAX MAC layer is to share efficiently the wireless channel and to provide an interface between the network layer and the PHY layer.

To provide functions such as QoS and security, the WiMAX MAC layer is divided into three sublayers, namely: service-specific convergence sublayer, MAC Common Part Sublayer (CPS) and Privacy Sub-layer (PS). These sublayers interact with each other via Service Access Points (SAPs), as shown in Figure 10.2. The service-specific convergence sublayer receives the external network data packets and forwards them to the CPS with

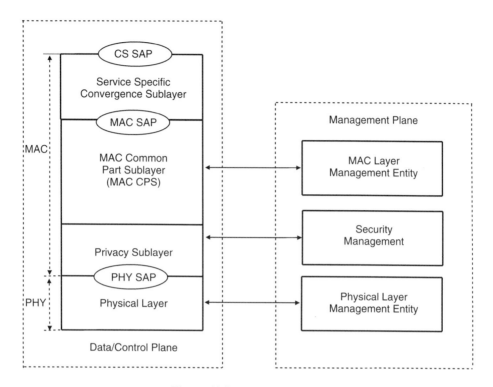

Figure 10.2 MAC sublayers.

the help of SAP. The MAC CPS, in turn, receives these packets (called MAC Service Data Units or MSDUs) and organizes them into MAC Protocol Data Units (MPDUs) for transmission. The privacy sublayer provides security features such as authentication, secure key exchange and encryption on the MPDUs and forwards them to the PHY layer.

The CPS acts as the core functional layer for providing bandwidth along with establishing and maintaining connections. WiMAX MAC is based on a connection oriented approach to provide service to SSs. Each connection is provided with a 16-bit connection identifier (CID) by the CPS. The 16-bit value constraints the maximum number of connections per BS to approximately 65,000.

The BS establishes transport connections along with three different bidirectional management connections on arrival of a new SS in its network. The three management connections are: basic connection, primary management connection and secondary management connection. The basic connections are established for short, time-critical MAC messages and radio link control messages. The primary management connection is for longer messages that can tolerate more delay and the secondary management connection is for standard-based messages, such as Simple Network Management Protocol (SNMP) and Trivial File Transfer Protocol (TFTP). The transport connections are unidirectional and facilitate different up-link and down-link QoS and traffic parameters. These transport connections can be mapped to application level connections, provided the applications have the same QoS and other requirements.

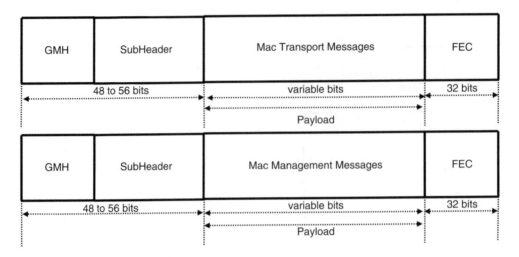

Figure 10.3   Generic frame format.

HT	EC	Type	Res	CI	ESK	Res	LEN	CID	HCS
1 bit	1 bit	6 bit	1 bit	1 bit	2 bit	1 bit	11 bit	16 bit	8 bit

6 Bytes

Figure 10.4   Generic MAC frame header.

WiMAX uses different MAC frame formats for uplink and downlink transmission. The generic MAC frame format is shown in Figure 10.3. Each frame consists of a 6-byte generic MAC frame header (GMH) which distinguishes between the uplink and downlink frame; optional sub-headers, payload (transport information or management information) and optional forward error correction codes (FEC). The 6-byte GMH includes other details such as header type (HT) bit, encryption control (EC) bit, encryption key sequence (EKS), cyclic redundancy check indicator as shown in Figure 10.4. Refer to [23, 24] for detailed information on GMH and other frame formats.

10.3.3   QoS Architecture for WiMAX

Quality of Service for a data packet entering into the WiMAX network is guaranteed by shaping, policing and prioritization at the subscriber station and the Base Station (BS). These data packets are associated with a QoS level based on the service flow QoS parameters. In the IEEE 802.16 standard, five scheduling service classes are defined:

10.3.3.1   Unsolicited Grant Service (UGS)

UGS supports constant bit rate (CBR) traffic, such as *audio streaming* without silence suppression. The QoS parameters defined for this service class are: the grant size to be allocated, the nominal interval between successive grants and the tolerated grant jitter. In UGS, BS provides fixed-size data grants periodically to UGS flows, which allow the SSs to transmit their packets without requesting bandwidth for each frame. Owing to bandwidth allocation without request contention, hard guarantees are provided in terms of bandwidth and access delay.

10.3.3.2 Real-Time Polling Service (rtPS)

For Variable Bit Rate (VBR) data, such as *MPEG streams*, the bandwidth requirements for the UGS grant interval cannot be determined at connection setup time. To handle such flows, rtPS service has been introduced. In rtPS service, the BS provides periodic transmission opportunities by means of a polling mechanism. The SS exploits these opportunities and asks for the bandwidth grants. The QoS parameters defined for this service class are the nominal polling interval between successive transmission opportunities and the tolerated poll jitter.

10.3.3.3 Extended Real-Time Polling Service (ertPS)

The ertPS service is similar to UGS with the only difference that in this service type the SS has the opportunity to request for a different bandwidth with change in the transmission rate. This service can be used for real time voice communication and avoids bandwidth wastage due to fixed allocation, as in UGS.

10.3.3.4 Non-Real Time Polling Service (nrtPS) and Best Effort (BE)

The nrtPS is similar to rtPS with the only difference that the polling interval is not guaranteed. In this, the polling interval depends on the network traffic load and hence this service class is suitable for variable packet size flows, such as large file transfers. In heavy load conditions, the BS cannot guarantee periodic unicast requests to nrtPS flows. Thus, the SS is required to use contention and piggybacking to send requests to the BS uplink scheduler.

BE service is the lowest priority service, introduced for traffic such as telnet and HTTP. In BE traffic, no periodic unicast requests are scheduled and thus there is no guarantee in terms of throughput or packet delay.

10.4 QoS Scheduling in WiMAX Networks

In order to provide efficient QoS support to the end users, the following set of protocol components are needed: admission control (controls the number of connections based on total available bandwidth and bandwidth requested by each connection) and packet scheduling (allocates time slots to the different connections based on QoS requirements), traffic policing (controls network traffic for conformity with a traffic contract) and traffic shaping (controls the volume of traffic being sent into a network in a specified period). Admission control is required to guarantee that the added traffic does not result in network overloading or service degradation for existing traffic. The dynamic nature of the

multimedia flows requires traffic shaping to conform to the pre-negotiated traffic specification and policing for conformance of user data traffic with the QoS attributes of the
corresponding service in order to ensure fair and efficient utilization of network resources.

The core component which determines network performance (QoS support) is the
scheduling algorithm. Since multimedia traffic is prone to delay, it is required to allocate
network resources to the traffic within a defined time duration. To achieve guaranteed
performance an efficient scheduling algorithm is used at the BS. The BS performs the
scheduling on per-connection basis. Thus, it requires information on the number of connections, number of pending connections, per connection throughput requirement and
queue status. This information is easily accessible as it is concerned with downlink connections which are established after the SSs send their bandwidth requests and queue
status to the BS, in uplink. For downlink scheduling, classical scheduling algorithms
such as Weighted Fair Queuing (WFQ) and Weighted Round Robin (WRR) can be used
since the scheduler has full knowledge of queue status. Scheduling uplink flows, on the
other hand, are much more complex owing to the location of queues in the SSs which is
separated from the BS.

For uplink scheduling, the BS receives the requests from the SSs and creates the uplink
map (UL-MAP) message of next UL frame and distributes the same to the SSs. Using the
UL-MAP, each SS knows the time and amount of bandwidth allocated for the next frame.
The traffic type affects the scheduling method requirement. For example, WFQ can be used
for UGS and rtPS, WRR for nrtPS class and FIFO (First In First Out) for non-real time
data (BE service class). Some of these scheduling algorithms [34] are described below:

10.4.1   Max-Min Weighted Fair Allocation

This algorithm works based on the available information of number of slots requested for
each flow (service type) by all the connections. BS examines the bandwidth requests of
each connection and calculates the total slots requested for each flow by all connections.
Each uplink flow is then allocated its percentage of bandwidth based on a normalized
weight. For example, UGS is given maximum weight followed by rtPS, nrtPS, and BE,
respectively. Further, the excess bandwidth allocated to any flow is distributed among
unsatisfied flows, again in proportion to their weight. This process continues until all the
uplink flows are satisfied or no bandwidth is available for allocation.

10.4.2   Deficit Fair Priority Queue

In this method, an active list of application services is maintained in the BS for scheduling.
The BS associates each service flow with a variable called 'deficit counter', D. Initially,
D is set to 0. The BS also maintains one more variable, $quantum$. The scheduler follows
a set of steps as given below:

1. Visits every non-empty queue and tries to serve one $quantum$.
2. For each visit, D is incremented by $quantum$.
3. If, the request size S at the head of the queue is greater than the $quantum$ size, then
   D is incremented by $quantum$.

4. Else if, $D + quantum$ is greater than S then the packet at the head of queue is scheduled and D is decremented by S.

The steps are repeated until D is 0 or the queue is empty (if so, D is set to zero). The scheduler then evaluates the next priority queue and if all the queues are empty the process stops; otherwise another pass is done for the unsatisfied flows in order of priority until either all flows are satisfied or no more uplink slots are available.

10.4.3   Weighted Fair Queuing [34]

In this algorithm, scheduling is done based on the amount of bandwidth requested by each connection. In each frame, the UGS connections are allocated the requested bandwidth. Other connections are allocated bandwidth based on their weight which is calculated as follows:

$$W_i = \frac{BW_i}{\sum_{j=1}^{N} BW_j} \tag{10.5}$$

where W_i is the weight of connection i, BW_i is the bandwidth requested by connection i and N is the number of connections.

Based on the calculated weight, bandwidth is allocated to a connection. The bandwidth allocated to each connection (BWA_i) is given by:

$$BWA_i = W_i * BW_{total} \tag{10.6}$$

where BW_{total} denotes the total uplink bandwidth available after satisfying the UGS connection requests. This algorithm suffers from the drawback that it does not consider the priorities of the service flows.

10.4.4   Weighted Fair Priority Queuing

The 'priority' term in WFPQ [34] eliminates the drawback of the WFQ algorithm by considering the priorities of the service flows. The algorithm works similarly to WFQ with the only difference that bandwidth is allocated on per service flow basis rather than per connection basis. The BS calculates the number of slots required by each service flow. After satisfying the UGS request the remaining slots are distributed to other service flows as follows: 50 % to rtPS flow, 30 % to nrtPS, and rest 20 % to BE. If some of the services get more than what is required then the collective excess bandwidth (all flows) is distributed among unsatisfied flows using WFQ. The unsatisfied service flows get the bandwidth in proportion to their need, given by:

$$BW_{add} = \frac{BW_{addReq}}{BW_{excess}} \tag{10.7}$$

where BW_{addReq} is the additional requirement of the service flow and BW_{excess} is the excess bandwidth available. The bandwidth allocated to each flow is distributed among all connections of that service flow depending on their requests.

10.5 Voice Traffic Scheduling in WiMAX

The uplink scheduling of packets is complex as compared to downlink scheduling. The complexity in uplink scheduling prompts us to discuss advanced uplink scheduling schemes for VoIP traffic designed to improve upon the network performance. The conventional scheduling methods for VoIP services in WiMAX have some limitations such as wastage of uplink resources (UGS algorithm), MAC overhead and access delay (rtPS algorithm), and thus cannot be used as such. Different scheduling algorithms have been proposed to improve the overall network performance and resource utilization.

10.5.1 Lee's Algorithm

This scheduling scheme proposed by Howon Lee *et al.* [22] is based on the modelling of voice traffic as an exponentially distributed ON/OFF model with mean on-time T_{ON} $(= \frac{1}{\lambda})$ and mean off-time T_{OFF} $(= \frac{1}{\mu})$. The system model can be represented by one-dimensional Markov chain considering N independent users, as shown in Figure 10.5.

Thus, based on the model the number of active voice users is given by:

$$\overline{N} = \sum_{n=1}^{N} n . P_N(n) \qquad (10.8)$$

where:

$$P_N(n) = \binom{N}{n} \left(\frac{T_{ON}}{T_{ON} + T_{OFF}} \right)^n \left(\frac{T_{OFF}}{T_{ON} + T_{OFF}} \right)^{N-n} \qquad (10.9)$$

Lee's algorithm works on the concept that the BS assigns resources to the SSs based on its knowledge of voice transition states of the SSs. The BS gets the voice status information from the SSs via the MAC header. The algorithm makes use of one reserved bit of MAC header called the Grant Me (GM) bit for informing the BS whether the voice state is "ON" or "OFF". It retrieves this information on the basis of the codec used. The BS, on receiving the MAC header, checks for the GM bit. If the GM bit is set to '0', it decreases the grant size by half until it is minimum; otherwise it increases it to the maximum grant size (equal to UGS grant size) which is sufficient to send voice packets. Owing to the use of the conventional MAC header, the MAC overhead (as seen in rtPS) is reduced leading to better throughput and acceptable access delay [22].

Lee's algorithm outperforms UGS and rtPS (even though UGS and rtPS are service classes, they are considered as scheduling methods due to the mechanism of slots allocation specified for each of them) in terms of throughput and required resources. For

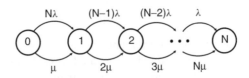

Figure 10.5 System model.

example, for $T_{ON} = 352$ ms, $T_{OFF} = 650$ ms, voice codec frame duration $= 20$ ms, information bit per voice codec frame $= 22$ bytes, and compressed header size $= 2$ bytes, the number of users serviced by Lee's algorithm is 77 in one frame duration with a throughput of 275 Kbps. For the same set of values, UGS serves 40 (225 Kbps) and rtPS serves maximum of 66 users (125 Kbps). In terms of used resources, Lee's algorithm uses 50 % of total resources for serving 40 users while rtPS and UGS use 60 % and 100 % of total resources, respectively.

Thus, Lee's algorithm provides around twice the improvement over UGS due to efficient utilization of resources by allocating bandwidth only when there is traffic. It shows 1.2 times improvement over rtPS owing to the reduced MAC overhead as the required traffic information is made available to BS through the use of a reserved bit. For further results refer to [22].

10.5.2 UGS with Activity Detection Scheduling (UGS-AD)

The UGS-AD algorithm [14] combines UGS and rtPS scheduling algorithms to eliminate the shortcomings of the two algorithms. This scheme works in two modes, that is, UGS mode and rtPS mode. At the initiation of the VoIP traffic, the rtPS mode is adapted by the algorithm. Further, with change in the bandwidth requests, the mode changes. For example, if a voice user requests a bandwidth of zero byte (silence) the mode remains in rtPS, otherwise it switches to UGS mode.

The UGS-AD algorithm works well with fixed data rate traffic but cannot be used efficiently for variable data rate traffic (such as traffic generated with enhanced variable rate codec) with silence suppression. In variable bit rate traffic, the waste of uplink resources occurs during the "ON" duration of the voice users.

UGS-AD supports more users as compared to conventional scheduling algorithm. For example, for a set of given parameters, frame symbols $= 36$, modulation $= $ QPSK 1/2, full, half, quarter and eight rates $= 290, 40 \ 70, 600$ ms, respectively; and compressed header $= 3$ bytes, the number of users supported by UGS-AD are 149 whereas rtPS and UGS saturates at 93 and 131, respectively. The reason for better performance is the same as that given for Lee's algorithm.

10.5.3 Extended-rtPS Scheduling

The Extended-rtPS scheduling algorithm [14] is proposed by Howon Lee *et al.* to solve the problems existing with conventional scheduling algorithms, UGS-AD and Lee's algorithm. The algorithm makes use of the grant management sub-header and bandwidth request header to inform the BS of its voice status information.

In case of decrease in voice packet size, extended piggyback request bits of grant management header are used for requesting the required bandwidth. The algorithm sets the most significant bit (MSB) of the piggyback request bit to 1 to distinguish it from the general request bit. The BS, on receiving the request, assigns resources as per the requested bandwidth size. In the case of increase in packet size, the SS sends the voice packet using bandwidth request bit of the bandwidth request header. The MSB of bandwidth request bit is set to 1 for distinguishing it from general request bit. The BS assigns uplink resources periodically according to the requested size until another request of different size comes to the BS.

This scheme works well with variable data rate traffic with silence suppression as the BS recognizes the grant management sub-header and bandwidth request header. Thus, if the user requests the bandwidth for sending voice packets, the BS changes its polling size according to the bandwidth size requested and keeps the changed polling size until the next request of different packet size arrives.

Extended rtPS appears to be the best amongst the conventional and Lee's algorithm. It supports a greater number of voice users that is 74 %, 24 %, and 9 % as compared to UGS, rtPS and UGS-AD/Lee's algorithm, respectively. For the same set of values as mentioned in the UGS-AD. Contradiction in the values for extended rtPS, that is, 74 % improvement over UGS as reported in [14] and Lee's algorithm, that is, 100 % improvement over UGS as reported in [22], is due to different simulation parameters. It outperforms the other algorithms owing to reduced wastage of uplink resources, reduced MAC overhead and reduced access delay. Refer to [14] for a detailed description of extended rtPS with performance results.

10.5.4   Multi-Tap Scheduling

Most of the scheduling schemes consider the ON-OFF model in studying the behaviour of VoIP traffic in the network. Although this model represents the nature of voice traffic, that is, periods of speech and silence, it does not consider the impact of transport layer protocols (TCP, UDP or SCTP) on the voice packet streams. Haghani *et al.* [13] proposed an extended distribution model called the multi-tap model to capture the exact behaviour of voice traffic which takes into account the packet size and inter-packet time.

The multi-tap scheduling scheme uses the information of the multi-tap traffic model to perform efficient scheduling. This scheme assumes that the SS has knowledge of the packet size and the parameters of VoIP traffic model, such as the inter-packet time matrix (Δ_{1XN}) and the probability matrix (P_{1XN}). Based on this information, the average inter-packet time ($\overline{\Delta}$) can be given by:

$$\overline{\Delta} = \sum_{i=1}^{N} P_i \Delta_i \tag{10.10}$$

The two main parameters used in the multi-tap scheduling scheme are the average bit-rate and the availability factor (ρ). The average bit rate (R_{avg}) of the VoIP traffic required by the SS to transmit to BS needs to be less than the maximum bit rate (R_{max}) the BS can allocate to the SS to keep the delay bounded. The R_{avg} value can be calculated based on the average inter-packet time and the packet size, which are known to the SS. Thus,

$$R_{avg}(bps) = \frac{S}{\overline{\Delta}} \tag{10.11}$$

The availability factor defines the available bandwidth for VoIP traffic and is given by the equation:

$$\rho = 1 - \frac{R_{avg}}{R_{max}} \tag{10.12}$$

It is assumed in the scheduling scheme that the SS has already transmitted the information of P, S and $\triangle$ to the BS. Further, the SS predicts the next packet generation from the time of previous packet generation based on the time intervals $\triangle_i$ where $i = 1, 2, \ldots, N$. The multi-tap scheduling is done in two phases: request phase at SS and grant phase at BS. The SS predicts the next packet arrival and piggybacks the index (i) in the packet. The SS transmits the packet to the BS and calculates the time difference (d) between the current time (t) and the last packet generation time (g) and finds the nearest $\triangle_i$ which is equal to or greater than d. If the value of d is equal to the $\triangle_i$ then it sets $index = i$ or it sets $index = N + 1$, if $d \geq \triangle_N$.

At the BS side, the BS reserves time slots based on the received *index* value and other network constraints such as the availability factor, ρ. BS assigns a transmission time at least at $\triangle_{index}$ value away from the previous packet transmission time. The delay increases with the decrease in ρ value and vice-versa.

This algorithm outperforms rtPS and UGS in terms of delay and bandwidth. For example, for $R_{avg} = 29$ Kbps, frame size $= 5$ ms, bits per time slot $= 192$, channel bandwidth $= 5$ MHz, modulation $= 16$ QAM, UL control slots $= 4$ and UL data symbols $= 21$, rtPS suffers a delay of more than 1.5 times the frame length, whereas multi-tap algorithm suffers a lesser delay, that is, 0.9 times of the frame length. In UGS, although the average delay is less, the BW wastage is more as compared to the multi-tap algorithm. The reason behind the performance of multi-tap algorithm is that it uses multi-tap model information. The multi-tap model converges very fast. This allows the user to estimate the model parameters in a short period of time and thus a new set of parameters can be derived if there is any change, such as IP address for the connection. For further detail the reader is referred to [10].

10.6 Video Traffic Scheduling in WiMAX

As described earlier, the traffic characteristics of video differ from voice in a variety of ways, such as data rate, delay and jitter tolerance. These characteristics present new challenges in video communication over WiMAX. Video traffic can be categorized into two forms: one way video (multicast; e.g. IPTV) and two-way video (e.g. video-conferencing). In our discussion, we will focus on each of these individually. For VoIP traffic, where the uplink scheduling is of main concern (due to ON/OFF period), for video it is not the same. As discussed in section 10.2.2, the main factor that decides video quality (one-way video) is the packet loss rate (packet drop probability). The packet drop probability should be minimized, without compromising on the goodput. This demands greater bandwidth in the downlink along with the use of Forward Error Correction (FEC) [23].

10.6.1 Opportunistic Scheduling

The opportunistic scheduling [2] mechanism exploits multiuser diversity to provide fairness along with QoS guarantees for both voice and video traffic. It is carried out in two phases: subcarrier allocation (WiMAX channel bandwidth is divided into a number of non-interfering bandwidth components; 256 subcarriers) and subcarrier assignment. The subcarrier allocation decides the number of subcarriers ($n_i(t)$) assigned to each user whereas subcarrier assignment determines which subcarriers have to be assigned to each user (δ_{ij}) to get maximum total rate.

10.6.1.1 Subcarrier Allocation Algorithm

The subcarrier allocation is decided based on three factors: (i) instantaneous subcarrier channel gains of active users, (ii) user's average rate and (iii) Head of Line (HoL) packet delays of these users.

In the first step, each active user is allocated subcarriers, $\acute{n}_i(t)$, so that users with better channels gets more subcarriers. The allocated number of subcarriers is given by:

$$\acute{n}_i(t) = \left\lceil \frac{r_i}{\frac{1}{|N_t|}\sum_{j\in N_t} r_j} \frac{\bar{\mu}_i(t)}{\frac{1}{|N_t|}\sum_{j\in N_t}\bar{\mu}_j(t)} \right\rceil \tag{10.13}$$

where N: Number of users; N_t: The number of active users at time t; r_i: Average traffic rate of ith user; and $\bar{\mu}_i(t)$: i^{th}user's average subcarrier capacity (if allocated all the subcarriers).

The value of $\bar{\mu}_i(t)$ can be calculated as follows:

$$\bar{\mu}_i(t) = \frac{1}{S}\sum_{j=1}^{S} \mu_{ij}(t) \tag{10.14}$$

where S is the total number of data subcarriers available and $\mu_{ij}(t)$ is the channel capacity of subcarrier j when allocated to user i. In equation 10.13, the first term, $\frac{r_i}{\frac{1}{|N_t|}\sum_{j\in N_t} r_j}$ weighs the allocation proportional to the user's average rate. It converges to one if the average traffic rate of all the users is same. In this case, the second factor, $\frac{\bar{\mu}_i(t)}{\frac{1}{|N_t|}\sum_{j\in N_t}\bar{\mu}_j(t)}$, decides the subcarrier allocation based on good or bad channel condition. The user with a good channel condition ($\bar{\mu}_i(t) > \frac{\sum_{j\in N_t}\bar{\mu}_j(t)}{|N_t|}$), gets more subcarriers and vice-versa. Three conditions can occur at the end of the allocation:

Case 1: All the data subcarriers are allocated to the set of users waiting for the service and the algorithm terminates.

Case 2: Some data subcarriers do not get allocated; in this case, the second step of the algorithm is invoked. The remaining subcarriers ($\acute{S}$) are calculated as:

$$\acute{S} = S - \sum_{N_t} \acute{n}_i(t) \tag{10.15}$$

Case 3: The number of allocated subcarriers exceeds S, in which case the third step of the algorithm is invoked.

The second step allocates the remaining subcarrier to the active users to minimize packet losses. It does so by allocating the biggest share of $\acute{S}$ to the user with the smallest deadline, $d_i(t)$, and the maximum number of violations it has suffered ($V_i(t)$). The history of previous assignments within a certain time window, that is, 1000 scheduling intervals is used to calculate this. Based on these factors, the number of subcarriers to be assigned to i^{th} active user is given by:

$$n_i(t) = \acute{n}_i(t) + \left\lceil \acute{S}\frac{\frac{max\{1,V_i(t)\}}{d_i(t)}}{\sum_{j\in N_t}\frac{max\{1,V_j(t)\}}{d_j(t)}} \right\rceil \tag{10.16}$$

At the end of this step, either Case 1 or Case 2 can be followed, based on the end condition.

In the third step of allocation algorithm, the number of subcarriers allocated to some of the users is decreased so as to satisfy the constraint of maximum subcarriers available, that is, S. The users are sorted based on the HoL packet time-to-expire in descending order. After sorting, the number of subcarriers is decreased by one in each iteration (in same order). If the total subcarriers allocated becomes equal to S then the step terminates. Otherwise, it iterates until termination.

10.6.1.2 Subcarrier Assignment

This phase enhances the fairness of scheduling algorithm. In this, the users with more packet drops are favoured. All the users are assigned a unity priority at the start of the time window and whenever a packet drops from a certain user queue, the priority gets incremented by one. The higher priority user followed by lower priority users are given the chance to select the best subcarriers. Once those subcarriers are assigned, they cannot be assigned to other users. In the case of a priority tie, the user with best channel quality is given precedence.

Opportunistic scheduling performs well overall in terms of better throughput, less packet dropping and fair delay distribution as compared to the conventional methods. The algorithm also requires less computation. The computation complexity of the allocation algorithm is given by $\leq O(N_t log(N_t))$ (for sorting) and the assignment algorithm is, $O(N_t S log(S))$. The interested reader is referred to [2] for detailed results and performance analysis.

10.6.2 Opportunistic DRR

O-DRR [32] is an uplink scheduling algorithm. It works on the basis of the polling mode operation at the WiMAX BS. In this mode, the BS polls the SSs to discover their bandwidth and QoS requirements. Before the algorithm starts, it requires the optimal value of polling interval, k for keeping a balance between efficiency and fairness. Rath et. al. Optimal calculation of k is explained in [32].

O-DRR works on the following assumptions:

1. Rayleigh fading model is used for the channel between BS and SS.
2. The coherence time (time interval within which wave's phase is predictable) of the frame is greater than the frame length (i.e. 5 ms).
3. The BS knows the signal-to-noise ratio (SINR) of all the channels.

The BS maintains a quantum size, $quantum_i$; a flag, $Flag_i$; and a deficit counter (called lead/lag counter, L_i) for all the SSs. The flag value denotes whether a SS has been assigned bandwidth in a given frame or not (1 or 0). The algorithm works as follows:

1. During the polling time, if $SINR_i$ is less than $SINR_{th}$ then SS_i is not scheduled. The BS distributes its $quantum_i$ to other SSs based on their weights, W_i, given by:

$$W_i = \frac{l_i \beta_i}{\sum_j l_j \beta_j} \qquad (10.17)$$

The BS then increments its L_i by $quantum_i$. The weight, W_i, is calculated based on the delay requirement and value of L_i. The delay count for the SS_i is given by the equation: $d_i = T_{d(i)}(j) - nT_f$ where T_f is the frame length, n is the number of frames elapsed since SS_i was scheduled and $T_{d(i)}(j)$ is the time delay of SS_i that belongs to j^{th} class of traffic (i.e. nrtPS, rtPS etc. to decide the delay bounds). In equation 10.17, the value of β_i is calculated as a reciprocal of delay and l_i is calculated as the scaled deficit counter value (sum of the magnitude of minimum deficit counter value among all SSs and L_i).

2. For SS_j that receives the extra quantum, the L_j value is decremented by the number of slots exceeding its $quantum_j$ value.

This algorithm achieves good performance compared to the DRR algorithm. It carries out the scheduling based on the delay requirements. The algorithm gives higher priority to the users with smaller delay counter value by assigning greater bandwidth to the user approaching towards its delay constraint. The results highlighted in [32] show that 91.5 % of the time, the delay requirements are met even under a heavy network load of 100 users with a frame length of 1 ms. This algorithm also allocates bandwidth depending on the class of traffic, making it more suited for multimedia applications.

10.6.3 Summary

The various algorithms explained in this section are summarized in Table 10.2.

Apart from scheduling schemes, the QoS of an application also depends on other mechanisms, such as admission control, fairness, congestion control, traffic shaping and policing. The reader is encouraged to go through these factors so as to understand the various issues related to multimedia and QoS in WiMAX. It is also worth reading [23] which explains the feedback based strategy for performance enhancement for streaming data, and [33] for video multicast over WiMAX networks and the other performance analysis for multimedia over WiMAX in [9] and [19].

This part of the chapter dealt with scheduling for multimedia traffic in WiMAX networks. One of the main reasons for lower QoS delivered to multimedia applications is weak signal strength, especially at nodes that are either far or moving away from the base station. A technique, called Mobile Multi-hop Relay (MMR), has been proposed for increasing network reach. The idea is to use intermediate relay stations that relay a user's transmissions to the base stations. This can help improve signal quality and thus deliver better QoS to the applications. The next sections deal with scheduling in WiMAX MMR networks.

10.7   Introduction to WiMAX MMR Networks

All wireless systems, including WiMAX, suffer from the challenging radio propagation characteristics of the wireless medium (see Figure 10.6 [5]). First, the achievable signal to noise ratio (SNR) and the resultant data rate decrease with increasing link distance. This results in low SNR at the cell border. Second, within a wireless network, there could be dead spots or coverage holes. These spots of poor connectivity are formed due to high path-loss and shadowing, because of the presence of obstacles such as large

Table 10.2 Summary of Scheduling algorithms, presented for Voice and Video support

Traffic Type	Algorithm	Type	Main Features	Main Results
Voice	Lee's Algorithm	Uplink	Based on information of ON-OFF Model	Good improvement over UGS. Resources utilized only in presence of traffic
Voice	UGS-AD	Uplink	Combined form of UGS and rtPS	Less MAC overhead and resource wastage as compared to rtPS and UGS, respectively
Voice	Extended rtPS	Uplink	Uses grant management subheader and bandwidth request header	Good improvement over UGS. Reduced bandwidth wastage and MAC overhead
Voice	Multi-tap	Uplink	Uses multi-tap model information, packet size and inter packet arrival time	Better than UGS and rtPS. Quickly adapts to changes, such as IP address change
Voice/Video	Opportunistic algorithm	Downlink	Subcarrier allocation and assignment	Reduces packet drops with fairness capability w.r.t conventional algorithms
Voice/Video	O-DRR	Uplink	Extended DRR based on the delay factor	Reduced delay; hence complies with the delay constraints

buildings, trees, etc. in the direct path between the base station and subscriber stations. In this chapter, the term *subscriber station* is used to refer to both *subscriber station* and *mobile station* (i.e. to refer to a user). The presence of these coverage holes within the network leads to non line-of-sight (NLOS) communication, which reduces the received signal quality [8, 15]. Also, it is required occasionally to provide wireless connectivity to an isolated area outside the reach of the nearest base station.

A simple solution to address these connectivity challenges is to deploy additional base stations. With more base stations in the network, the distance between a subscriber station and a base station decreases thereby improving the SNR at the subscriber stations. Increasing the number of base stations also reduces the probability of shadowing since a subscriber station not 'covered' by one base station could be 'covered' by another base station. However, owing to the cost of WiMAX base stations, such a solution could be prohibitively expensive.

A cost-effective alternative is to use WiMAX relay stations. WiMAX relay stations are low-cost counterparts of WiMAX base stations. They implement the minimal functionality necessary for relaying signals between the base station and subscriber stations. The introduction of relay stations into a WiMAX network can significantly enhance the quality of wireless links leading to throughput enhancements and extended network coverage

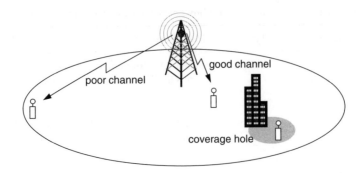

Figure 10.6 Channel conditions in a wireless network.

[12]. A basic WiMAX network (with only base station and subscriber stations) together with relay stations is referred to as a *WiMAX Mobile Multi-hop Relay (MMR)* network. The IEEE 802.16j working group has been created to design specifications for WiMAX MMR networks. As at the time of writing of this chapter the IEEE 802.16j standard is still at a draft stage.

10.7.1 How WiMAX MMR Networks Work

In a WiMAX MMR network, signals from subscriber stations with weak direct connectivity to the base station will take a multi-hop route through one or more relay stations. Figure 10.7 illustrates an example WiMAX MMR network. In this figure, each of the shaded ovals indicates the range of communication of the base station or the relay station positioned at the centre of the area. Subscriber station 1 is in direct range of the WiMAX base station and can be serviced directly by the base station. Subscriber station

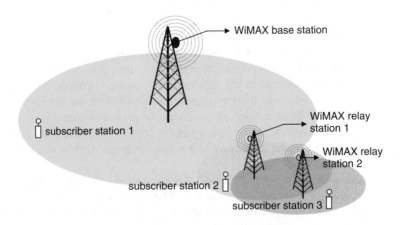

Figure 10.7 Illustration of a WiMAX mobile multi-hop relay network.

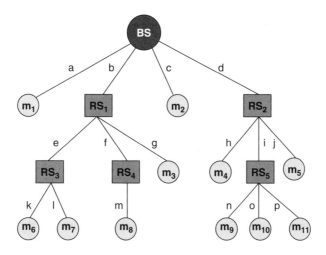

Figure 10.8   Schematic of a WiMAX MMR network. Reprinted with permission © IEEE 2009.

2 and subscriber station 3 are not directly within the range of the WiMAX base station. WiMAX base station serves these users by going over multiple hops through WiMAX relay stations. Subscriber station 2 is served through relay station 1 and subscriber station 3 is served through relay stations 1 and 2.

A schematic of a WiMAX MMR network is shown in Figure 10.8. The relay stations generally operate in *decode-and-forward* mode, although this is not stated explicitly in the standard [31]. In *decode-and-forward* mode, a relay station, on receiving an incoming signal, decodes the signal to extract data, interprets the data according to the packet format and performs local error detection/correction before re-encoding the signal and transmitting it on to the next hop.

Relay stations can be classified as either *transparent* or *non-transparent*. In the case of *transparent* relay stations, the subscriber stations are not aware of the presence of relay stations, whereas in the case of *non-transparent* relay stations, the subscriber stations are aware of the presence of relay stations as these subscriber stations synchronize and collect control information from them. However, in the case of *non-transparent* relay stations, the subscriber stations have an implicit 'understanding' that a *non-transparent* relay station is actually a base station [31].

In general:

1. Each relay station knows the subscriber stations it is serving directly.
2. The base station knows the relay station that it has to transmit to, to serve a particular subscriber station.
3. The base station schedules for all subscriber stations, even those that are serviced by relay stations, in the case of centrally controlled relay stations, as noted in section 10.7.3.
4. The base station sends the schedule to the relay stations and the relay stations simply obey the schedule (for the case above).

5. At any relay station, at any point in time, at most, only the data sent by the base station in the current frame is available (i.e. queued). That is, relay stations do not queue user data from previous scheduling frames.

10.7.2 Performance Impact

The performance of potential range extension of a WiMAX network using relay stations has been studied in [8]. The authors show that compared to a no-relay scenario (i.e. only base station and subscriber stations) in a cell with 1 km radius, five relays are sufficient to extend the cell radius by 20 % while providing network coverage to 95 % of subscriber stations and seven relays are sufficient to extend the cell-radius by 60 % while providing network coverage to 90 % of subscriber stations. However, with the increase in the number of relay stations, the number of hops that packets have to take from the base station to the subscriber station increases, increasing the probability of error. This, together with the fact that not all subscriber stations get network coverage when using the above mentioned numbers of relay stations, brings down the achievable system throughput. In the case of extending network coverage by using five relays, the mean throughput reduces by 11 % from the no-relay scenario, and by 36 % when using seven relays. These results show that an operator can deploy relay stations in an incremental manner depending upon cost-benefit tradeoffs, achieving progressively 100 % coverage.

Relay stations, as noted earlier, can also be used to increase the system throughput. Figure 10.9 illustrates a scenario where relay stations are deployed to reduce coverage holes within a cell. The impact of increased capacity has also been studied in [8] using 36 randomly generated topologies, in a cell with 1 km radius. Not all the 36 cases showed significant increase in network capacity when relay stations are used. In more than 50 % of these 36 cases, the median throughput increased by at least 15 % and in 78 % of these

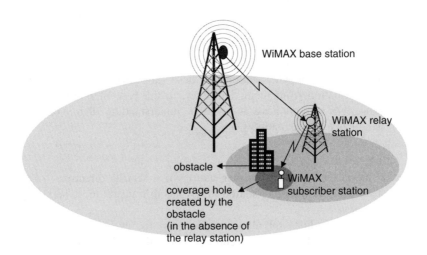

Figure 10.9 Illustration of deployment of relay stations for increasing network capacity. Reprinted with permission © IEEE 2009.

36 cases, median throughput increased by at least 5 %. Further, in more than 30 % of the 36 cases, the mean throughput increased by at least 10 % and in more than 82 % of the 36 cases, mean throughput increased by at least 5 %.

The results in [8] show that although relaying does not result in significant benefits for every random topology, there are several topologies where subscriber stations suffer from coverage holes due to shadowing and high path loss, and using relay stations for these topologies will increase the network capacity significantly, justifying the deployment of relay stations to fill these coverage holes.

10.7.3 Radio Resource Management Strategies

In WiMAX MMR networks, there are two approaches to radio resource management: centrally controlled and de-centrally controlled relay stations. In the former scenario, the base station controls directly all subscriber stations and all relay stations. The base station schedules all transmissions in the cell. The relay stations forward the relevant subset of the control information to the subscriber stations that they serve. The relay stations behave according to the base station's schedule, that is, they receive and transmit during time-slots (and in sub-channels) allotted to them by the base station. This leads to a simpler (and hence low cost) relay station design [12]. In the de-centrally controlled relay station scenario, each relay station has full control over the subscriber stations that are associated with it. The entire functionality that is required for the multi-hop operation is encapsulated in relay stations. The base station is not affected. For the base station, a relay station appears like an ordinary subscriber station, and for subscriber stations, a relay station appears to be a regular base station [15]. We assume the use of centrally controlled relay stations for the rest of the discussion in this chapter.

This chapter focuses on the WirelessMAN-OFDM physical layer interface. In an OFDM system, to make simultaneous data transmission possible at any given moment in time, the available channel bandwidth is divided into a number of orthogonal (i.e. non-interfering) smaller bandwidth components called sub-carriers. The WirelessMAN-OFDM physical layer interface provides for 256 orthogonal subcarriers [16, 17]. These subcarriers are further grouped into multiple subchannels. For example, the 256 subcarriers can be grouped into four subchannels of 64 sub-carriers each or they can be grouped into five subchannels of 51 subcarriers each. The subcarriers in a subchannel are classified as noted below [16]: Data subcarriers for data transmission; Pilot subcarriers for various estimation purposes; and Null sub-carriers for guard bands, etc.

In Figure 10.8, for example, with four subchannels, at a particular moment in time, the base station could be simultaneously transmitting to subscriber stations m_1 and m_2 and relay stations RS_1 and RS_2, using a different subchannel for each of m_1, m_2, RS_1 and RS_2. As per the standard draft, the same subchannel cannot be used to transmit different flows even when these flows are on different hops. We should also note here that not all the subcarriers in a subchannel can be used for data transmission.

For the discussion in the rest of this chapter, we make the following assumptions:

1. Orthogonal allocation of time/frequency resources: Over the multi-hop communication links (both in the downlink and uplink directions) between the base station and subscriber stations, only a single node (base station, relay station or subscriber station)

can transmit to another node at a given cellular time and in a given sub-channel. This assumption is required to totally avoid intracell interference [28].

2. Multi-hop route selection: Capacity-optimal routes (i.e. which subscriber station is served by which relay station or by the base station) are predetermined by using a routing algorithm (e.g. DSDV).

3. Availability of route metrics: The base station has complete information of the end-to-end route metrics of all users over all subchannels. The base station can use this information for opportunistic scheduling by assigning subchannels to users based on their route qualities.

In a multi-hop network, one can use different strategies to allocate subchannels to competing users as noted below:

1. *Same subchannel on all hops*: In this approach, each user is allocated a particular subchannel (so as to optimize on one or more of the objectives listed in section 10.8). Once a subchannel is allocated to a user, this same subchannel is used over all hops in the routing path from the base station to the subscriber station, to transmit data to the subscriber station [28]. This approach leads to low-complexity centralized opportunistic scheduling algorithms. But, it does not exploit the frequency selectivity fully and as a consequence may not achieve the best possible system throughput. This strategy of allocating the same subchannel on all the hops of the multi-hop path from the base station to the subscriber station, is the essential characteristic of a resource allocation policy called $OFDM^2A$ (Orthogonal Frequency Division Multihop Multiple-Access) [28].

2. *Different subchannels on different hops*: In this approach, each user can be allocated a different subchannel on different hops so as to optimize on one or more of the objectives listed in section 10.8. This approach has the advantage of providing best possible system throughput by exploiting frequency selectivity to the greatest extent. However, this approach has the disadvantage of increased runtime complexity.

10.8   Scheduling in WiMAX MMR Networks

In an OFDM system, scheduling is the process of determining which user should be serviced at a given moment in time and on a given subchannel. Figure 10.10 shows a schematic of a scheduling frame. Each small box in the scheduling frame in Figure 10.10 represents a (time-slot, sub-channel) pair and is referred to as a *tile*. Scheduling involves filling up each tile of a scheduling frame (Figure 10.10 [5]) with a subscriber station (i.e. user) that should be serviced in the time-slot and using the subchannel associated with that tile.

10.8.1   Objectives of Scheduling

The basic driving principle behind any scheduling algorithm is to use effectively the available system resources so as to optimize one or more desired objective(s). Some of these desirable objectives are:

- satisfying QoS guarantees provided to end users;

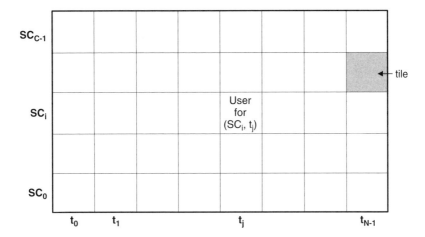

Figure 10.10 Basic Structure of an OFDM scheduling frame. Reprinted with permission © IEEE 2009.

- maximization of system throughput;
- providing fairness to end users;
- maximization of operator's revenue;
- Simple and fast implementation.

It should be noted that some of these desirable objectives are contradictory to each other. For example, maximizing operator's revenue may imply resource allocation that is inherently unfair to the end users, and vice versa. Also, one could strive to optimize a combination of the desirable objectives. For example, one could try to design a scheduling algorithm that optimizes both operator's revenue and fairness to end users at the same time. While in such a system, neither of the objectives are fully optimized individually, they provide a compromise that does not manifest the disadvantages of optimizing the objectives individually.

10.8.2 Constraints on Scheduling

While a scheduling algorithm tries to optimize its desired objective function, it may also have to work within the constraints imposed by the system. Some of these constraints in a WiMAX MMR network are [8]:

1. *Decode-and-Forward Relay (DFR)* Constraint: All the data that a relay station receives in one scheduling frame is also sent out during the course of the same scheduling frame.
2. *Transmit-Receive (TR)* Constraint: If a relay station has a single transceiver (as is the case of WiMAX MMR networks), it cannot transmit and receive concurrently. This constraint requires that a relay station cannot be transmitting on any subchannel over any of its child links while it is receiving a packet on some subchannel over its parent link.

3. *Spectrum Sharing (SS)* Constraint: In a given time slot, each subchannel can be used by only one link.

4. *Single Transmitting Node on all Sub-channels of a Time Slot (STS)* Constraint: In a given time slot, only one of the relay stations (or the base station) can transmit on all the subchannels.

5. *Low Runtime Complexity (LRC)* Constraint: In WiMAX MMR networks, a scheduling frame is typically constructed once every 5 ms. The scheduling algorithm's runtime complexity should be accordingly low.

The SS constraint means that there is no spatial reuse within the same sector. This is essential as the relay stations will lie within the same sector of same cell, when relays are deployed for capacity enhancement. We note that even though this assumption may not be valid when relay stations are deployed to extend the network coverage, it would be difficult to use different algorithms depending on whether relays are being used for capacity enhancement or for extending the network coverage, and in some situations relay stations could be used for both capacity enhancement and for extending network coverage. The STS constraint aims to reduce wasted bandwidth due to a relay station having to switch from receive mode to transmit mode (and vice-versa) frequently. This constraint also simplifies the complexity of the scheduling decision.

10.8.3 Diversity Gains

A scheduling algorithm that exploits the available diversities (different users and subchannels) is called an *opportunistic* scheduler. For example, when the objective is to service users in a fair manner, the scheduling algorithm can choose from among all the available users, the user that best matches the desired fairness metric. When the objective is to maximize the system throughput, the scheduling algorithm could choose the subchannel that maximizes the possible data rate during that time slot. The choice to exploit these diversities comes at the cost of higher complexity given the several possible combinations of diversities. Hence, we often need to restrict the amount of diversity under consideration, so as to bound the runtime of the scheduling decision.

10.9 Basic Wireless Scheduling Algorithms

In this section, we look at some of the conventional scheduling algorithms and how they can be extended for use in a multi-hop scenario such as a WiMAX MMR network. The multi-hop versions of the algorithms discussed in this section are as much applicable to generic multi-hop networks as they are to in WiMAX MMR networks.

10.9.1 Round Robin Scheduling

This is the simplest scheduling algorithm. To fill up each tile (c,t) in the scheduling frame, this algorithm chooses users in a round-robin fashion. The algorithm does not consider the channel conditions of the users. It is simple and potentially fair, but may not achieve high system throughput compared to an opportunistic algorithm. It is easy to see that the runtime complexity of this algorithm is $O(CN)$, where C is the number of sub-channels and N is the number of time slots in a scheduling frame. This algorithm does not need any special modifications for use in the multi-hop case.

In the following sections, we will describe more complex algorithms.

10.9.2 Max-SINR Scheduling

The Maximum Signal to Interference plus Noise Ratio (Max-SINR) scheduling algorithm chooses the user with the best channel quality at the given instant t when using sub-channel c. That is, to fill up each tile (c,t) this algorithm chooses user j such that

$$j = \arg\max_i r_i(l, c, t)$$

where $r_i(l, c, t)$ is the maximum data rate for user i, possible on link l (the link that should be used to transmit to user i) using subchannel c at time t. Even though this algorithm maximizes the achieved system throughput, it is inherently unfair to users with poor channel quality, serving them with arbitrarily large delays [28]. The runtime complexity of this algorithm is $O(CMN)$, where C is the number of subchannels, M is the number of users in the system that are simultaneously contending for a transmission opportunity and N is the number of time slots in a scheduling frame. This algorithm is also known as *MaxCap* (Maximum Capacity) [7].

10.9.3 Extension for Multi-Hop Case

For the multi-hop case, the Max-SINR algorithm can be extended as discussed below:

1. At time t, for each user, calculate the minimum of the possible data rates over all the hops (i.e. minimum of the data rates on the designated link on each of the hops) from the base station to that user, using subchannel c on all the hops. The rate at which this user can be serviced at time t using subchannel c is upper bounded by this value.
2. Among all the users, select the user with the maximum of the value calculated in point 1 above.

That is, to fill up each tile (c,t) in the scheduling frame, this extension chooses user j such that

$$j = \arg\max_i \min_{n=1..h} r_i(l_n, c, t)$$

where $r_i(l_n, c, t)$ is the data rate for user i, possible on link l_n (the link that should be used on the nth hop to transmit to user i) using subchannel c at time t; h is the number of hops from the base station to user i. The above extension is also referred to as *Max-Route* algorithm [28]. This algorithm also has a runtime complexity of $O(CMN)$, since the maximum number of hops in a network is bounded by a small number that can be considered as a constant.

10.9.4 Proportional Fair Scheduling

This is one of the widely used scheduling algorithms for wireless systems [3]. This algorithm is designed to achieve multi-user diversity considering fairness. This algorithm chooses the user with the maximum value of:

$$\frac{r_i(l, c, t)}{R_i(t)}$$

, where $r_i(l, c, t)$ is the maximum achievable data rate on link l using sub-channel c, for user i at time t; and $R_i(t)$ is the long term average service rate for user i at time t. In the above expression, l is the link over which data has to be sent to serve user i. $R_i(t)$ is updated as:

$$R_i(t+1) = \begin{cases} (1-\tau)R_i(t) + \tau r_i(t) & \text{if user } i \text{ is served at time } t \\ (1-\tau)R_i(t) & \text{otherwise} \end{cases} \tag{10.18}$$

for some time constant τ $(0 \leq \tau \leq 1)$. This algorithm gives priority to users with a high instantaneous channel rate $(r_i(l, c, t))$ and a low average service rate $(R_i(t))$.

The Proportional Fair algorithm maximizes, over all feasible schedules, the metric $\sum_i \log R_i$, also known as the Proportional Fair metric. This metric ensures that no user is starved (since the metric will evaluate to $-\infty$ even if one user is starved). The algorithm thus tries to optimize the two objectives of fairness to individual users and the achieved system throughput. It has a runtime complexity of O($CMN + CN$), where CMN is the cost associated with first selecting one of the M users for filling up each of the CN tiles in the scheduling frame and the second term CN is the cost associated with updating $R_i(t+1)$ value of the user selected for each of the CN tiles.

10.9.5 Extension for Multi-Hop Case

For the multi-hop case, to fill up each tile (c,t), the Proportional Fair algorithm can be extended as discussed below:

1. At time t, for each user, calculate the minimum of the possible data rates over all the hops (i.e. minimum of the data rates on the designated link on each of the hops) from the base station to that user, using sub-channel c on all the hops. The rate at which this user can be serviced at time t using sub-channel c is upper bounded by this value.
2. Calculate the value $\dfrac{r_i(c, t)}{R_i(t)}$, where $r_i(c, t)$ is the value calculated in step 1 above, and $R_i(t)$ is the long-term average service rate of user i.
3. Among all the users, select the user with the maximum of the value calculated in point 2 above.

That is, to fill up each tile (c,t) in the scheduling frame, the multi-hop extension of the Proportional Fair algorithm chooses user j such that

$$j = \arg\max_i \frac{1}{R_i(t)} \min_{n=1..h} r_i(l_n, c, t)$$

where $r_i(l_n, c, t)$ is the maximum data rate for user i, possible on link l_n (the link that should be used on the nth hop to serve user i) using subchannel c at time t; h is the number of hops from the base station to user i. This extension also has a runtime complexity of $O(CMN + CN)$.

10.9.6 Performance Comparison

The performance of Round Robin (multi-hop version), Proportional Fair (multi-hop version) and Max-Route algorithms, using the $OFDM^2A$ resource allocation policy has been studied in [28]. The spectral efficiencies (in bits/Hz) of the algorithms (with $\tau = 0.25$ for Proportional Fair algorithm) are compared fixing the target outage probability at 10%. For a system with 10 users and maximum 2 hops, the spectral efficiencies were 0.4, 1.6 and 2.0 respectively. For further details, the reader is referred to [28]. The PF algorithm was able to exploit multi-hop and multi-user diversity but had lower capacity compared to Max-Route algorithm due to inherent fairness constraints.

10.9.7 The PFMR Scheduling Algorithm

Even though the Proportional Fair scheduling algorithm maximizes the Proportional Fairness metric, it does not provide any absolute user level service rate guarantees. Some applications, such as streaming video, need a minimum bandwidth for an acceptable level of performance. Also, in some cases, we may want to restrict the amount of service that any individual user receives, perhaps to encourage the user to upgrade to a more expensive service. Therefore, we would like to bound the average service rate R_i that a user receives by a minimum rate R_i^{min} and a maximum rate R_i^{max}. That is, we would like to maximize the value $\sum_i \log R_i$ subject to $R_i^{min} \leq R_i \leq R_i^{max}$ [3].

An algorithm for this problem called Proportional Fair with Minimum/ Maximum Rate Constraints (PFMR) is described in [4]. The algorithm maintains a token counter $T_i(t)$ for each user i. The role of this token counter is to enforce the rate constraints. It is updated based upon:

$$T_i(t+1) = \begin{cases} T_i(t) + R_i^{token} - r_i(t) & \text{if user } i \text{ is served at time } t \\ T_i(t) + R_i^{token} & \text{otherwise} \end{cases}$$

where $R_i^{token} = R_i^{min}$ if $T_i(t) \geqslant 0$ and $R_i^{token} = R_i^{max}$ if $T_i(t) < 0$. To fill up each tile (c,t) in the scheduling frame, the PFMR scheduling algorithm chooses the user with the maximum value of:

$$\frac{r_i(l, c, t)}{R_i(t)} e^{a_i T_i(t)}$$

where $r_i(l, c, t)$ is the maximum achievable data rate on link l using sub-channel c, for user i at time t, and $R_i(t)$ is the long term average service rate for user i at time t. $R_i(t)$ is updated according to the formula noted in equation 10.18. In the above formula, l is the link over which data has to be sent to serve user i and a_i is a parameter that determines the time scale over which the rate constraints are satisfied. If the average service rate to user i is less than R_i^{min}, then $T_i(t)$ becomes positive and so we are more likely to serve user i. If the average service rate to user i is more than R_i^{max}, then $T_i(t)$ becomes negative and so we are less likely to server user i [3]. The PFMR algorithm only increases the probability of satisfying the minimum rate and the maximum rate QoS constraints. It, however, does not guarantee these QoS constraints. This algorithm has a runtime complexity of O($CMN + CN$), where CMN is the cost associated with first selecting one of the M users for filling up each of the CN tiles in the scheduling frame (C is the number of sub-channels and N is the number of time slots in a scheduling frame) and the second term CN is the cost associated with updating $R_i(t+1)$ and $T_i(t+1)$ values of the user selected for each of the CN tiles. The extension of this algorithm to the multi-hop case is essentially the same as multi-hop extension of the Proportional Fair algorithm as detailed in section 10.9.4.

10.10   Scheduling Algorithms for WiMAX MMR Networks

In section 10.9, we presented conventional scheduling algorithms and discussed how they could be extended to the multi-hop case. Even though the multi-hop extensions presented earlier do not consider the constraints mentioned in section 10.8, these algorithms can be adapted for use in WiMAX MMR networks. In this section, we present scheduling algorithms that have been specifically designed for multi-hop networks in general and WiMAX MMR networks, in particular. These algorithms try to optimize system throughput and fairness. Please note that all of the algorithms discussed in this section assume the use of *infinitely backlogged* model where each of the users waiting for a transmission opportunity, have an infinite number of packets to transmit (i.e. they do not consider the packet arrival process).

10.10.1   The Scheduling Problem

In [8], the authors present three scheduling algorithms–*GenArgMax*, *TreeTraversingScheduler* and *FastHeuristic16j*, that solve the problem they call *PSOR* (Proportional Fair Scheduling for OFDMA Relay networks) stated below.
   Given:

- a tree topology with the base station as the root and the relay nodes as the intermediate links;
- the sustainable data rates $r(l, c)$ (in bits per time-slot) over each of the links l for every sub-channel c;
- the long-term average data rate R_m that each user m has received till the previous scheduling frame

Find: a complete schedule (in the form of a filled-up scheduling frame), subject to DFR, TR, SS and LRC constraints noted in section 10.8, such that we maximize the objective function

$$F = \sum_{m \in \mathcal{M}} \frac{d_m}{R_m} \tag{10.19}$$

where d_m is the sum of the rates at which data transfer is scheduled for user m (by the base station or the relay station that the user is directly connected to) on any of the subchannels in any of the time slots in the current scheduling frame. $\mathcal{M}$ is the set of users in the network that are waiting to be serviced.

TreeTraversingScheduler and *FastHeuristic16j* algorithms take into account the STS constraint in addition to the DFR, TR, SS and LRC constraints. The authors in [8] call the resultant problem as *16jPSOR*. So, while *GenArgMax* algorithm solves the PSOR problem (actually, a simplified version of PSOR problem, as we will see in a little while), *TreeTraversingScheduler* and *FastHeuristic16j* algorithms solve a simplified version of the 16jPSOR problem.

In [8], the authors note that this problem is NP-hard. As a result, we need to simplify this problem to make it solvable within acceptable time limits.

10.10.2   The GenArgMax Scheduling Algorithm

The *GenArgMax* algorithm uses the following two heuristics to simplify the PSOR problem:

1. The time slots in a scheduling frame are divided into multiple segments, so that the links of the multi-hop path from base station to the user are in different segments. That is, all links in the first hop of the routing tree are in segment 1, all links in the second hop of the routing tree are in second segment and so on. This heuristic is a way to simplify the PSOR problem while satisfying the TR constraint.
2. The different sub-channels on different links are assigned to users by considering users in the ascending order of the number of tiles in the scheduling frame that need to be used in serving this user for a unit increment in the objective function. Smaller the number of tiles required by a user for a unit increment in the objective function, higher will be the priority for that user.

The number of tiles required to be used in serving user i so as to achieve a unit increment in the objective function is given by $\frac{R_i(t)}{r_i(c,t)}$, where $r_i(c,t)$ can be calculated as $\min_{l \in P_m} r_i(l,c,t)$ where P_m is the set of links from user m to the base station, and c is the sub-channel (on link l) that has the largest data rate.

GenArgMax is essentially a four-step algorithm, as noted below:

Step 1 – Divide slots in a scheduling frame, into multiple segments: In this step, the time slots in the scheduling frame are split into H segments, where H is the height of the routing tree. The number of time-slots reserved for each segment is proportional to the number of users that need to use links in that hop. For example, in Figure 10.8, all the 11 users would have to use links in the first hop, nine users (m_3 to m_{11}) would have to use links in the second hop and only six users (m_6 to m_{11}) would have to use links in the third hop. So, the slots in the scheduling frame are divided into three segments in the ratio 11/26, 9/26 and 6/26. If the scheduling frame has nine time slots, four time slots would be reserved for segment 1, three slots for segments 2 and another two slots for segment 3. Figure 10.11 illustrates this segmentation done on a scheduling frame with nine time slots and three sub-channels.

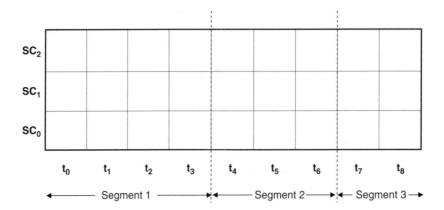

Figure 10.11   Illustration of segmenting of a scheduling frame. Reproduced with permission ©
IEEE 2009.

Step 2 – Select eligible users: In this step, all eligible users are considered for scheduling. An user m is considered eligible if, on the last hop link to this user, this user has the largest value of $\frac{r_i(c,t)}{R_i(t)}$ for some subchannel c, among all its siblings. That is, among all its directly serviced users, each relay station and base station chooses only one user to service for each subchannel. This is an important step that makes *GenArgMax* scalable. This step, regardless of the number of users contending for a transmission opportunity, fixes the number of users that are considered for scheduling at any time to be not more than $((R+1)C)$, where R is the number of relay stations and C is the number of sub-channels. Call this list of eligible users $\mathcal{M}_c$.

Step 3 – Select the most eligible user from among the eligible users: In this step, from among the eligible users (i.e. the users in $\mathcal{M}_c$), the most eligible user is selected as follows:

> **Step 3a:**   For each user, for each link in the path from the user to the base station, calculate the number of tiles required to be used on this link (using the best available subchannel), so as to increment the objective function (F) by 1 unit, by only serving this user. Note down the sub-channel used on this link to achieve this.

> **Step 3b:**   Take the maximum of the value calculated above in step 3a. This serves as the number of slots required to service this user so as to increment F by one unit.

> **Step 3c:**   Choose the user with the minimum value calculated in step 3b.

Step 4 – Allocate slots to the selected user: For the user chosen in step *3c*, allot the maximum number of tiles possible, on all the segments corresponding to the hops on which the data transmitted by the base station has to travel to reach this user (using hop-specific sub-channels noted in step 3a for this user).

Step 5 – Termination: In this step, we repeat steps 3 and 4 till there is some tile available in all the segments of the scheduling frame.

The *GenArgMax* algorithm has a runtime complexity of O(LC) where L is the number of links in the routing tree and C is the number of subchannels [8].

Generally, *line-of-sight* (LOS) communication is possible for communication between the base station and a relay station and for communication between two relay stations. As a result, the links between the base station and relay stations and between two relay stations have higher capacity (due to low path-loss and near-zero shadowing). However, communication between a subscriber station and a relay station (or the base station) that directly serves it, is *non line-of-sight* (NLOS) communication. NLOS links are subject to higher path-loss and shadowing (and so have lower capacities) compared to LOS links. Because of this reason, to support one tile's worth of communication on the last-hop NLOS link, we may only require to use a fraction of a tile on the segments corresponding to the multi-hop LOS path from the base station to this user. Consider the NLOS link k for user m_6 in Figure 10.8. Assume that the capacity of NLOS links is taken to be 0.75 of the capacity of LOS links. To support one tile on link k, we only need to allocate 0.75 tiles on the LOS links e and b.

10.10.3   The TreeTraversingScheduler Algorithm

The authors in [8] present another algorithm called *TreeTraversingScheduler* that also takes into account the STS constraint. To satisfy the LRC constraint, this algorithm restricts the amount of diversity under consideration by making the assumption that a subchannel at a relay station is dedicated for transmission to only one of the child nodes (either a subscriber station or another relay station). The algorithm traverses the routing tree in a bottom up manner and computes for every relay station/base station u, the fraction of time-slots to be assigned to relay stations in the subtrees rooted at each of its child relay stations, out of every one time-slot allocated to the subtree rooted at u. Note that the subtree rooted at u also includes node u. For every relay station/base station u, the algorithm computes three quantities i_u, c_u and t_u defined below:

- i_u: For every one time-slot allocated to the entire subtree rooted at u, the increase in the objective function F, due to data transmitted to subscriber stations in the subtree rooted at u. This term indicates the increase in F (due to data transmitted to subscriber stations in the subtree rooted at u) in a time period equal to one time-slot (that is allocated to entire subtree rooted at u).
- c_u: The total data per time-slot that has to be transmitted to the subscriber stations in the subtree rooted at u, for incrementing the objective function F by i_u units. This term indicates the rate (data per time-slot) at which we transmit data to subscriber stations in the subtree rooted at u.
- t_u: The fraction of transmission time allocated to relay stations in the subtree rooted at u for every one time-slot allocated to the subtree rooted at the parent of u.

TreeTraversingScheduler is a four-step algorithm, as described below:

Step 1 – Compute i_u and c_u for each of the leaf relay stations u (i.e. relay stations that have only subscriber stations as their children):   In section 10.10.2, we saw that for subchannel c, at time t, the number of tiles required by subscriber station m for a unit increment in the objective function to be equal to $\frac{R_m(t)}{r_m(c,t)}$. For every leaf relay node, the algorithm assigns each subchannel c to the subscriber station m that requires the least number of tiles for unit increment in the objective function F. If m_c is the subscriber station selected for subchannel c, then from definitions of c_u and i_u, we have

$$c_u = \sum_c r(l_{m_c}, c, t), \quad i_u = \sum_c \frac{r(l_{m_c}, c, t)}{R_{m_c}(t)}$$

where l_{m_c} is the link between subscriber station m_c and its parent.

Step 2 – For each intermediate relay station u, find the node (relay station or subscriber station) to which u transmits: In this step, the algorithm traverses the entire routing tree in a bottom-up manner and decides for each intermediate relay station, the child relay station/subscriber station that it has to transmit to. This decision is based on the number of tiles used up for a unit increment in the objective function. The one that takes up the least number of tiles for a unit increment is selected. The number of tiles used by a relay station (in serving the subscriber stations in the subtree rooted at this relay station) for a unit increment in the objective function is then calculated as follows.

Consider an intermediate relay node v whose parent is u. From the definition of c_v and i_v, $\frac{c_v}{i_v}$ is the amount of data required to be transmitted to v for a unit increment in the objective function. To transmit this amount of data using sub-channel c alone, relay station u needs to transmit to relay station v over subchannel c for $(c_v/i_v)/r(l_v, c, t)$ (or $\frac{c_v}{i_v r(l_v, c, t)}$) tiles, where l_v is the link between relay stations u and v. Note that this is the number of tiles required, to transmit to the subscriber stations in the subtree rooted at v. This would also incur an overhead for transmitting data from relay station u to relay station v (so as serve subscriber stations in subtree rooted at v). From the definition of i_v, this overhead is $\frac{1}{i_v}$ time slots. Because of the STS constraint, relay station u is constrained to transmit to relay station v on all the subchannels. Therefore, the overhead can be written as $\frac{C}{i_v}$, where C is the number of subchannels.

For a relay station u, the total number of tiles used by the nodes in its subtree for a unit increment in the objective function if u transmits to relay station v over subchannel c is $t_v = \frac{c_v}{i_v r(l_v, c, t)} + \frac{C}{i_v}$. As we noted previously, the number of tiles used up by a subscriber station for a unit increment of the objective function is $t_m = \frac{R_m(t)}{r_m(c, t)}$. Relay station u chooses to transmit to a relay station v or a directly associated subscriber station m, whichever takes the least number of tiles for unit increment of the objective function (i.e. least of t_v and t_m).

Step 3 – Compute i_u, c_u and t_v for each of the intermediate (non-leaf) relay stations u and for each relay station v that is a child of relay station u: Consider an intermediate relay node v whose parent is u. The data rate at which u transmits to v is given by

$$r_v = \sum_c r(l_v, c, t)$$

if u transmits to v at time t. l_v is the link between relay stations u and v. If t_v be the fraction of time allocated to the subtree rooted at v, for every unit time slot allocated to the relay nodes in the subtree rooted at u, and if t'_u be the fraction of time for which u transmits for every unit time allocated to the relay nodes in the subtree rooted at u, then, at node v, by conservation of flows, we have

$$r_v t'_u = c_v t_v \qquad (10.20)$$

Also, we have,

$$t'_u + \sum_v t_v = 1 \tag{10.21}$$

From 10.20 and 10.21, we have

$$t'_u = \frac{1}{1 + \sum_v \frac{r_v}{c_v}} \tag{10.22}$$

Consider an intermediate relay station u. From the definition of c_u, we have

$$c_u = t'_u \sum_c r(l_{n_c}, c, t) \tag{10.23}$$

where l_{n_c} is the link between relay station u and the relay station or subscriber station that u has chosen to transmit to, on subchannel c, at time t. Also, from the definition of i_u, we have

$$i_u = t'_u \left(\sum_{c:n_c \text{ is a relay station}} \frac{r(l_{n_c}, c, t) \, i_{n_c}}{c_{n_c}} + \sum_{c:n_c \text{ is a subscriber station}} \frac{r(l_{n_c}, c, t)}{R_{n_c}} \right) \tag{10.24}$$

where the first part of the sum in the above equation accounts for the increment in the objective function for a unit time slot allocated to a relay station n_c that was chosen by u to transmit to, on subchannel c, at time t, and the second part accounts for the increment in the objective function for a unit time slot allocated to a subscriber station n_c that was chosen by u to transmit to, on subchannel c, at time t. Note that the component $\frac{C}{i_v}$ noted in step 2 is not present in the first part of the sum in equation 10.24 as this component is needed only for supporting transmission to the child relay station and therefore does not directly contribute towards increment in the objective function.

Step 4 – Compute the time allocations for every relay station u: In steps 2 and 3, the entire routing tree was traversed and the value t_u was calculated for every relay station u. With this, and with the knowledge that all the N time-slots in a scheduling frame are available to the tree rooted at the base station, the entire routing tree can be traversed in a top-down manner to calculate the exact time-allocations for each of the relay stations.

This algorithm has a runtime complexity of $O(LC)$ where L is the number of links in the routing tree and C is the number of sub-channels [8].

10.10.4 The FastHeuristic16j Scheduling Algorithm

The authors in [8], define another scheduling algorithm called *FastHeuristic16j*, that solves a simplified version of the 16jPSOR problem. This algorithm is suitable when there is little or no frequency selectivity for the base station to relay station and relay station to relay station links. The algorithm has two steps: (i) Step 1: Solve the LP corresponding to 16jPSOR problem under two simplifying yet realistic assumptions noted below; (ii) Step 2: Round the LP solution without violating constraints of the 16jPSOR problem.

The assumptions used for Step 1 are as follows:

- For any relay node u, at time t, transmissions over sub-channel c to subscriber stations directly associated with u, happen only to the subscriber station i for which the ratio $\dfrac{r_i(l, c, t)}{R_i}$ is the maximum.
- Time allotted to each relay station/base station is partitioned into multiple segments–one segment for transmitting to subscriber stations that are directly serviced by this base station/relay station and one segment each for transmitting to each of the child relay stations. We note that because of the STS constraint, the selected relay station transmits on all the sub-channels of a particular time slot.

This algorithm has a runtime complexity of $O(LC)$ where L is the number of links in the routing tree and C is the number of sub-channels. The details of the LP formulation for 16jPSOR and a detailed performance study of the above three algorithms are available in [8].

10.10.5 Improved Hop-Specific Scheduling Algorithms

In some cases, algorithm *GenArgMax* wastes free tiles in a scheduling frame by not using them in serving any user. A study of such wastage and mechanisms to reduce this were originally presented in [25] and are summarized in this section.

Figure 10.12 shows an instance of a scheduling frame when *GenArgMax* terminates even though there are some free tiles. In this case, there is no free tile in segment 1, but there is at least one free tile in at least one of the other segments. As a result, no user can be scheduled (because to schedule second hop and third hop users on tiles that belong to segment 2 and segment 3, we require free tiles in segment 1 also). Algorithm *GenArgMax* fails to reclaim these free tiles since it considers users at all hops of the multi-hop network while constructing a scheduling frame. This can be remedied by only considering the users at a specific hop of the multi-hop network, while constructing a scheduling frame.

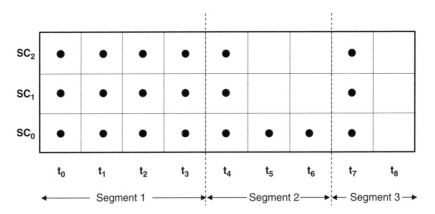

Figure 10.12   GenArgMax wastes free tiles.

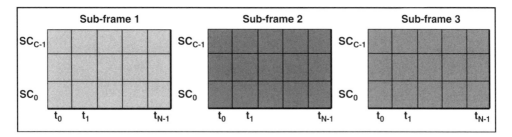

Figure 10.13 Virtual super frame for a three-hop network. Reproduced with permission © IEEE 2009.

10.10.5.1 Algorithm SuperFrame

In this algorithm [25], a virtual super frame consisting of H sub-frames (H is the height of the routing tree), is constructed. In each subframe only the users belonging to a particular hop are scheduled. For example, in the first sub-frame, only the first hop users are served, in the second subframe, only the second hop users are served and so on. The structure of this virtual super frame is shown in Figure 10.13. Algorithm *GenArgMax* is used to compute each of the sub-frames in the virtual super frame. The computed subframes are scheduled one after the other in a Round Robin fashion.

10.10.5.2 Algorithm RandomOrdered-SuperFrame

In algorithm *SuperFrame*, one subframe is constructed for scheduling users at each hop and the subframes are served in a Round Robin manner. In algorithm *RandomOrdered-SuperFrame* (*RO-SuperFrame*, for short) [25], a subframe is randomly chosen. To do this, a random number is generated in the range $1..H$ with a probability that is distributed in the proportion of the number of users at each hop. For example, in a three-hop MMR network with 40 users distributed as 14, 12 and 14 at hops 1, 2, and 3 respectively, sub-frames 1, 2, and 3 are randomly chosen with probabilities 0.35, 0.3 and 0.35 respectively.

10.10.5.3 Algorithm Cost Adjusted Proportional Fair-SuperFrame

Even though algorithm RO-SuperFrame provides fairness to users, its non-deterministic nature may not be suitable for applications that require QoS guarantees. Further, this algorithm does not consider variations in users' channel quality and therefore it does not provide proportional fairness. In algorithm *Cost Adjusted Proportional Fair-SuperFrame* (CAPF-SuperFrame) [25], the subframes within a super frame are ordered such that proportional fairness is maintained across hops. To provide this fairness, the subframes within a super frame are ordered using the following heuristic.

Step 1: In this step, the users that were serviced in the previous subframe for that hop, are chosen as representatives of all users in that hop. This is shown in Table 10.3. This table shows that for hop 1, in the previous instance of the scheduling frame for hop 1 (i.e., in subframe 1), the set of users $U_{(0,1)}$ were served on subchannel C_0, the set

Table 10.3   Choosing representatives for each hop

	C_0	C_1	C_2
Sub-frame 1	$U_{(0,1)}$	$U_{(1,1)}$	$U_{(2,1)}$
Sub-frame 2	$U_{(0,2)}$	$U_{(1,2)}$	$U_{(2,2)}$
Sub-frame 3	$U_{(0,3)}$	$U_{(1,3)}$	$U_{(2,3)}$

of users $U_{(1,1)}$ were served on subchannel C_1, and so on. These users are chosen as representatives of all users at hop 1.

Step 2:   In this step, for each user $i_{(c,k)}$ in the set of users $U_{(c,k)}$ selected for each subframe k and each sub-channel c, the modified proportional fairness value $effHopGain_{i_{(c,k)}} * r_{i_{(c,k)}}(l,c,t)/R_{i_{(c,k)}}(t)$ is calculated. The term $effHopGain$, explained below, compensates the overhead of scheduling users at hops further away from the base station. For each subframe k and for each subchannel c, the user $i'_{(c,k)}$ that has the largest value of the modified proportional fairness value, among all the users $i_{(c,k)}$ in set $U_{(c,k)}$, is then chosen.

Step 3:   In this step, for each subchannel c, the sub-frame k' is chosen such that the user $i'_{(c,k')}$ has the largest value of the modified proportional fairness value, among all users chosen for subchannel c, in any of the subframes. This is repeated for all sub-channels. Finally, the sub-frame k' that was selected for a majority of the subchannels is chosen. This is the subframe that we consider next for scheduling.

An important point to note is that the cost of scheduling users that are n hops away from the base station, increases with the value of n. This is because, to serve a user at the nth hop, tiles need to be reserved in segments $n-1, n-2, \ldots, 1$ (for supporting the multi-hop communication), in addition to reserving tiles in segment n (for the actual communication). Therefore, gain in system throughput when considering only single-hop users is higher than when considering only two-hop users (which is higher than when considering only three-hop users) and so on.

To accommodate this variation in gain at different hops, the factor $effHopGain_{i_{(c,k)}}$ is used in Step 3 of algorithm *CAPF-SuperFrame*. The factor $effHopGain_{i_{(c,k)}}$ for user $i_{(c,k)}$ in $U(c,k)$, is calculated as follows: let $TilesGained$ = average capacity of LOS links/$r_{i_{(c,k)}}(l,c,t)$. To support $TilesGained$ worth of communication to user $i_{(c,k)}$ at hop k, one tile would have to be used in each of the $k-1$ hops. Therefore, let $TilesLost = k-1$. Then, $effHopGain_{i_{(c,k)}} = TilesGained/(TilesGained + TilesLost)$. The factor $effHopGain_{i_{(c,k)}}$ adjusts the proportional fairness value of user i_k considering the cost associated with transmitting to this user i_k which is located at hop k.

As with algorithm *GenArgMax*, algorithms *SuperFrame*, *RO-SuperFrame* and *CAPF-SuperFrame* have runtime complexity of $(O|\mathcal{L}| |\mathcal{C}|)$ [25].

10.10.6   Performance Evaluation

The authors in [8] have studied the performance of *GenArgMax*, *TreeTraversingScheduler* and *FastHeuristic16j* scheduling algorithms in the setting of a 120° sector of radius 1 km consisting of three relay stations that are placed equidistant from each other along an arc of radius 0.8 km. Their analysis showed that *GenArgMax*, *TreeTraversingScheduler* and *FastHeuristic16j* algorithms perform close to the optimum (off by less than 0.5 %) [8]. Further, the authors in [8] note that the STS constraint imposed by IEEE 802.16j framework does not result in significant performance degradation. The running time of these algorithms, as measured on a Intel Centrino Core 2 Duo machine running at 2 GHz, with 1 GB RAM, was of the order of microseconds and so the deadline of 5 ms can be easily met [8].

Figure 10.14 [25] presents the system throughput and the proportional fairness metric for the hop-specific scheduling algorithms discussed in section 10.10.5, for two-hop generic OFDM relay networks. With reference to the average values, we note that the system throughput obtained in case of algorithms *SuperFrame*, *RO-SuperFrame* and *CAPF-SuperFrame* are 0.73 %, 8.19 % and 8.94 % higher respectively, than the system

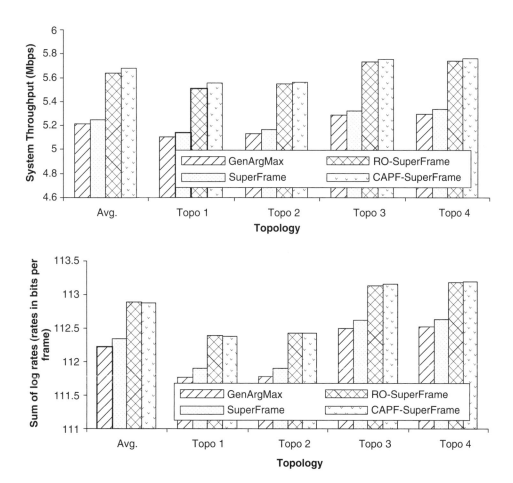

Figure 10.14 System throughput and proportional fairness metric for two-hop generic OFDM relay networks. Reproduced with permission © IEEE 2009.

Table 10.4 Run times of the hop-specific scheduling algorithms for two-hop generic OFDM relay networks

Algorithm	Average runtime (ms)
GenArgMax	0.029
SuperFrame	0.017
RO-SuperFrame	0.013
CAPF-SuperFrame	0.021

Reprinted with permission © IEEE 2009

throughput for algorithm *GenArgMax*; and that the proportional fairness metric for algorithms *SuperFrame*, *RO-SuperFrame* and *CAPF-SuperFrame* are 0.11 %, 0.59 %, and 0.59 % higher respectively, than the proportional fairness metric for algorithm *GenArgMax*. The average running time of these algorithms is summarized in Table 10.4. Additional results are available in [25].

10.11   Further Reading

A survey of scheduling theory as applicable to wireless data networks is presented in [3]. In this survey, the authors describe some of the models that have been proposed for modeling wireless data networks and analyze the performance of several scheduling algorithms that use these models. Some of the models considered are infinitely backlogged queues and stochastic arrival process (for modelling arrival times) and stationary stochastic process and worst-case adversary model (for modeling channel conditions). In [5], the authors study the performance of many variants of the *Max-Weight* scheduling algorithm in a situation where finite queues are fed by a data arrival process. The *Max-Weight* algorithm always serves the user that maximizes $Q_i^s(t)r_i(t)$ at each time step t, where $Q_i^s(t)$ is the queue size of user i at the beginning of time slot t [5]. These variants extend the *Max-Weight* algorithm (which works in a single-carrier setting) to the multi-carrier setting.

In this chapter we have assumed the availability of routing information (i.e. information on which relay station to transmit to, to reach a particular subscriber station) at the base station. In [7], the authors propose two routing algorithms (*Fixed-hop-count routing* and *Opportunistic-hop-count routing*) for routing between the base station and subscriber stations, and analyze the system-level performance of multi-cellular multi-hop networks in the presence of co-channel interference. The authors model statistically the co-channel interference in a downlink multi-cellular multi-hop communication setting accounting for random transmissions from multiple (possibly sectorized) base stations and omni-directional relay stations, and study the total obtainable capacity under different multi-user scheduling algorithms such as MaxCap, Proportional Fair, and Round Robin, in the multi-hop cellular network based on this co-channel interference model.

In [26], the authors consider the design of multi-hop wireless backhaul networks with delay guarantees. The authors propose a generalized link activation framework (called *even-odd* framework) that reduces interference and maps a wireless backhaul to a half-idle wireline network. The authors also propose optimal and heuristic backhaul routing algorithms and show that when a multi-hop wireline scheduler with worst case delay bounds (such as WFQ or Coordinated EDF) is implemented over a wireless backhaul, the even-odd framework guarantees approximately twice the delay compared to the corresponding wireline topology.

In this chapter, we have only considered scheduling for the downlink. In [20], the authors consider a traffic adaptive uplink scheduling scheme for relay stations in a WiMAX MMR network. The authors argue that signalling overhead (and the latency due to this) in conventional uplink scheduling is very high, and propose new uplink scheduling algorithms to minimize this signalling overhead. The authors propose a technique where relay stations preallocate bandwidth for the relay station to base station uplink communication and the relay station uses this to forward data to the base station as soon as it receives data from subscriber station. To avoid wastage of pre-allocated bandwidth in the case that subscriber station to relay station communication fails, the authors propose traffic-dependent uplink scheduling algorithm that can both avoid resource wastage and minimize the delay and the signaling overhead. The authors propose two different strategies—one for real-time traffic and the other for non real-time traffic.

10.12 Summary

This chapter presented a survey of multimedia traffic scheduling and multi-hop relay based scheduling in WiMAX networks. The characteristics and requirements of multimedia traffic were first presented. This was followed by a description of scheduling algorithms that support QoS in WiMAX networks. The chapter then described the IEEE 802.16j standards related WiMAX mobile multi-hop relay (MMR) networks and related scheduling algorithms. Further work is necessary in the context of designing and evaluating scheduling algorithms that consider other traffic types such as mobile TV, games and also relay networks that consider peer-to-peer relays and mobile relays.

References

[1] D.P. Agrawal and Q. Zeng, *Introduction to Wireless and Mobile System*. Thomson, Brooks/Cole, 2003.

[2] Ahmed K.F. Khattab and K.M.F. Elsayed, Opportunistic Scheduling of Delay Sensitive Traffic in OFDMA-Based Wireless. In *WOWMOM '06: IEEE Proceedings of the 2006 International Symposium on World of Wireless, Mobile and Multimedia Networks*, pp. 279–88, 2006.

[3] M. Andrews, A Survey of Scheduling Theory in Wireless Data Networks. In *Proceedings of the 2005 IMA summer workshop on wireless communications*, 2005.

[4] M. Andrews, L. Qian and A. Stolyar, Optimal Utility Based Multi-user Throughput Allocation Subject to Throughput Constraints. In *Proceedings of IEEE INFOCOM*, Vol. 4, pp. 2415–24, March 2005.

[5] M. Andrews and L. Zhang, Scheduling Algorithms for Multi-Carrier Wireless Data Systems. In *Proceedings of MobiCom '07: Proceedings of the 13th annual ACM international conference on Mobile computing and networking*, pp. 3–14, New York, NY, USA, 2007. ACM.

[6] J.L. Burbank and W.T. Kasch, Chapter 1: WiMAX Past, Present, and Future: An Evolutionary Look at the History and Future of Standardized Broadband Wireless Access. In Syed Ahson and Mohammad Ilyas, editors, *WiMAX: Applications*, pp. 1–14. CRC Press (Taylor and Francis Group), 2008.

[7] M. Charafeddine, O. Oyman and S. Sandhu, System-Level Performance of Cellular Multihop Relaying with Multiuser Scheduling. In *Proceedings of Annual Conference on Information Sciences and Systems (CISS)*, pp. 631–6, March 2007.

[8] S. Deb, Vi. Mhatre and Ve. Ramaiyan, WiMAX Relay Networks: Opportunistic Scheduling to Exploit Multiuser Diversity and Frequency Selectivity. In *Proceedings of the 14th ACM international conference on Mobile computing and networking (MobiCom)*, pp. 163–74, New York, NY, USA, 2008. ACM.

[9] Do. Zhao and X. Shen, Performance of Packet Voice Transmission using IEEE 802.16 Protocol. *IEEE Wireless Communications* **14**(1): 44–51, Feb. 2007.

[10] E. Haghani and N. Ansari, VoIP Traffic Scheduling in WiMAX Networks. In *Proceedings of IEEE GLOBECOM*, pp. 1–5, December 2008.

[11] F. Ohrtman, *How WiMAX Works*. McGraw Hill Communications, 2005.

[12] R. Ganesh, S.L. Kota, K. Pahlavan and R. Agusti, Chapter 5: Fixed Relays for Next Generation Wireless Systems. In Norbert Esseling, B.H. Walke and R. Pabst, editors, *Emerging Location Aware Broadband Wireless Ad Hoc Networks*, pp. 72–93. Springer US, 2005.

[13] E. Haghani, S. De and N. Ansari, On Modeling VoIP Traffic in Broadband Networks. *Proceedings of IEEE Global Telecommunications Conference (GLOBECOM)*, pp. 1922–26, Nov. 2007.

[14] H. Lee T. Kwon and D.-H. Cho, Extended-rtPS Algorithm for VoIP Services in IEEE 802.16 Systems. In *Proceedings of IEEE International Conference on Communications (ICC)*, Vol. 5, pp. 2060–65, June 2006.

[15] C. Hoyman, K. Klagges and M. Schinnenburg, Multihop Communication in Relay Enhanced IEEE 802.16 Networks. In *Proceedings of IEEE International Symposium on Personal, Indoor and Mobile Radio Communications (PIMRC)*, pp. 1–4, 2006.

[16] IEEE 802.16-2004–IEEE Standard for Local and Metropolitan Area Networks Part 16: Air Interface for Fixed Broadband Wireless Access Systems, 2004.

[17] IEEE 802.16e-2005–IEEE Standard for Local and Metropolitan Area Networks Part 16: Air Interface for Fixed and Mobile Broadband Wireless Access Systems Amendment for Physical and Medium Access Control Layers for Combined Fixed and Mobile Operation in Licensed Bands, 2005.

[18] IEEE 802.16 Task Group m (TGm). http://wirelessman.org/tgm/, July 2009.

[19] J. Woo So, Performance Analysis of Uplink Scheduling Algorithms for VoIP Services in the IEEE 802.16e OFDMA System. *International Journal on Wireless Communication Networks* **47**(2): 247–63, 2008.

[20] O. Jo and D.-H. Cho, Traffic Adaptive Uplink Scheduling Scheme for Relay Station in IEEE 802.16 Based Multi-Hop System. In *Proceedings of IEEE VTC Fall*, pp. 1679–83, 2007.

[21] L.E.G. Richardson, *The MPEG-4 and H.264 Standards*. John Wiley & Sons Ltd, 2003.

[22] H. Lee, T. Kwon and D.-H. Cho, An Efficient Uplink Scheduling Algorithm for VoIP Services in IEEE 802.16 BWA Systems. *In Proceedings of IEEE Vehicular Technology Conference-Fall*, **5**: 3070–4, Sept. 2004.

[23] M. Chatterjee, S. Sengupta and S. Ganguly, Feedback-based Real-time Streaming over WiMAX. *IEEE Wireless Communications*. **14**(1): 64–71, Feb. 2007.

[24] M. Chatterjee and Sh. Sengupta, Chapter 4: VoIP over WiMAX. In S. Ahson and M. Ilyas, editor, *WiMAX Applications*, pp. 60–5. CRC Press, 2008.

[25] Sr. Narasimha, K.M. Sivalingam, Opportunistic Scheduling Algorithms for WiMAX Mobile Multihop Relay Networks. M. Tech. Project, Indian Institute of Technology Madras, Chennai, India, June 2009.

[26] G. Narlikar, G. Wilfong and L. Zhang, Designing Multihop Wireless Backhaul Networks with Delay Guarantees. In *Proceedings of IEEE INFOCOM 2006*, pp. 1–12, April 2006.

[27] L. Nuaymi, *WiMAX: Technology for Broadband Wireless Access*. John Wiley & Sons, 2007.

[28] O. Oyman, OFDM2A: A Centralized Resource Allocation Policy for Cellular Multi-hop Networks. *Proceedings of Fortieth Asilomar Conference on Signals, Systems and Computers (ACSSC)*, pp. 656–60, Nov 2006.

[29] F. Ohrtman, *WiMAX Handbook–Building 802.16 Wireless Networks*. McGraw-Hill Communications, 2005.

[30] O. Hersent, D. Gurle and J.-P. Petit, *IP Telephony: Packet-Based Multimedia Communications Systems*. Addison-Wesley, Washington, DC, USA, 2000.

[31] S.W. Peter and R.W. Heath, The Future of WiMAX: Multihop Relaying with IEEE 802.16j. *IEEE Communications Magazine* **47**(1): 104–11, January 2009.

[32] H.K. Rath, A. Bhorkar and Vishal Sharma, An Opportunistic Uplink Scheduling Scheme to Achieve Bandwidth Fairness and Delay for Multiclass Traffic in WiMAX (IEEE 802.16) Broadband Wireless Networks. In *Proceedings of IEEE Global Telecommunications Conference (GLOBECOM)*, pp. 1–5, December 2006.

[33] S. Deb, S. Jaiswal and K. Nagaraj, Real-Time Video Multicast in WiMAX Networks. *INFOCOM 2008. IEEE 27th Conference on Computer Communications*, pp. 1579–87, April 2008.

[34] K. Sai Suhas, M. Sai Rupak, K.V. Sridharan and K.M. Sivalingam, Scheduling Algorithms for WiMAX Networks: Simulator Development and Performance Study. In *Emerging Wireless LANs, Wireless PANs, and Wireless MANs*, Edited by Y. Xiao and Y. Pan, pp. 108–117. John Wiley and Sons, Inc., New York, NY, USA, 2009.

[35] S. Sengupta, M. Chatterjee, S. Ganguly and R. Izmailov, Improving R-Score of VoIP Streams over WiMAX. *IEEE International Conference on Communications (ICC)*. **2**: 66–71, June 2006.

[36] WiMAX Forum, http://www.wimaxforum.org, July 2009.

11

On the Integration of WiFi and WiMAX Networks

Tara Ali Yahiya and Hakima Chaouchi
Computer Science Laboratory, Paris-Sud 11 University, France
Telecom and Management Sud Paris, Evry cedex, France

11.1 Introduction

The next generation of networks will be seen as a new initiative designed to bring together all heterogeneous wireless and wired systems under the same framework, to provide connectivity anytime and anywhere using any available technology. Network convergence is therefore regarded as the next major challenge in the evolution of telecommunications technologies and the integration of the computer and communications.

During recent years, IEEE802.11 Wireless Local Area Networks have been deployed widely and 802.11 access points (APs) are able to cover areas of a few thousand square meters, making them suitable for enterprise networks and public hot spot scenarios such as airports and hotels. Recently, WiMAX using the IEEE802.16e standard received a great deal of attention because of the high rate of data support, its intrinsic QoS and mobility capabilities and a much wider area of coverage enabling ubiquitous connectivity. An interworking between these technologies has been considered as a viable option for the realization of the 4G scenario. However, this interoperation raises several challenges especially when seamless session continuity is required for, for example, media calls such as VoIP or video telephony.

Since the WiMAX and the WiFi networks have different protocol architectures and QoS support mechanisms, protocol adaptation would be required for their interworking. For example, with a layer 2 approach, adaptation would be required in the medium access control (MAC) layer for the WiMAX BS and WiFi nodes. With a layer 3 approach, the adaptation would be performed at the IP layer and a WiFi user would interact only with the corresponding WiFi AP/router (as in Figure 11.1). This layer 3 approach is preferred for

WiMAX Security and Quality of Service: An End-to-End Perspective Edited by Seok-Yee Tang,
Peter Müller and Hamid Sharif

the WiMAX/WiFi integrated network, since WiFi APs/routers can fully control bandwidth allocation among the nodes. Since a WiFi AP/router is responsible for protocol adaptation up to the IP layer, modifications of WiFi user equipment and the WiMAX BS (in hardware and/or software) are not required.

The deployment of an architecture that allows users to switch seamlessly between these two types of network would provide several advantages to both users and service providers. By offering integrated WiFi/WiMAX services, users would benefit from enhanced performance and the high data rate of such a combined service. For the providers, it could capitalize on their investment, attract a wider user base and ultimately facilitate the ubiquitous introduction of high speed wireless data. The required WiFi access network may be owned either by the WiMAX operator or by any other party, which requires proper rules and SLAs set up for smooth interworking on the basis of business and roaming agreements between the WiFi and WiMAX operators. Ongoing efforts are being made in IEEE802.21 WG in order to integrate different types of network by introducing MIH (media independent handover) which aims to achieve a seamless handoff among different wireless networks regardless of the type of technology [1].

We begin this chapter with an outline of the design tenets for an interworking architecture between both WiFi and WiMAX technologies. We then define the various functional entities and their interconnections. Next, we discuss end-to-end protocol layering in the interworking architecture, network selection and discovery and IP address allocation. We then describe in more detail the functional architecture and processes associated with security, QoS and mobility management.

11.2 General Design Principles of the Interworking Architecture

The development of the WiFi/WiMAX interworking architecture followed several design tenets, most of which are based on 3GPP. 3GPP2 works with loosely and tightly coupled architectures. However, some of the important design principles that guided the development of interworking architecture include the following.

11.2.1 Functional Decomposition

The interworking architecture will be based on functional decomposition principles, where the required features are broken down into functional entities.

11.2.2 Deployment Modularity and Flexibility

The internetworking architecture will be modular and flexible enough so as not to preclude a broad range of implementation and deployment options. The access network for both networks may be broken down in many ways and multiple types of decomposition topologies may coexist within a single access network. The architecture will range from a single operator with a single base station to a large-scale deployment by multiple operators with roaming agreements.

11.2.3 Support for Variety of Usage Models

The interworking architecture will support the coexistence of fixed, nomadic, portable and mobile usage including all of the versions of IEEE 802.16e and IEEE 802.11. The interworking architecture will also support seamless handover for different levels of mobility and end-to-end QoS and security support.

11.2.4 Extensive use of IETF Protocols

The network-layer procedures and protocols used across the architecture will be based on the appropriate IETF RFCs. End-to-end security, QoS, mobility, management, provisioning and other functions will rely as far as possible on existing IETF protocols. Extensions may be made to existing RFCs, if necessary.

11.3 WiFi/Mobile WiMAX Interworking Architecture

Figure 11.1 shows the interworking architecture of WiFi/WiMAX which is based on loosely coupled architecture. The necessary changes in both WiFi and Mobile WiMAX systems are rather limited as they will integrate both systems at the IP layer and rely on the IP protocol to handle mobility between access networks.

The main characteristic of this architecture is to assume two overlapped cells of a Mobile WiMAX and a WiFi, where both cells are served by a BS and an Access Point (AP) respectively. We assume that the AP is connected to the WiFi access network which can have a dedicated gateway to the Mobile WiMAX. Traffic from Mobile WiMAX to WiFi or vice versa will be routed through this gateway. The MN has dual interfaces: WiMAX and WiFi.

As shown in Figure 11.1, the Mobile WiMAX supports access to a variety of IP multimedia services via WiMAX radio access technologies which are called the Access Service network (ASN) [2]. The ASN is owned by a Network Access Provider (NAP) and comprises one or more BS and one or more ASN gateways (ASN-GW) that form the radio access network. Access control and traffic routing for MSs in Mobile WiMAX is handled entirely by the Connectivity Service Network (CSN), which is owned by a Network Service Provider (NSP) and provides IP connectivity and all the IP core network functions. The WiFi access network may be owned either by the NAP or by any other part (e.g. public WiFi operator or an airport authority), in which case the interworking is enabled and governed by the appropriate business and roaming agreement.

For the purpose of enabling the interworking of WiFi/Mobile WiMAX, the Mobile WiMAX CSN core network incorporates four new functional elements: the Mobile WiMAX AAA server, the CSN-GW, the WiFi access gateway (WAG) and the Packet Data Gateway (PDG). The WiFi must also support a similar interworking functionality so as to meet the access control and routing enforcement requirements. The Mobile WiMAX AAA server in the Mobile WiMAX domain terminates all AAA signalling originated in the WiFi that pertains to HMS. This signalling is typically based on Radius

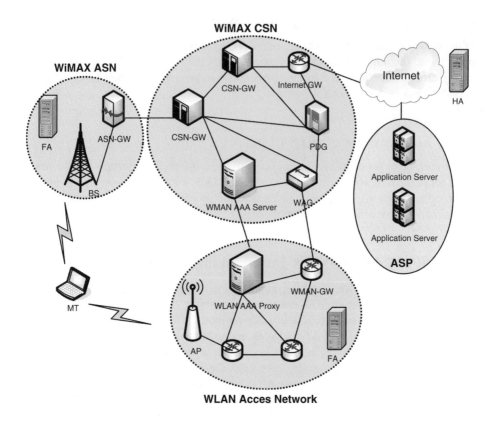

Figure 11.1 WiFi/WiMAX interworking architecture.

[3] or Diameter [4]. The Mobile WiMAX AAA server interfaces with other Mobile WiMAX components, such as the WAG and PDG. The Mobile WiMAX AAA server can also route AAA signalling to or from another Mobile WiMAX AAA server, in which case it serves as a proxy and is referred to as the Mobile WiMAX AAA proxy. To support mobility, the Foreign Agents (FA) located in ASN Gateway are considered as the local FAs in the interworking architecture. However, for enabling vertical handover, a Mobile IP Home Agent (HA) has been added to the architecture. While the HA may be local by either network, it must be accessible by both networks.

As shown in Figure 11.1, traffic from HMS is routed to the WAG and finally to the PDG. This routing is enforced by establishing appropriate traffic tunnels after a successful access control procedure. The PDG function is much like a CSN-GW in Mobile WiMAX domain. It routes the user data traffic between the HMS and an external packet data network (in our case, the IP multimedia network) and serves as an anchor point that hides the mobility of the HMSs within the WiFi domain. The WAG functions mainly as a route policy element, ensuring that user data traffic from authorized HMS is routed to the appropriate PDGs, located in either the same or a foreign Mobile WiMAX.

11.4   Network Discovery and Selection

The interworking architecture is required to support automatic selection of the appropriate network, based on MT preference. It is assumed that an MT will operate in an environment in which multiple networks are available for it to connect to and where multiple service providers are offering services over the available networks. To facilitate such an operation, the following principles have been identified regarding multi-access network selection (between WiFi and WiMAX) and discovery when both access networks are available:

- The interworking architecture may provide the mobile terminal with assistance data/policies about available access to allow the mobile terminal to scan for and select access.
- The interworking architecture allows the home and visited operator to influence the access that the mobile terminal will hand off (when in active mode) or re-select (when in idle mode).
- Multi-access network discovery and selection works for both single-radio and multiple-radio terminals.
- No architectural impact is foreseen for network selection – upon initial network attachment.

Figure 11.2 shows the architecture for Access Network Discovery Support Functions (ANDSF) which may be used for access network discovery and selection [5]. The ANDSF contains the data management and control functionality necessary to provide network discovery and selection assistance data as per operators' policies. The ANDSF is able to initiate data transfer to the MT, based on network triggers, and respond to requests from the MT.

A part of the network selection process is the IP address assignment for the MT when it moves from one network to another. Usually, the Dynamic Host Control Protocol (DHCP)

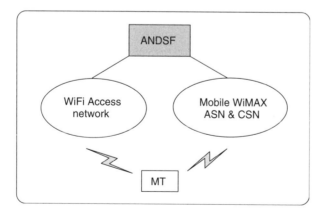

Figure 11.2   Architecture for network discovery.

is used as the primary mechanism to allocate a dynamic point-of-attachment (PoA) IP address to the MT. The DCHP server can reside in any part of the network, ASN, CSN, etc.

11.5 Authentication and Security Architecture

As for any network access connectivity, authentication is the first communication exchange within the network as soon as the physical layer of a terminal is connected to the access media. In fact the network has to identify the device/terminal and then authorize it, based on the user's contract, to use network resources and finally account for the used resources so as to bill the user. In wireless connectivity, it will also be mandatory to provide link layer encryption for every packet sent on the air (radio) since radio can easily be eavesdropped upon. So both authentication and confidentiality completed by integrity control based schemas will have to be deployed in both WiFi and WiMAX networks and based on the integration level of these two networks, the interaction between the security framework has to be adapted. In the following we'll describe the classical AAA (Authentication, Authorisation, Accounting) architecture followed by a brief description of WiFi and WiMAX security solutions and by security considerations in the integrated WiFi-WiMAX integrated architecture.

11.5.1 General Network Access Control Architecture

Within the deployment of charged network services, the network operator puts in place an architecture known as AAA (Authentication, Authorization, Accounting). Authentication identifies the user requesting access to network services. Authorization limits the user's access to permitted services only. Accounting calculates the network resources that are consumed by the user.

The AAA architecture creates interactions between three entities, as shown in Figure 11.3: the user terminal, the AAA client installed at the access router of the operator and the AAA server installed in the operator's network.

The terminal interacts with the access router. In the case where a terminal connects from a switched network (PSTN, ISDN, GSM), the access router becomes a NAS (Network Access Server) gateway that ensures connectivity between the switched network and IP network. Once it is physically connected to the network, the user terminal is authenticated. At the beginning of a communication between the terminal and the network, only those packets belonging to the authentication protocol and addressed to the AAA server are authorized and relayed by the NAS. Upon a successful authentication, the NAS authorizes other packets coming from the user terminal to go through. This is made possible by the configuration of two ports at the NAS: a controlled port and an uncontrolled port. During the authentication phase, the traffic is going through the controlled port which recognizes the authentication traffic and lets it go through. After user authentication, the traffic goes through the second port.

From the operator's point of view, the AAA client located on the NAS captures the authentication messages (e.g. EAP: Extensible Authentication Protocol) coming from the terminal, encapsulates them into AAA messages and sends AAA messages to the AAA

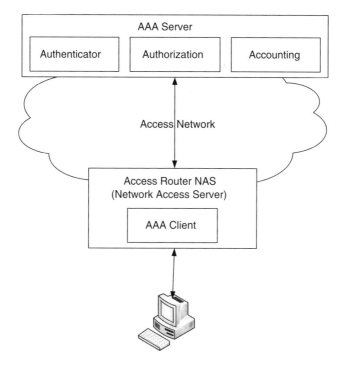

Figure 11.3   AAA architecture.

server. The AAA server accesses a database that stores all the information relative to the users and necessary for authentication. In general, the AAA server and the terminal share a secret that allows the AAA server to authenticate the user. Other authentication methods are also possible.

In the context of roaming, the AAA architecture defines domains of authentication. Each domain has its AAA server. A mobile user is registered with its home AAA (AAAH) server where it subscribed, and can be authenticated by any visited network or domain through an inter-domain AAA protocol where a roaming contract has previously been signed. This inter-domain authentication is conducted by AAA broker, as presented in Figure 11.4, running an inter-domain AAA protocol.

The IETF has standardized protocols designed to implement AAA functions for both inter-domain and intra-domain situations:

- The interface terminal-NAS: two protocols are now envisaged for the transport of link layer authentication messages, namely 802.1X and PANA (Protocol for carrying Authentication for Network Access).
- The interface NAS-AAA server for intra-domain which is provided by the RADIUS protocol.
- The interface between AAA servers for inter-domain which is implemented by the Diameter protocol.

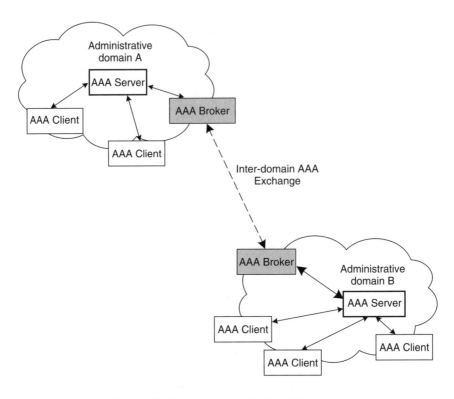

Figure 11.4   Inter-domain AAA architecture.

11.5.2   EAP and PANA

The AAA service usually requires a link layer protocol between the terminal and the access network. EAP (Extensible Authentication Protocol) is one of the most commonly used link layer authentication protocols which replaced the existing PAP (PPP Authentication Protocol) and CHAP (Challenge based Authentication Protocol) link layer authentication. It is used to authenticate the terminal before it gets an IP address, triggering authentication at the link layer through a controlled port at the NAS point (access network) and blocking IP address allocation until the authentication is successful. In WiFi networks, IEEE 802.1X is the authentication scheme standard, which applies the EAP protocol combined with the RADIUS protocol to the AAA server. An alternative to the link layer protocol is PANA (Protocol for carrying Authentication for Network Access). It works over UDP and needs to obtain an IP address before proceeding with the authentication of the terminal. The PANA protocol encapsulates the EAP protocol, like 802.1X, but unlike 802.1X, PANA is applicable to any type of network access (WiFi, WiMAX ...) when an IP connection can be mounted. However, it is necessary to ensure that the access network accepts only PANA messages at the beginning of the connection until the terminal is authenticated successfully. This is not straightforward because, unlike EAP which proceeds with the authentication before getting the IP address, PANA runs over IP and therefore cannot block other application messages at the entrance of the network unless a special filter is

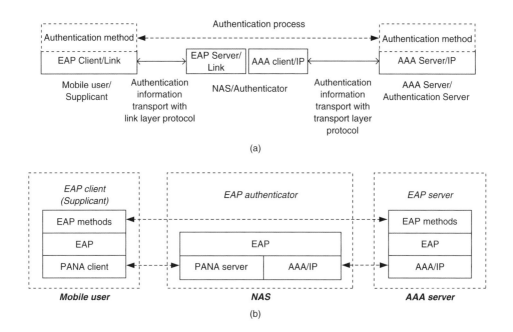

Figure 11.5   (a) EAP architecture (b) EAP architecture with PANA at network access.

installed to allow PANA packets during the authentication and block any other packet, allowing all the packets when the user is authenticated. In EAP, no IP address is allocated during the authentication phase, so no application packets can go through the network until the authentication is finished.

As shown in Figures 11.5.a and 11.5.b, the architecture of EAP at the link and network layers [10, 11] involves an authentication (Authenticator) at the NAS that communicates with the supplicant entity at the terminal using the EAP protocol. The server sends an authentication request to the terminal. The request depends on the authentication method. The identity of the user is known as NAI (Network Access Identifier) send by the terminal during the first connection to the network and based on this NAI, the AAA server can choose the authentication method stored in the AAA database where the user is already registered. In this architecture, the authentication server or NAS acts as a bridge between the terminal and the AAA server during the phase of user authentication. This is mainly to avoid, for security reasons, direct communication between terminals and the AAA server. Once the authentication is completed successfully, the terminal gets an IP address and is authorized to issue traffic to the network.

11.5.3   RADIUS and Diameter

RADIUS (Remote Authentication Dial In User Service) has been designed for intra-domain AAA service [12]. It uses IPsec between its various entities: the RADIUS client and RADIUS server. The RADIUS client in the NAS receives the request to connect to the network, initiates the process of authentication and transfers authentication messages between the terminal and the RADIUS server. The RADIUS server stores the information needed to authenticate the user. Different authentication algorithms can be used.

With the mobility of users, the Diameter protocol was developed by the IETF to handle the AAA inter-domain authentication scheme since the Radius message format doesn't allow the transport of inter-domain messages. The RADIUS protocol is limited technically to intra-domain authentication and Diameter can be seen as an enhanced and scalable version of RADIUS [4]. The support of inter-domain mobility, the support for quality of service and the extension to accounting are some of the extensions implemented in Diameter.

Note that RADIUS and Diameter protocols are authentication and authorization protocols but are limited somewhat in the support of authorization compared to other protocols such as the COPS protocol (Common Open Policy Service) where different authorization policies may be expressed. [13] provides an AAA protocol evaluation as between RADIUS, Diameter and COPS.

11.6   Security in WiFi and WiMAX Networks

11.6.1   Security in WiFi

The first generation of WiFi came with a simple security schema called WEP (Wired Equivalent Privacy) where all the communication is encrypted at the link layer with some semi static key and RC4 (Rivest Cipher 4) cipher algorithm using mainly a simple XOR operation between a message and the WEP built key. The weakness of this schema, which is still used in many WiFi communications today, lies in the key semi dynamicity. In fact the key used in WEP protection has only a small part which is dynamic and owing to the power of calculation of current processors it is possible after a certain number of packets to guess the key and thus corrupt the WEP protection. The available key size is 128 or 256 bits [16].

In the second generation of WiFi security (in the absence of the standard of security in IEEE 802.11i which was not yet ready at that time) in order to strengthen the authentication architecture the IEEE standardized the process under the IEEE 802.1x framework – where link layer authentication is carried with the EAP protocol and where the WiFi Access point acts similar to a NAS and runs the AAA client to communicate through radius in order to proceed with the AAA authentication server. More precisely, in 2002, before the 802.11i standard was fully ratified, a new protocol was introduced. The WiFi Protected Access (WPA) method implements stronger encryption algorithms and provides two usage levels. WPA-Personal is used in situations where there is no server for authentication. A 'pre-shared key' (PSK) is created in order to authorize contact. This PSK is a phrase, eight to 63 characters long, or a hexadecimal string up to 64 characters long, which is shared manually between access point and client. When an authentication server is available, WPA utilizes the 802.1X protocol to communicate with the server and assigns dynamically a different key for each attached device. A RADIUS server can be used to handle verification of those requesting access and enforce policies for access. WPA provides greater security by changing the key often. Therefore, if a key is discovered by an outside entity, that key provides only a limited window of access to the network.

To do this, WPA uses the Temporal Key Integrity Protocol (TKIP) to resets the 128-bit keys periodically. Optionally, the protocol provides support for AES-CCMP (Advanced Encryption Standard-Counter mode for the Cipher Block Chaining/CBC based MAC Protocol), a very strong security protocol that handles four security aspects (authentication, confidentiality, replay protection and integrity) [14].

Finally, the security standard of WiFi was ratified in June 2004. IEEE 802.11i is last generation of security in WiFi under the IEEE 802.11i standard and is named WPA2 offering not only the dynamicity of the key, but also a new strong encryption standard aka AES (Advanced Encryption Standard). So as long as AES remains robust; it is predicted to continue to be strong until 2050 depending on the evolution of the speed of processors. So until then we may consider that the security offered in Wifi is robust. WPA2 implements the required elements of the final ratification of the 802.11i specification that occurred in June of 2004. The primary difference between WPA and WPA2 is that the support of AES-CCMP protocol, optional in WPA, is now required in WPA2. Interestingly enough, WPA2 is not backwards compatible, meaning that clients and access points must be reconfigured in order to switch between the two versions. The final security device, known in WiFi communication, is the use of smart cards and USB tokens. Most of these devices carry strong forms of encryption that combine two or more types of authentication, such as biometrics and a password. Although this form of security is considered by some to be the strongest and safest, a drawback might be that it can be quite expensive. The higher cost is due to the need for purchasing the physical devices for each employee and member of the team as well as supporting the authentication methods that are selected.

In the context of mobility in WiFi, the IEEE 802.11r working group proposed a fast-based station security process where the key is generated in the previous access point before moving from the current one. This is an amendment to the IEEE 802.11 standard to allow fast authentication and minimize the delay of re-authentication before attachment to the next access point in the case of a handover.

11.6.2 Security in WiMAX

WiMAX was designed as a solution for the 'last mile' of a Wireless Metropolitan Area Network (WMAN) that would bring internet access to an entire metropolitan area. Even though the network architecture is different, WiMAX concepts are deeply rooted in the IEEE 802.14 project (cable-TV access method and physical layer specification) now given up, which started in 1996. This suggested defining a MAC protocol, based on the ATM infrastructure and dedicated to the TV broadcasting via cables. On the one hand, the headend is connected to an operator network. On the other, it is connected to a group of users, who are kitted out with *cable modem* (CM). The security of the exchanges between subscribers and headend is based on several parameters: a cookie, a cryptographic key computed via a Diffie Hellman procedure and two random numbers generated by each entity [15].

From the start, WiMAX was designed with security in mind. At the lower-edge of the Media Access Control sublayer of TCP/IP, a privacy sublayer was defined in the official 802.16e-2005 specification to handle encryption of packets and key management. To handle authentication, the specification relies on the already existing Extensible Authentication Protocol (EAP) [10] similar to WiFi.

There are two schemes for data encryption, which are supported in the 802.16 standard, the Advanced Encryption Standard (AES) and Triple Data Encryption Standard (3DES). Both of these schemes are block ciphers, which are security algorithms which operate on one chunk (or block) of data at a time as opposed to stream ciphers which can act on a single byte. AES handles a 128-bit block of data at a time, and has been shown to be very fast in both software and hardware implementations. Both because of its speed and because of its ease to implement, AES has become the algorithm of choice for WiMAX just like WiFi. During the authentication process, a 128- or 256-bit key is created and that is used in conjunction with the cipher. Additionally, it is recreated at intervals for optimal security; the robustness of a security solution relies also on the dynamicity of the key where the best solution would be to use one time key. The 802.16e-2005 amendment specifies Privacy and Key Management Protocol Version 2 as the key management implementation [14]. This system handles the transfer of keys between the base station and the subscriber station by using X.509 digital certificates; based on asymmetric encryption using RSA public-key algorithm. Additional security is provided by refreshing the keys and connections at frequent intervals. If long keys (1024 bits) are used, the RSA algorithm is considered to be secure.

User and device authentication for WiMAX consists of certificate support using (IETF) Extensible Authentication Protocol. EAP is a structure designed to perform authentication through the use of functions that can negotiate with many different possible procedures. There are around 40 different procedures, called EAP methods, including some defined within the IETF standard and others that have been developed by outside entities. Some of the types of credentials that WiMAX can use for authentication purposes are digital certificates, smart cards and user name/password [14]. In terminal devices, a X.509 digital certificate with both the MAC address and public key can perform device authentication as needed. Adding both user authentication and device authentication creates an additional layer of security.

Another authentication method used with WiMAX is support for control messages. This type of handshake is used to assure both the message authenticity and the integrity of the data that the message contains. CMAC (Cipher-based Message Authentication Code) uses a block cipher algorithm while HMAC (keyed-Hash Message Authentication Code) uses a hash function to combine with the secret key. Both of these types of scheme are supported by WiMAX [14]. Regarding the handover the key exchange process might be accelerated to help in minimizing the delay of the handover.

11.6.3 Security Consideration in WiFi-WiMAX

WiFi and WiMAX use different physical and data layers. As a result, security attacks can differ depending on which scheme is in place. WiFi being the older, more prevalent wireless standard, it has long been assailed by security attacks from all sides. Security researchers note increasing instances of so-called 'evil twin' attacks, in which a malicious user sets up an open WiFi network and monitors traffic in order to intercept private data [14]. Some of the other types of security threats that have been used on WiFi networks are identity theft in the form of MAC spoofing, man-in-the-middle attacks, Denial-of-Service

(DoS) attacks and network injection attacks where intruders inject commands into the network to re-configure it.

In WiMAX, jamming and packet scrambling are the general kinds of attacks that can most affect WiMAX's physical layer. Signals in the lower frequencies that cross or are in close proximity to the WiMAX antenna can produce second and third harmonic waves that interfere and can overload the WiMAX signal. For example, if we take a 850 MHz signal, we will find a second harmonic, although not as strong, at 1700 MHz (2 × 850). A third harmonic, much weaker, will be located at 2550 MHz (3 × 850). Because WiMAX is transmitted over frequency bands that are licensed, unintentional jamming is rare. Taking a spectrum analysis at intervals can mitigate constant jamming, whether malicious or not [14]. Within the Data Link Layer of the network stack, digital certificates work very well for establishing the identity of a mobile station to a base station. However, a simple one-way authentication could provide an opportunity for intruders to create a rogue base station and snoop on traffic. Authentication using EAP-TLS will enable both the base station and the mobile station to use X.509 certificates to establish their legitimacy [14].

Integrating WiFi and WiMAX access to allow smooth handover from one technology to another will inherit the security weaknesses of both technologies. From the security architecture point of view, the interaction between the security entities such as the AAA system will be different based on the integration/coupling scenarios.

11.6.4 WiFi-WiMAX Interworking Scenarios

For effective interworking between available Radio Access Technologies a variety of approaches can be taken, depending on the level of integration that is required or deemed necessary. The main requirements for interworking that need to be taken into consideration are as follows [17]:

- Mobility support (Handover WiMAX WiFi); the user should be notified of service derogation during handover.
- Partnership or roaming agreements between a WiMAX network operator and a WiFi network; the operator should give the user the same benefits as if the interworking was handled within one network operator.
- Subscriber billing and accounting between roaming partners must be handled.
- Subscriber identification should be such that it can be used both in a pure WiMAX/WiFi environment.
- The subscriber database could either be shared or could be separate for the two networks while sharing the subscribers' security association. The subscriber database could be the HLR/HSS (3GPP terminology: Home Locator Registrar/Home Subscriber Server) or an AAA server (IETF terminology).

If the integration between different technologies is close, the provisioning of the service is more efficient and the choice of the mode in order to find the best radio access as the well as the handover procedure is faster. However, a high level of integration requires considerable effort in the definition of interfaces and mechanisms able to support the necessary exchange of data and signalling between different radio access networks.

Based on these trade-off considerations, different types of coupling and therefore different integration approaches can be classified:

- Open coupling.
- Loose coupling.
- Tight coupling.

11.6.4.1 Open Coupling

Open coupling means that there is no effective integration between two or more radio access technologies. As reported in [17], in an open coupling situation, two access networks, for example WLAN and WMAN, are considered in an independent way, with only a Billing system being shared between them. Separate authentication procedures are used for each access network and no vertical handovers take place, an ongoing session is simply lost and has to be reinitiated by the user manually at the new access point. In this case, there is only an interaction between the billing management systems of each network technology; however there is no interaction between the control procedures related to the QoS and mobility management, as shown in Figure 11.6.

11.6.4.2 Loose Coupling

ETSI defined loose coupling as a complementary integration of generic radio access technology networks such as WiFi with 3G access networks without any user plane interface, thus avoiding the servicing and gateway nodes of the packet switching part of the network. Operators are still be able to make use of the existing subscriber database for 3G clients and generic radio access technologies' clients, allowing centralized billing and maintenance for different technologies. In case of WiMAX integration with WiFi, we will follow the same logic since WiMAX is a licenced radio technology, as a 3G radio access.

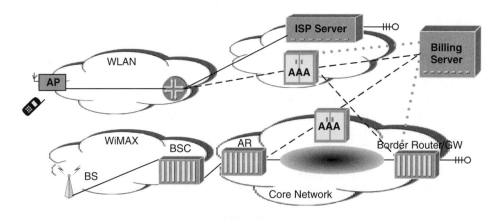

Figure 11.6   Open coupling.

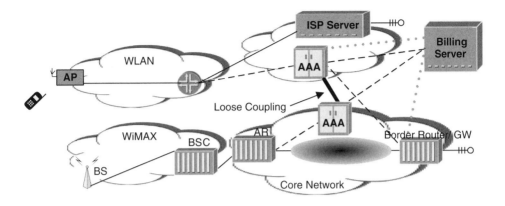

Figure 11.7   Loose coupling.

The main consequence of this kind of coupling is that during the switch over between the two radio access technologies, the service in progress is dropped, therefore no seamless vertical handover is available; however there is an interaction with the AAA procedures. In this case, there is an interaction between the billing management systems of each operator. In addition, there is an interaction between the control planes of each operator regarding the authentication procedure. This is similar to the inter-domain authentication as Diameter between the AAA server of the WiFi network and the AAA server of the WiMAX server.

Figure 11.7 shows a WiMAX-WiFi loose coupling scenario. The core network here coordinates subnetworks during the interworking. For authentication and billing, one customer database and procedure is used and a new link between the WiFi hotspot and the WiMAX network is provided. It means that the user has to perform a unique subscription if the network provider is the same for both networks, or alternatively, the user has to perform a unique subscription to a certain service that will be available for both access networks.

11.6.4.3   Tight Coupling

For tight coupling the WiFi network is connected to the rest of the WiMAX network of the same operator. This means that the access router of the operator will handle the controller of the WiFi and the WiMAX access points. In Figure 11.8, vertical handover will be possible between the two technologies as moving from a WiFi to a WiMAX access network is possible under the same access network hosting both technologies. This main characteristic allows for seamless handover between WiFi and WiMAX to take place. As compared to loose coupling this provides improved handover performance; however the WiFi and WiMAX access networks will be required to expect users coming from the other access technology. This type of coupling may occur if one operator is running both networks. In this case, the control plane as well as the management plane of each network technology interacts closely with each other. Mobility, the AAA and QoS management can be supported by the same core network. However, in this coupling it is difficult to support

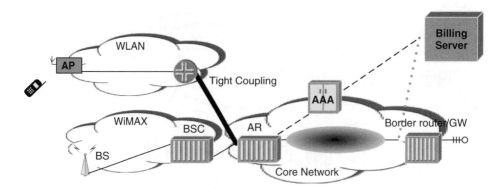

Figure 11.8   Tight coupling.

a seamless handover between the different technologies for all applications since QoS interpretation and support from each technology is different. The AAA service is the same, meaning that the authentication protocol is the same in a network such as Radius; intra domain, and link layer authentication will be adapted to WiFi or WiMAX connectivity. The acceleration of the authentication process will be achieved by more interaction at access between the WiFi and WiMAX access point. For example, by using some additional layer such as IEEE 802.21 at the mobile terminal and the APs to allow the exchange of some information related to the key management process between both technologies.

Finally, in very tight coupling [17], with the introduction of reconfigurable terminals, a very tight interworking between different networks will make it possible for a terminal to reconfigure its interface very quickly and connect to the available access point from a different technology. This will allow very fast handover by accelerating the security procedure at the radio part between the terminal and the access network.

11.7   Mobility Management

The mobility management procedures specified to handle mobility between WiFi and WiMAX networks should include mechanisms to minimize service interruption during handover and where possible support bidirectional service continuity. Mobility management is typically triggered when the MS moves across base stations based on radio conditions.

- This applies to mobile terminals supporting either single or dual radio capability.
- The mobility management procedures should minimize any performance impact on the mobile terminal and the respective accesses; for example, mobile terminal battery consumption and network throughput.
- The mobility management procedures should minimize the coupling between the different accesses allowing independent protocol evolution in each access.

11.7.1 Handover Support

Handover in WiFi is almost all the time hard, which means (break-then-make approach). This is because the WiFi station cannot be serviced in parallel by more than one AP and therefore has to break its communication with its current AP before establishing a connection with a new one. Thus, considerable transmission disruptions may occur that result in QoS degradation. Moreover, handovers in IEEE 802.11 are controlled solely by the WiFi AP, so the WiFi infrastructure cannot provide tight control of QoS provisioning.

However, in IEEE 802.16e, there are three types of handover: (i) hard handover, (ii) Fast Base Station Switching (FBSS) and (iii) Macro Diversity Handover (MDHO) [6]. In the last two methods, the MS maintains a valid connection simultaneously with more than one BS. In the case of FBSS, the MS maintains a list of the BSs involved, called the active set. The MS monitors the active set continuously, does ranging and maintains a valid connection ID with each of them. The MS, however, communicates with only one BS, called the anchor BS. When a change of anchor BS is required, the connection is switched from one base station to another without having to perform handover signalling explicitly. The MS simply reports the selected anchor BS on the Channel Quality Indicator Channel (CQICH). Macro diversity handover is similar to FBSS, except that the MS communicates on the downlink and the uplink with all the base stations in the active set simultaneously – called a diversity set here. In the downlink, multiple copies received at the MS are combined using any of the well-known diversity-combining techniques. In the uplink, where the MS sends data to multiple base stations, selection diversity is performed to pick the best uplink.

11.7.2 Cell Selection

Cell selection is an important step in the handover process since the performance of the whole process depends on the success of the selection. In Mobile WiMAX, the cell selection process begins with the decision for the MS to migrate its connections from the serving BS to a new target BS. This decision can be taken by the MS, the BS or some other external entity in the WiMAX network and is dependent on implementation. While in WiFi, the cell selection decision is limited to the WLAN AP only.

Handover decision criteria assist the determination of which access network or cell should be chosen. Traditionally, handoff occurs when there is a deterioration of signal strength received at the mobile terminal from the Access Point/Base Station in WiFi and mobile WiMAX respectively. Decision-based signal strength can be very useful in the case of horizontal handoff, that is, WiFi to WiFi and WiMAX to WiMAX. However, in vertical handoff between WiFi and mobile WiMAX, there is no comparable signal strength available to aid the decision as in horizontal handoff because the received signal strength sample from WiFi and mobile WiMAX are heterogeneous quantities that cannot be compared directly. Thus, additional criteria should be evaluated such as monetary cost, offered services, network conditions, terminal capabilities (velocity, battery power, location information, QoS) and user preferences.

11.7.3 IP for Mobility Management

Mobile IP (v4 and v6) being the standard for IP layer support of mobility, in the WiFi-WiMAX mobility scenario, the IP layer will have to rely on such protocol to repair the connectivity at the IP layer if the IP address has changed. So this will depend on the integration layer of both technologies. Also, in order to avoid implementing Mobile IP in WiFi or WiMAX devices, the IETF decided on a network assisted IP mobility approach. Proxy Mobile IP (PMIP) is an embodiment of the standard MIP framework wherein an instance of the MIP stack is run (ASN in the case of WiMAX and Access router in WiFi case) on behalf of an MS that is not MIP capable or MIP aware [7]. Using proxy MIP does not involve a change in the IP address of the MS when the user moves and obviates the need for the MS to implement a MIP client stack.

Similar to PMIP, the IETF has also proposed an IP mobility solution where mobility management is done by the network and not by the end node. This approach avoids adding a Mobile IP stack in the terminal and would allow a rapid deployment of such solutions. Obviously the network operator has to deploy new entities/functionalities in order to allow such mobility.

The IETF NetLMM Working group (Network-based Localized Mobility Management) works on this solution with a micro-mobility based approach [19].

When the MN enters in a NetLMM network, it obtains an IPv6 address that it will keep during its movements in the NetLMM domain. An entity called the *Local Mobility Anchor* (LMA) is responsible for redirecting packets for the MN towards the AR in charge of this MN and is called the *Mobile Access Gateway* (MAG). Since the MN enters in a subnetwork managed by a MAG, the latter informs the LMA. The interface between an MN and a MAG is described in the document while the mechanism used between a MAG and a LMA is described in RFC 5213 [18].

The main security problem is the discovery of the MN's arrival by a MAG. Indeed, this discovery is based on a neighbour discovery mechanism. Thus, it is recommended that MNs compatible with NetLMM use the protection mechanism of SEND/CGA in order to guarantee the integrity and uniqueness of their IPv6 address.

Moreover, it is necessary to secure the information exchanged between MAGs and the LMA of a NetLMM network. For this, the use of IPsec is recommended.

In the case of WiFi-WiMAX integration, it would be interesting to support fast mobility at the IP layer by IP mobility improved schema such as Fast Mobile IP as well as anticipated authentication to better support smooth and seamless handover such as mobile VoIP which is presently only wireless VoIP as in case of VoWiFi (Voice over WiFi) where the session disconnects from one AP to another one due the non efficient mobility support at the IP layer. For this objective, an information layer such as IEEE 802.21 or ANDSF would be necessary to make these technologies exchange information at the lower layers and better inform the IP and upper layers about the connectivity changes at the physical layer of the multimode terminal.

11.7.4 Session Initiation Protocol for Mobility Management

The IETF has developed a signalling protocol SIP [20], Session Initiation Protocol, which can also be used to support so-called personal mobility. Personal mobility allows a user to change terminal and recover its session. Unlike Mobile IP, SIP acts at the transport layer and not at the network layer of TCP/IP model [9]. SIP is independent of transport protocol (UDP, TCP, ATM ...). It uses a logical address instead of IP addresses. It controls a multimedia session with two or more participants. It is a lightweight protocol and not complex with little load on the network. SIP was accepted by the 3GPP as the signalling protocol in November 2000 and is a permanent element of the next generation network IMS (IP Multimedia Subsystem). SIP terminals are already on the market for applications such as voice over IP. Several conversation clients also use SIP (Windows Messenger, AOL Instant Messenger ...) via the Internet.

SIP proposes mainly adding a 'user agent' in the terminal user who plays the role of SIP client, a registrar or registration server. It keeps the location information provided by the 'user agent' and a proxy between two 'user agents' that can relay SIP requests and asks the right 'registrar' to locate the corresponding 'user agent'. These components are separated logically and not necessarily physically. SIP can operate in peer to peer mode, but in the context of deployment of public services, registration servers and proxies are necessary.

SIP is a text protocol and shares similar response codes with HTTP. However, SIP differs from HTTP as a SIP agent is at the same time a client and a server. Figure 11.9 depicts SIP functionality. In general, SIP is composed of the following elements:

- *User Agent (UA)*: We may find it in all SIP phones or any other SIP-based applications. A communication between two SIP agents is established based on a *URI (Uniform Resource Identifier)* that is similar to an e-mail address.

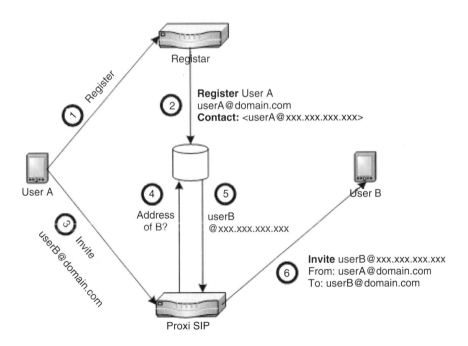

Figure 11.9   SIP functionality [23].

- *Registar*: As we obviously need to know the IP address of the target SIP UA to establish a communication, the *Registar* is in charge of registering and maintaining that IP address in a database that will then link it with the target URI.
- *Proxy*: A SIP proxy has a middleman role between two SIP UAs in order to obtain their respective IP addresses. The SIP Proxy retrieves the destination IP address from the database and then contacts the destination SIP UA. Data traffic never transits through a SIP Proxy but is exchanged directly between two SIP UAs.
- *Redirect Server*: A SIP redirect server receives requests from a SIP UA and is in charge of returning a redirection response indicating where the request should be retrieved.
- *Session Border Controller (SBC)*: A SIP-ready intelligent firewall. When a SIP UA initiates a SIP session, two connections are built, one for signalling and one for data transmission. Although this process does not pose any problems when both SIP UAs are located within the same subnetwork, firewalls or NAT separating different networks may not be aware of the relationship between these two connections. They could therefore reject traffic to a subscriber in its subnetwork even if signalling established that connection successfully. NATs also generate address translation problems between multiple temporary addresses established by ISPs and their visibility on the Internet. In order to resolve these issues, it has therefore been proposed to create Session Border Controller (SBC) acting as an application-layer gateway and guaranteeing correct address translation and assisting network administrators in managing the flow of sessions transiting into their subnetworks.

In the case of WiFi-WiMAX integration, the SIP protocol deals with the application oriented reachability of the user whether it is connected to a WiFi or WiMAX access network. This is similar to IMS (IP Multimedia SubSystem) where a multimode terminal can benefit from accessing a service (e.g. Telephony) no matter which access network it is connected to. SIP will then handle the signalling part of the session establishment to the correct point of attachment where the node is located at that time.

11.7.5 Identity Based Mobility

In today's Internet architecture, IP addresses are used both as locators and identifiers. This dual role poses several problems. Firstly, IPv4 is still more widely used than IPv6, so the address space of IPv4 becomes insufficient owing to increasing Internet usage and the number of hosts. Furthermore, as the mobility of devices increase, the dual role of IP addresses make mobility management complicated.

In order to solve these problems the Host Identity Protocol/HIP [21] is proposed by the IETF (Internet Engineering Task Force) and IRTF (Internet Research Task Force). It proposes separating locators and identifiers. In HIP, IP addresses act only as locators while host identities identify themselves. This situation requires adding a new layer in the TCP/IP stack between the transport layer and the IP layer. The role of this layer is to compensate host identities with upper layer protocols.

One of the issues defined in HIP is that the Host Identity (HI) is the public key within a public/private key pair. HIP is illustrated in Figure 11.10. This key can be represented by the Host Identity Tag (HIT), a 128-bit hash of the HI, and has to be globally unique in

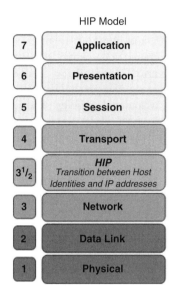

HIP Model

Figure 11.10   Host identity protocol in ISO layers.

the entire Internet universe. Another representation of the HI is the Local Scope Identity (LSI) which is 32-bits size and can only be used for local purposes.

The HIP Base Exchange is a cryptographic key-exchange procedure performed at the beginning of the establishment of HIP communication. The HIP Base Exchange is built around a classic authenticated Diffie-Hellman key exchange. The BE is four-way packet exchange between the Initiator (I) and the Responder (R). The initial IP address of a HIP host should be stored in order to make the host reachable. Traditionally, the DNS is used for storing this information. The problem with the DNS system is the latency; updating the location information each time the MN moves, the update is not fast enough. The Rendezvous Mechanism is designed to solve this problem. The Rendezvous Server (RVS) keeps all the information of HIP communication. The location information of RVS is stored in DNS. If a MN wants to communicate with other MNs, all nodes have to register with their RVS.

The HIP enabled Responder (R) should register with the RVS its HIT and current IP address. When the Initiator (I) wants to establish a connection with R, it first sends the I1 packet to one of R's rendezvous servers or to one of IP addresses (if it can be learned via DNS). I gets the IP address of R's RVS from DNS and send the I1 packet to the RVS for Base Exchange. RVS checks whether it has the HIT of the I1 packet. If the HIT belongs to itself, it sends the I1 packet to the relevant IP address. R sends the R1 packet directly to I without RVS. It is proposed to accelerate the registration process of HIP by early update through the previous connecting access point [22]. In the case of WiFi-WiMAX integration it means that each access point would serve the other to proceed with the acceleration of registration; it means that a level of trust is already established between these access points of different technologies and perhaps different network administration.

11.8   Quality of Service Architecture

The interworking between WiFi and WiMAX requires the development of an end-to-end QoS architecture that integrates both networks within one QoS framework. This framework should define the various QoS-related functional entities in both networks and the mechanisms for provisioning and managing the various service flows as well as the access categories and their associated policies. The QoS framework should support the simultaneous use of a diverse set of IP services, such as differentiated levels of QoS per user, per service flow or per access category; admission control; and bandwidth management. The QoS framework calls for the use of standard IETF mechanisms for managing policy decisions and policy enforcement between operators.

11.8.1   End-to-End QoS Interworking Framework

Figure 11.11 shows the proposed QoS functional framework that may form a possible interworking QoS architecture for both networks. A traffic flow may be admitted into the heterogeneous system only if it can guarantee the requested end-to-end QoS of this flow.

The important functional entities in the architecture which concern WiMAX are as follows:

1. **Policy function.** The policy function (PF) and a database reside in the home NSP. The PF contains the general and application-dependent policy rules of the NSP. The PF database may optionally be provisioned by an AAA server with user-related QoS profiles and policies. The PF is responsible for evaluating a service request it receives

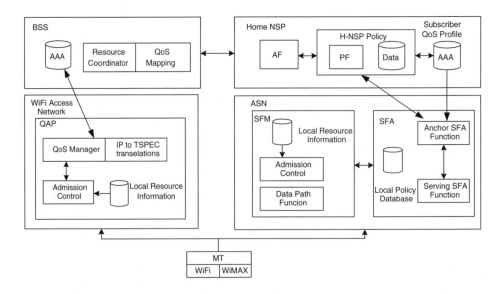

Figure 11.11   QoS end-to-end interworking architecture.

against the provisions. Service requests to the PF may come from the service flow authorization (SFA) function or from an application function (AF), depending on how the service flows are triggered.

2. **AAA server.** The user QoS profiles and associated policy rules are stored in the AAA server. User QoS information is downloaded to an SFA at network entry as part of the authentication and authorization procedure. The SFA then evaluates incoming service requests as against this downloaded user profile to determine handling. Alternatively, the AAA server can provision the PF with subscriber-related QoS information. In this case, the home PF determines how incoming service flows are handled.

3. **Service flow management.** Service flow management (SFM) is a logical entity in the BS that is responsible for the creation, admission, activation, modification and deletion of 802.16e service flows. The SFM manages local resource information and performs the admission control (AC) function, which decides whether a new service flow can be admitted into the network.

4. **Service flow authorization.** The SFA is a logical entity in the ASN. A user QoS profile may be downloaded into the SFA during network entry. If this happens, the SFA evaluates the incoming service request against the user QoS profile and decides whether to allow the flow. If the user QoS profile is not with the SFA, it simply forwards the service flow request to the PF for decision making. For each MS, one SFA is assigned as the anchor SFA for a given session and is responsible for communication with PF. Additional SFAs may exist in a NAP that relays QoS-related primitives and applies QoS policy for that MS. The relay SFA that communicates directly with the SFM is called the serving SFA. The SFAs may also perform ASN-level policy enforcement, using a local policy function (LPF) and database. The LPF may also be used for local admission control enforcement.

5. **Application function.** The AF is an entity that can initiate service flow creation on behalf of a user. An example of an AF is a SIP proxy client.

6. **WLAN QoS Manager.** The WLAN QoS manager will map the QoS parameters in the MT request within appropriate 802.11e TSPEC parameters. This mapping is achieved both using automatic mapping to 802.11e TSPEC parameters with appropriate algorithms wherever possible and using user configurable values for the rest of the 802.11e TSPEC parameters. The QoS manager will also interface with the authorization and authentication server (such as the Diameter or Radius server) to validate the request. If the request is authenticated properly and authorized to access the service, the QoS manager will use the mapped TSPECs to set up a traffic stream between the end point MT and the AP.

7. **Resource Coordinator.** This is the entity carrying out the main role within the framework. The aim of this module is to coordinate the management of QoS across the two network segments by coordinating the service offered over the WiMAX radio interface with that offered over the WiFi. To perform its task, the resource coordinator collects and exchanges information from both the WiFi and WiMAX sides. The resource coordinator plays a key role in the end-to-end QoS management as its main tasks are (i) mapping between the different QoS classes and their QoS parameters for both networks, (ii) cooperation with the collocated admission control module in both WiMAX and WiFi and (iii) flow management.

11.8.2 QoS Considerations

One vital component for the provision of seamless multimedia session continuity is QoS consistency across the WiFi and Mobile WiMAX networks. This is vital because without QoS consistency the multimedia sessions will experience different QoS levels in both network domains, and thus seamless continuity will not be achievable. It is unfortunate, however, that the WiFi and Mobile WiMAX specifications were based on different sets of requirements, and ended up supporting different sets of QoS features. Consequently, QoS consistency turns out to be a very challenging issue. To provide greater insight on this issue, we discuss a list of WiFi QoS deficiencies with respect to Mobile WiMAX QoS. When we target multimedia session continuity across WiFi and Mobile WiMAX networks, we should take these deficiencies into consideration and understand their impact. The discussion is based on the assumption that the WiFi Medium Access Control (MAC) layer complies with IEEE 802.11 plus the amendments of IEEE 802.11e [802.11e], and the physical layer complies with IEEE 802.11g [802.16g]; while mobile WiMAX is based on IEEE 802.16e the mobile version of IEEE 802.16d.

11.8.2.1 QoS Support and Classes

The IEEE 802.11 standard is intended to support only the best effort service; however IEEE 802.11e introduced basic QoS support by defining four different access categories (ACs), namely AC_VO (voice) with highest priority, AC_VI (video), AC_BE (best effort), and AC_BK (background) with lowest priority [8] (The ACs are depicted in Table 11.1).

In IEEE 802.16e the QoS is represented by five SFs: Unsolicited Grant Service (UGS) for (VoIP without silence suppression), Real-time polling service (rtPS) for (video), Extended Real-time polling service (ertPS) for (voice with silence suppression), Non Real-time polling service (nrtPS) for (FTP) and best effort (see Table 11.2). From the difference in the supported QoS classes, it becomes clear that a vertical handover from Mobile WiMAX to WiFi needs to involve a QoS mapping procedure.

Therefore, we have to define for each WiFi QoS class or AC which Mobile WiMAX SF type we assign and additionally we have to provide a mapping from TSPEC (Traffic Specification) negotiated in WiFi in Action.ADDT request to Dynamic Service Addition Request (DSA-REQ) negotiated in Mobile WiMAX. The mapping is illustrated in Table 11.3.

Table 11.1 IEEE 802.11e access category and user priority mapping

User priority	Category(AC)	Informative
1	AC_BK	Background
2	AC_BK	Background
0	AC_BE	Best Effort
3	AC_BE	Video
4	AC_VI	Video
5	AC_VI	Video
6	AC_VO	Voice
7	AC_VO	Voice

Table 11.2 The five service flow (SF) types defined by IEEE 802.16e

Service Flow Designation	QoS Parameters	Application Examples
Unsolicited grant services (UGS)	Maximum sustained rate Maximum latency tolerance Jitter tolerance	Voice over IP (VoIP) without silence suppression

Real-time Polling service (rtPS)	Minimum reserved rate
	    Maximum sustained rate
	    Maximum latency tolerance
	Traffic priority
		Streaming audio and video, MPEG (Motion Picture Experts Group) encoded

Non-real-time Polling service (nrtPS)	Minimum reserved rate Maximum sustained rate Traffic priority	File Transfer Protocol (FTP)
Best-effort service (BE)	Maximum sustained rate Traffic priority	Web browsing, data transfer

Table 11.3 QoS mapping between IEEE802.11e and IEEE802.16e classes

802.11e Access categories	802.16 Service Flows	Application
AC_VO	UGS, ertPS	Voice
AC_VI	rtPS	Video
AC_BE (high load)	nrtPS	FTP(high load)
AC_BE (medium load)	BE	FTP(medium load), Web browsing
AC_BK (low load)	BE	FTP(low load), Email

11.8.2.2 Mechanisms of Channel Access

An important aspect to consider is that the basic support for QoS differs significantly between WiFi and Mobile WiMAX owing to their different PHY and MAC layers design. While access to the channel in Mobile WiMAX is completely centralized, it can be distributed or centralized in WiFi based 802.11e. This can be explained as follows.

11.8.2.3 WiFi Access Methods

It is wort describing the mandatory access method in 802.11 by comparing it with 802.11e. The access method is based on Distributed Coordination Function (DCF). The basic DCF uses a Carrier Sense Multiple Access with Collision Avoidance (CSMA/CA) mechanism to regulate access to the shared wireless medium. Before initiating a transmission, each WS is required to identify the medium and perform a binary exponential back off. If the medium has been identified as idle for a time interval called DCF Interframe Space (DIFS), the WS enters a back off procedure. A slotted back off time is generated randomly from a Contention Window (CW): back off time = rand[0, CW] x slot time. In DCF, only a best effort service is provided. Time-bounded multimedia applications (e.g. voice over

IP, videoconferencing) require certain bandwidth, delay and jitter guarantees. The point is that with DCF, all the WSs compete for the channel with the same priority. There is no differentiation mechanism to provide a better service for real-time multimedia traffic than for data applications. This is the reason behind introducing the hybrid coordination function in IEEE 802.11e which consists of two different methods of medium access and uses the concepts of Traffic Opportunity (TXOP), which refers to a time duration during which a WS is allowed to transmit a burst of data frames [8].

1. **The Enhanced Distributed Channel Access (EDCA)** method in which each AC behaves as a single DCF contending entity with its own contention parameters (CWmin, CWmax, AIFS and TXOP), which are announced by the AP periodically in beacon frames. Basically, the smaller the values of CWmin, CWmax, and AIFS[AC], the shorter the channel access delay for the corresponding AC and the higher the priority for access to the medium. In EDCA a new type of IFS is introduced, the Arbitrary IFS (AIFS), instead of DIFS in DCF. Each AIFS is an IFS interval with arbitrary length as follows: AIFS = SIFS + AIFSN × slot time, where AIFSN is called the arbitration IFS number. After sensing that the medium is idle for a time interval of AIFS[AC], each AC calculates its own random back off time (CWmin[AC] ≤ back off time ≤ CWmax[AC]). The purpose of using different contention parameters for different queues is to give a low priority class a longer waiting time than a high priority class, so the high priority class is likely to access the medium earlier than the low priority class.
2. **The polling based HCF Controlled Channel Access (HCCA)** method in which different traffic classes called traffic streams (TSs) are introduced. Before any data transmission, a traffic stream (TS) is first established, and each WS is allowed to have no more than eight TSs with different priorities. In order to initiate a TS connection, a WS sends a QoS request frame containing a traffic specification (TSPEC) to the AP. A TSPEC describes the QoS requirements of TS, such as mean/peak data rate, mean/maximum frame size, delay bound, and maximum Required Service Interval (RSI). On receiving all these QoS requests, the AP scheduler computes the corresponding HCCA-TXOP values for different WSs by using their QoS requests in TSPECs (TXOP1, TXOP2, etc.) and polls them sequentially.

11.8.2.4   Mobile WiMAX Access Method

In Mobile WiMAX, the MAC layer at the base station is responsible for allocating bandwidth to all users, in both the uplink and the downlink. The only time the MS has some control over bandwidth allocation is when it has multiple sessions or connections with the BS. Depending on the particular QoS and traffic parameters associated with a service, one or more of these mechanisms may be used by the MS. The BS allocates dedicated or shared resources periodically to each MS, which it can use to request bandwidth. This process is called polling. Polling may be done either individually (unicast) or in groups (multicast). Multicast polling is done when there is insufficient bandwidth to poll each MS individually. When polling is done in multicast, the allocated slot for making bandwidth requests is a shared slot which every polled MS attempts to use. Mobile WiMAX defines a contention access and resolution mechanism for the case when more than one

MS attempts to use the shared slot. If it already has an allocation for sending traffic, the MS is not polled. Instead, it is allowed to request more bandwidth by (1) transmitting a stand-alone bandwidth request MPDU, (2) sending a bandwidth request using the ranging channel, or (3) piggybacking a bandwidth request on generic MAC packets.

11.9 Summary

This chapter presented an overview of the interworking architecture between WiMAX/WiFi networks, functional entities that enable the seamless handover between these two different systems. The interworking architecture provides flexibility for implementation while at the same time providing a mechanism for interoperability.

- The 3GPP has developed interworking models that provide flexibility for implementation while at the same time providing a mechanism for interoperability.
- The interworking architecture provides a unified model for fixed, nomadic and mobile usage scenarios.
- The WiMAX/WiFi interworking architecture defines various QoS-related functional entities and mechanisms to implement the QoS features supported by IEEE 802.16e and IEEE 802.11e.
- The WiMAX/WiFi interworking architecture supports both layer 2 and layer 3 mobility. Layer 3 mobility is based on mobile IP and can be implemented without the need for a mobile IP client.
- WiMMAX/WiFi security architecture relies on the security level of both WiMAX and WiFi and could interwork in some functionalities as the authentication by sharing the AAA server in the case of tight coupling, or interwork as inter-domain AAA service similar to roaming AAA service. Regarding data confidentiality, the key management is different between the two technologies; however if a standardized information exchange layer such as IEEE 802.21 is deployed, then exchanging information through this layer would link the key management process of both access technologies.

References

[1] IEEE Standard P802.21/D02.00, IEEE Standard for Local and Metropolitan Area Networks: Media Independent Handover Services, September 2006.
[2] J.G. Andrews, A. Ghosh and R. Muhamed, *Fundamentals of WiMAX Understanding Broadband Wireless Networking*, Pearson Education, Inc., 2007.
[3] C. Rigney, A. Rubens, W. Simpson and S. Willens, 'Remote Authentication Dial in User Services (RADIUS)', Internet Engineering Task force RFC 2138, Sept. 2003.
[4] P. Calhoun, J. Loughney, E. Guttman, G. Zorn and J. Arkko, 'Diameter Base Protocol', Internet Engineering Task force RFC 3588, Sept. 2003.
[5] Third Generation Partnership Project, 'Technical Specification Group Services and System Aspects; Architecture enhancements for non-3GPP accesses (Release 8)', 3GPP TS 23.402, 2007.
[6] IEEE Standard 802.16e, IEEE Standard for Local and Metropolitan Area Networks – Part 16: Air Interface for Fixed and Mobile Broadband Wireless Access Systems, February 2006.
[7] R. Wakikawa, 'IPv4 Support for Proxy Mobile IPv6 draft-ietf-netlmm-pmip6-ipv4-support-00.txt'. IETF 2007.
[8] IEEE 802.11e/D6.0, EEE Draft Standard Medium Access Control (MAC) Quality of Service (QoS) Enhancements, November 2003.

[9] E. Wedlun and H. Schulzrinne, 'Mobility Support Using SIP', in Proceedings of ACM international workshop on Wireless mobile multimedia, Vol. 1, pp. 719–37, 1999.

[10] B. Aboba, L. Blunk, J. Vollbrecht, J. Carlson and H. Levkowetz (ed), 'Extensible Authentication Protocol (EAP)', Internet RFC3748.

[11] P. Jayarama, R. Lopez, Y. Ohba (Ed.), M. Parthasarathy and A. Yegin, 'Protocol for Carrying Authentication for Network Access (PANA) Framework', Internet RFC 5193.

[12] C. Rigney, S. Willens, A. Rubens and W. Simpson, 'Remote Authentication Dial In User Service (RADIUS)' Internet RFC2865

[13] D. Mitton, M. St.Johns, S. Barkley, D. Nelson, B. Patil, M. Stevens and B. Wolff, 'Authentication, Authorization, Accounting: Protocol Evaluation' Internet RFC3127

[14] http://software.intel.com/en-us/articles/wi-fi-and-wimax-protocols-of-security/.

[15] P. Urien 'WiMAX Security', in H. Chaouchi and M. Laurent, *Security of Wireless and Mobile Networks*, ISTE/John Wiley & Sons Ltd, May 2009.

[16] G. Pujolle 'Wifi Security', in H. Chaouchi and M. Laurent, *Security of Wireless and Mobile Networks*. ISTE/John Wiley & Sons Ltd, May 2009.

[17] V. Friderikos, A. Shah Jahan, H. Chaouchi, G. Pujolle and H. Aghvami, 'QoS Challenges in All-IP based Core and Synergetic Wireless Access Networks', IEC Annual Review of Communications, Vol. 56, November 2003.

[18] S. Gundavelli (Ed.), K. Leung, V. Devarapalli, K. Chowdhury and B. Patil, 'Proxy Mobile IPv6', Internet Engineering Task Force RFC 5213, August 2008.

[19] J. Kempf *et al.*, 'Problem Statement for Network-Based Localized Mobility Management (NETLMM)', Internet RFC 4830.

[20] J. Rosenberg *et al.*, 'Session Initiation Protocol' Internet RFC 3261.

[21] R. Moskovitz *et al.*, 'Host identity protocol: HIP' Internet RFC 4423.

[22] Z. AYdin and H. Chaouchi, 'eHIP : Early update HIP', ACM Mobility 2009.

[23] J. Haari, C. Bonnet "Security in next generation mobile networks", book chapter, ISTE/Willey "Wireless and Mobile Network Security" 2009, ISBN: 9781848211179.

12

QoS Simulation and An Enhanced Solution of Cell Selection for WiMAX Network

Xinbing Wang, Shen Gu, Yuan Wu and Jiajing Wang
Department of Electronic Engineering, Shanghai Jiaotong University, Shanghai, China

12.1 Introduction

Over recent years wireless network infrastructure has been expanding, driven rapidly by the development of broadband wireless technology and huge market need. Amongst these new wireless technologies, WiMAX represents a next generation broadband wireless access (BWA) solution. While it is no longer uncommon that a mobile device user finds itself in an area covered by multiple wireless cells, the base station (BS) of a wireless cell will more often work at a higher data transmission rate, or even reach its maximum capacity in order to accommodate the increasing number of subscribers and provide high speed Internet and multimedia data services such as HDTV. How to and 'smartly' select and switch promptly among the accessible wireless cells has become a very interesting issue. WiMAX is also affected by this problem. In this chapter, we first perform the QoS simulation of WiMAX and then present our study on an enhanced cell selection solution designed for WiMAX networks and analyze how it would improve overall network performance.

The research in [1]–[3] has been performed with regard to related problems. Much research on WiMAX cell switch has focused on the different handover mechanisms, which usually use signal power as a threshold trigger and the decision factor. A general cell selection algorithm based on several additional decision criteria was proposed in [3], with a non-implementable assumption that the mobile subscriber station (MS) is constantly and accurately aware of the BS parameters. We hereby propose an enhanced cell selection

WiMAX Security and Quality of Service: An End-to-End Perspective   Edited by Seok-Yee Tang, Peter Müller and Hamid Sharif
© 2010 John Wiley & Sons, Ltd

method for WiMAX networks by using the following measurements as handover triggers and as target cell decision factors: signal power, estimated effective idle capacity of the BS and requested dataflow rate. Specifically, we have addressed the problems of how to estimate the effective capacity of a WiMAX BS in line with the physical layer definition of the IEEE 802.16 standard [4]–[5] and how WiMAX BSs could update MSs dynamically with capacity information by using the 802.16 MOB_NBR_ADV advertisement messages. Our work is adapted to the WiMAX/IEEE 802.16 standard and is implementation-ready.

The rest of the chapter is organized as follows. We first present a brief overview of the major WiMAX network simulation tools. Then in section 12.3 we perform our simulation for QoS of WiMAX network in several scenarios. In section 12.4 a brief analysis for the simulation result is presented. We explain in detail the system model of the proposed cell selection algorithm and the simulation results in the environment of a NS-2 network simulator in section 12.5. We give our summary in section 12.6.

12.2   WiMAX Simulation Tools – Overview

The IEEE 802.16 workgroup published two milestone standards, 802.16d [4] and 802.16e [5], in 2004 and 2005. The 802.16d standard specified the MAC and PHY interface for fixed BWA systems within local and metropolitan area networks and the 802.16e standard serves as an amendment to the 802.16d standard with support for mobile BWA. WiMAX is a non-profit, industry-led forum promoting the IEEE 802.16 standards. Nowadays, WiMAX has become synonymous with the IEEE 802.16 standard and we use it as such in this chapter.

Network simulators are useful tools for the study of WiMAX QoS performance. Here we introduce three major simulation tools suitable for WiMAX networks.

12.2.1   NS2

NS2 is a well-known discrete event simulator for network simulation. It began as a variant of the REAL network simulator in 1989. The design implementation of NS2 includes two kind of language, C++ and Otcl. NS2 integrates many kinds of network protocols, services, routing algorithms and queuing management mechanisms. It can be used for the simulation of many kinds of networks such as fixed network, wireless network, satellite network and hybrid network. The major characteristics of NS2 are open-sourced, good scalability and efficiency in design.

The ns-2 WiMAX PMP module was designed and developed by Chang Gung University. The implemented module comprises the fundamental functions of the service-specific convergence sublayer (CS), the MAC common part sublayer (CPS), and the PHY layer. A simple call admission control (CAC) mechanism and the scheduler are also included in this module.

12.2.2   OPNet Modeler

The OPNet Modeler is a product of OPNet Technologies, Inc. The modelling methodology of OPNET is organized in a hierarchical structure. It has a three-layer model mechanism, which consist of the Process Model, Node Model and Network Model. Modeler incorporates a broad suite of protocols and technologies, such as VoIP, TCP, MPLS, etc. The statistical and analysis capability of Modeler are very strong so that Modeler can collect the performance statistic data of each common network layer and generate simulation reports. The OPNET WiMAX Specialized Model supports the IEEE 802.16-2004 and IEEE 802.16e-2005 standards. It can be used to evaluate custom scheduling algorithms for WiMAX base and subscriber stations, optimizing application performance by leveraging WiMAX QoS policies and predicting network performance for different MAC and PHY layer profiles. Compared with NS2, Modeler is easier to use, and has a uniform interface. But it is not easy for users to create new modules with a specific function.

12.2.3 QualNet

QualNet is a commercial product of SNT (Scalable Network Technologies) which is derived from the GloMoSim project of UCLA. QualNet is based on a Parsec parallel simulation core and each node has the capability of independent calculation. It also contains many module libraries including Developer Library, Wireless Library and Advanced Wireless Library, to name just a few. QualNet's WiMAX channel model incorporates co-channel interference, urban path loss, fading, shadowing and mobility effects. It supports different type of QoS priority flow including UGS, rtPS, nrtPS, ertPS and BE. Compared with the other two simulators, QualNet is easier to learn and has a richer module library. Therefore, it is widely accepted from a scientific point of view

12.3   QoS Simulation of WiMAX Network

Here the simulation was performed in the framework of WiMAX with a transport mode of a PMP (Point-to-Muiltipoint), NLOS (Non-Line-of-Sight) environment, a frequency of 2–11 GHz, the maximum data transmission rate being 75 Mbps. We employed the tool gawk to process the trace files for data analysis and performance testing and we used Gnuplot to present the figures.

12.3.1 Performance Comparison Between Different Services

12.3.1.1 Setup of Simulation Scenario

Figure 12.1 shows the network topology in NS2 simulation.

Parameter Setup of Simulation Scenario

Table 12.1 Parameter setup of simulation scenario

Simulation duration (s)	BS coverage (m)	Modulation
40.0	1000	OFDM

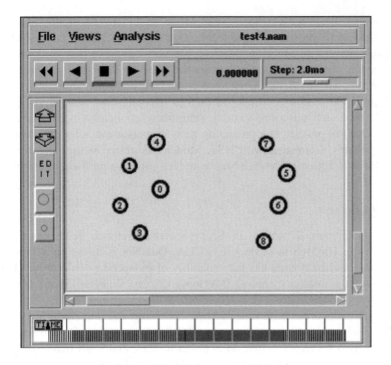

Figure 12.1   Network topology.

Relationship Between Nodes

Table 12.2   Relationship between nodes

node	node0	node1	node2	node3	node4	node 5	node6	node7	node8
station	BS	SS1	SS2	SS3	SS4	SS5	SS6	SS7	SS8

Datalink Setup Between Nodes

Table 12.3   Datalink setup between nodes

Send	Receive	Service Flow	Start	Stop	Packet Size	Rate
SS1	BS	UGS	0.5 s	40.0 s	1500	512
SS2	BS	RtPs	1.0 s	40.0 s	1500	512
SS3	BS	NrtPs	1.5 s	40.0 s	(512,1024)	
SS4	BS	BE	2.0 s	40.0 s	(512,1024)	
BS	SS5	UGS	0.5 s	40.0 s	1500	512
BS	SS6	RtPS	1.0 s	40.0 s	1500	512
BS	SS7	NrtPS	1.5 s	40.0 s	(512,1024)	
BS	SS8	BE	2.0 s	40.0 s	(512,1024)	

12.3.1.2   Simulation Result

Initialization

---

num_nodes is set 9
channel.cc:sendUp - Calc highestAntennaZ_ and distCST_
highestAntennaZ_ $= 1.5$, distCST_ $= 550.0$
SS 1 is sending RNGREQ to BS
SS 4 is sending RNGREQ to BS
SS 2 is sending RNGREQ to BS
SS 3 is sending RNGREQ to BS
BS is sending RNGRSP to SS 1
SS_X: 265.000000 SS_Y: 550.000000
BS_X: 425.000000 BS_Y: 450.000000 distance:188.679623 64 QAM
BS is sending RNGRSP to SS 2
SS_X: 385.000000 SS_Y: 652.000000
BS_X: 425.000000 BS_Y: 450.000000 distance:205.922315 64 QAM
BS is sending RNGRSP to SS 3
SS_X: 225.000000 SS_Y: 380.000000
BS_X: 425.000000 BS_Y: 450.000000 distance:211.896201 64 QAM
BS is sending RNGRSP to SS 4
SS_X: 310.000000 SS_Y: 265.000000
BS_X: 425.000000 BS_Y: 450.000000 distance:217.830209 64 QAM
SS 8 is sending RNGREQ to BS
SS 7 is sending RNGREQ to BS
SS 6 is sending RNGREQ to BS
SS 5 is sending RNGREQ to BS
BS is sending RNGRSP to SS 5
SS_X: 850.000000 SS_Y: 284.000000
BS_X: 425.000000 BS_Y: 450.000000 distance:456.268561 64 QAM
BS is sending RNGRSP to SS 6
SS_X: 860.000000 SS_Y: 642.000000
BS_X: 425.000000 BS_Y: 450.000000 distance:475.488170 64 QAM
BS is sending RNGRSP to SS 7
SS_X: 900.000000 SS_Y: 352.000000
BS_X: 425.000000 BS_Y: 450.000000 distance:485.004124 64 QAM
BS is sending RNGRSP to SS 8
SS_X: 925.000000 SS_Y: 515.000000
BS_X: 425.000000 BS_Y: 450.000000 distance:504.207299 64 QAM

Signal Transmission Start

At 0.5 s SS1 established UL connection with BS, SS5 established DL connection with BS, the UGS service began. (Refer to Figure 12.2)

At 1.0 s SS2 established UL connection with BS, SS6 established DL connection with BS, the rtPS service began.

At 1.5 s SS3 established UL connection with BS, SS7 established DL connection with BS, the nrtPS service began.

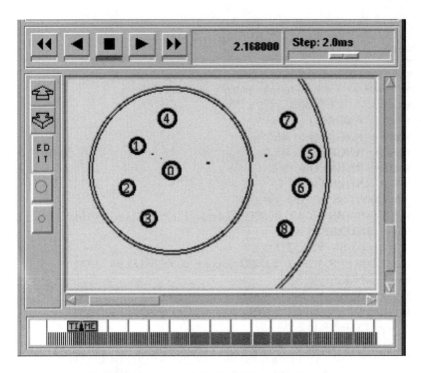

Figure 12.2   At t = 2.0 s the UGS service began.

At 2.0 s SS4 established UL connection with BS, SS8 established DL connection with BS, the BE service began.

12.3.1.3   Performance Comparison

Throughput
Figure 12.3 shows the throughput of four different services:

- The rtPS service, signal transmission started around 1.0 s, the throughput remained around 1.8 Mbps.
- The UGS service, signal transmission started around 0.5 s, the throughput remained around 370 kbps.
- The nrtPS service, signal transmission started around 1.5 s, the throughput remained around 100 kbps.
- The BE service, signal transmission started around 2.0 s, the throughput remained around 40 kbps.

Delay
System MAC Delay
As shown in Figure 12.4, the data transmission started at 0.5 s, the real-time delay at MAC layer stabilized around 0.006 s.

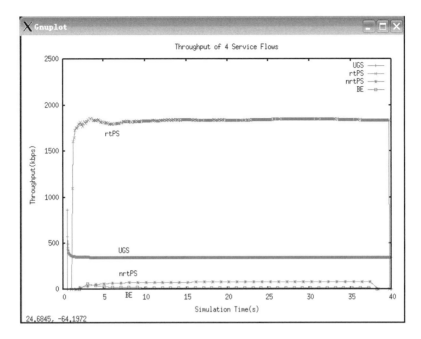

Figure 12.3   Throughput comparisons of different services.

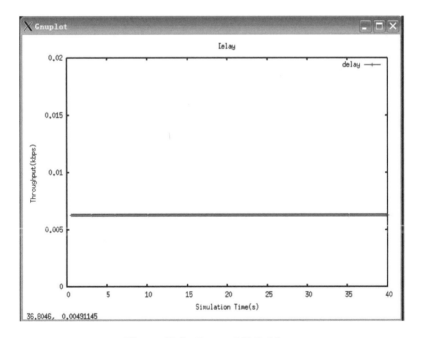

Figure 12.4   System MAC delay.

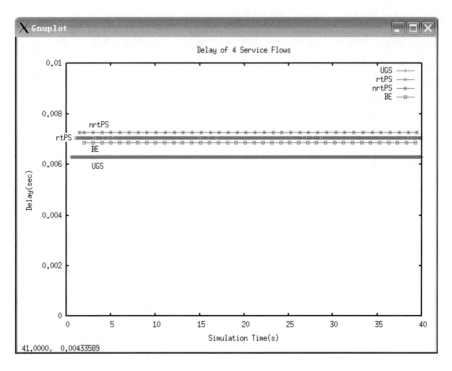

Figure 12.5   Delay comparison of different services.

Delay of Different Services
As shown in Figure 12.5, the delay for the 4 services UGS, rtPS, nrtPS, BE all remained between 0.006 s-0.008 s.
   And

$$delay_{UGS} < delay_{BE} < delay_{rtPS} < delay_{nrtPS}$$

Jitter
We used the following formula for jitter calculation:

$$Jitter = \{[(receivetime(j) - sendtime(j)] - (receivetime(i) - sendtime(i)]\}/(j - i)$$

   Figure 12.6 shows the system MAC jitter and the respective jitter for the four services. As can be observed from the figure, the jitter basically remained 0.

12.3.2   Mobility Support

12.3.2.1   Setup of Simulation Scenario

Setup of Fixed SS
The scenario setup was the same as the scenario setup in subsection 12.3.1.1.

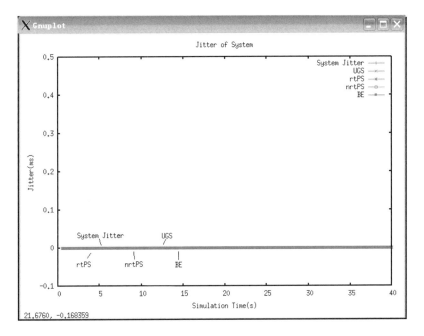

Figure 12.6 Jitter of system.

Setup of Mobile SS

Table 12.4 Mobility setup of SSs

Node	1	2	3	4	5	6	7	8
Original Position	265, 550	225, 380	310, 265	385, 652	925, 515	900, 352	860, 642	850, 284
Destination Position	661, 88	958, 234	960, 42	141, 174	777, 408	325, 839	1357, 762	1331, 187
Moving Speed	2 m/s	2 m/s	2 m/s	2 m/s	2 m/s	2 m/s	2 m/s	2 m/s
	5 m/s	5 m/s	5 m/s	5 m/s	5 m/s	5 m/s	5 m/s	5 m/s
	10 m/s	10 m/s	10 m/s	10 m/s	10 m/s	10 m/s	10 m/s	10 m/s

Simulation Result
Fixed SS
The simulation result remained the same as depicted in subsection 12.3.1.3.

Mobile SS
At t = 0 SSs started moving towards the destination positions (refer to Figure 12.7).

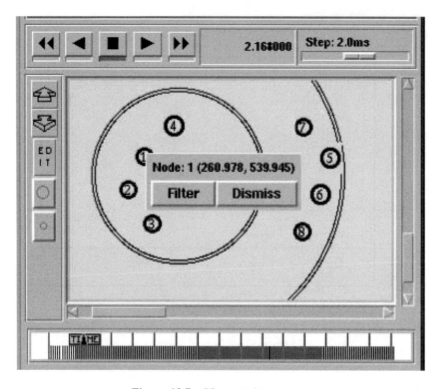

Figure 12.7 SSs started to move.

At t = 0.5 s SS1 began transmitting information – SS5 began receiving information...

Performance Comparison
Throughput
UGS Service
The throughput of the UGS service with SSs moving at a speed of 2 m/s, 5 m/s and 10 m/s
are shown in Figure 12.8. The throughput was almost identical and we had approximately
the same results as for fixed SSs. As can be observed from the figure, the variation of
moving speed has little impact on the throughput of UGS.

rtPS Service
The throughput of the rtPS service with SSs moving at a speed of 2 m/s, 5 m/s and 10 m/s
are shown in Figure 12.9. The throughput was almost identical and we had approximately
the same results as for fixed SSs. As can be observed from the figure, the variation of
moving speed has little impact on the throughput of rtPS.

nrtPS Service
The throughput of the nrtPS service with SSs moving at a speed of 2 m/s, 5 m/s and 10 m/s
are shown in Figure 12.10. The throughput was almost identical and we had approximately

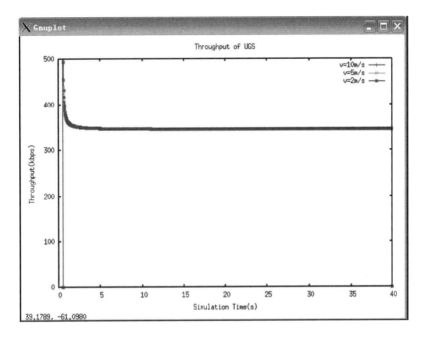

Figure 12.8   Throughput variation of UGS.

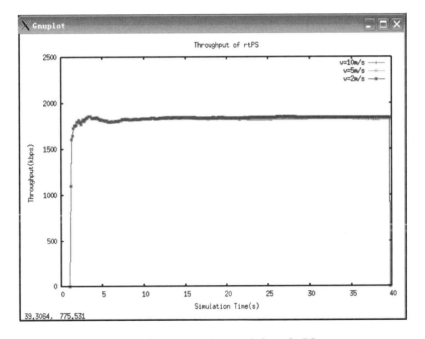

Figure 12.9   Throughput variation of rtPS.

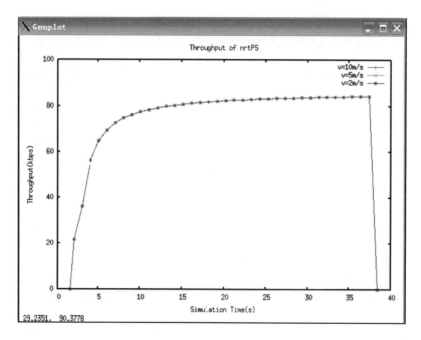

Figure 12.10   Throughput variation of nrtPS.

the same results as for fixed SSs. As can be observed from the figure, the variation of
moving speed has little impact on the throughput of nrtPS.

BE Service
The throughput of the BE service with SSs moving at a speed of 2 m/s, 5 m/s and 10 m/s
are shown in Figure 12.11. The throughput was almost identical and we had approximately
the same results as for fixed SSs. As can be observed from the figure, the variation of
moving speed has little impact on the throughput of BE.

Delay
MAC Layer Delay
As shown in Figure 12.12:
   During an interval of 0–20 s, since the SSs were moving towards the BS, the delay
diminished and the delay decreased at a quicker rate as when the node moved faster.
   During an interval of 20–40 s, the system's MAC delay continued to decrease for SSs
moving at v = 2 m/s and 5 m/s; as for SSs with v = 10 m/s, since they probably arrived
at the destination point and remained static, the system's MAC delay started to increase.

UGS Service Delay
As shown by Figure 12.13, the variation in the UGS service delay was similar to the
average MAC delay and the curves were relatively smooth.

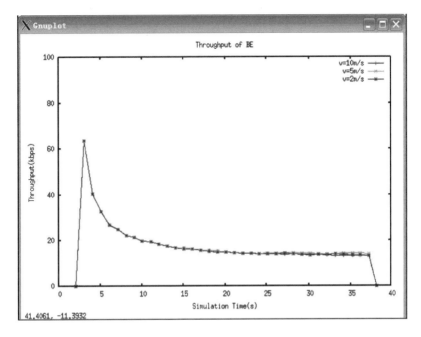

Figure 12.11   Throughput variation of BE.

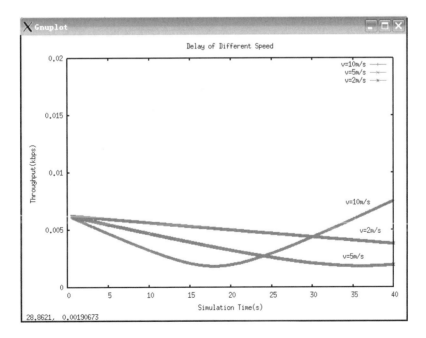

Figure 12.12   MAC layer delay variation.

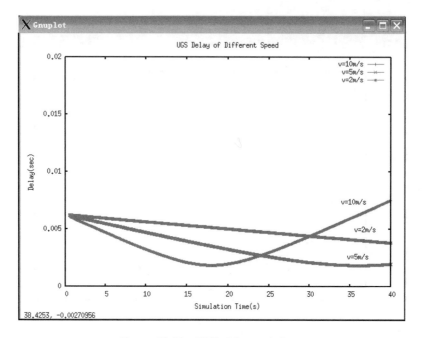

Figure 12.13   UGS delay variation.

rtPS Service Delay
As shown by Figure 12.14, the variation in rtPS service delay was similar to the average MAC delay, but the real-time delay was slightly higher than the system delay and UGS delay.

nrtPS Service Delay
As shown by Figure 12.15, nrtPS service delay was increasing with the movement of SSs; while the increase rate became higher when the speed of movement was greater. In particular, for SSs with a speed of $v = 10$ m/s, the delay reached 13 ms.

BE Service Delay
As shown by Figure 12.16, the variation in BE service delay was similar to the average MAC delay.

Jitter
MAC Layer Jitter
As shown by Figure 12.17, the jitter remained basically 0.

Jitter of UGS
As shown by Figure 12.18, the jitter of UGS was similar to MAC jitter, which remains around 0–0.01 ms.

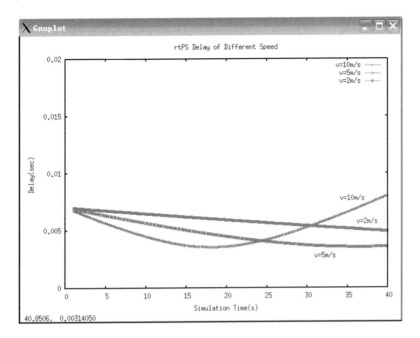

Figure 12.14   rtPS delay variation.

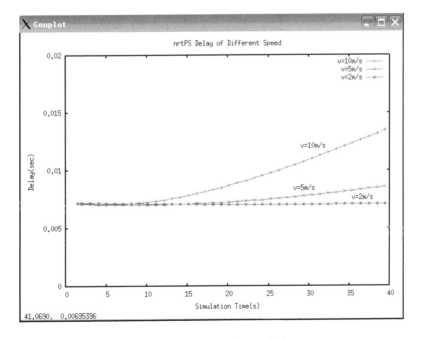

Figure 12.15   nrtPS delay variation.

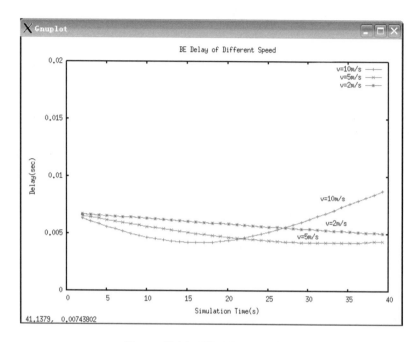

Figure 12.16   BE delay variation.

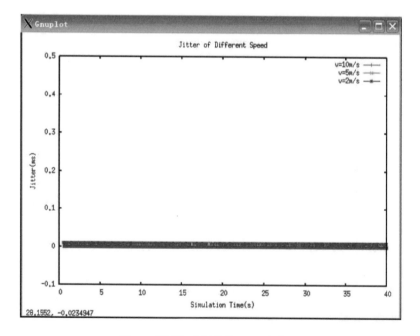

Figure 12.17   MAC jitter variation.

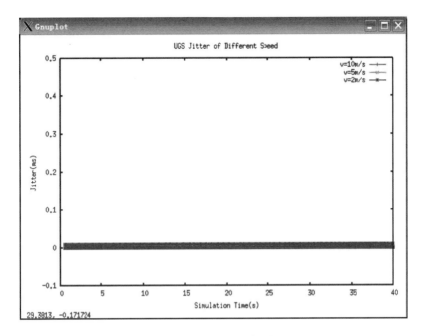

Figure 12.18   Jitter of UGS.

Jitter of rtPS
As shown by Figure 12.19, the jitter of rtPS was around 0–0.04 ms, slightly greater than that of the UGS service.

Jitter of nrtPS
As shown by Figure 12.20, when the moving speed is $v = 2$ m/s, the jitter remained zero; as the speed increased, the jitter increased accordingly and the rate of increase was proportional to the speed of movement.

Jitter of BE
As shown by Figure 12.21, the jitter was higher at the beginning of the service.

When the speed was $v = 2$ m/s, the jitter decreased in accordance with the time, which remained around 0.05 ms. When the speed increased, the jitter of BE represented greater fluctuation without a particular pattern.

12.4 Analysis of QoS Simulation Results

12.4.1 Fixed SSs

12.4.1.1 Throughput

$$Throughput_{BE} < Throughput_{nrtPS} < Throughput_{UGS} < Throughput_{rtPS}$$

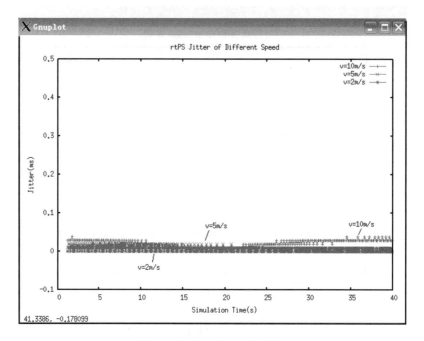

Figure 12.19 Jitter of rtPS.

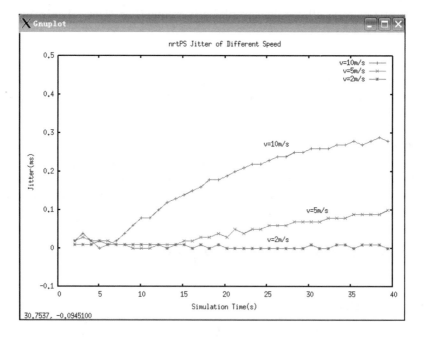

Figure 12.20 Jitter of nrtPS.

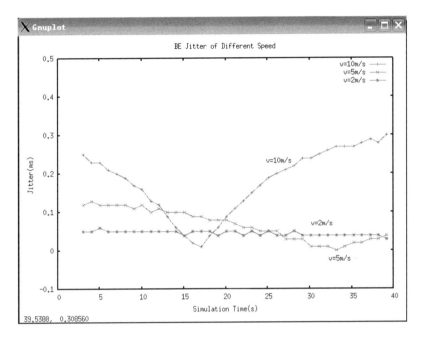

Figure 12.21 Jitter of BE.

The rtPS service had a periodic variable-sized packet. It had the highest throughput at around 1.8 Mbps, which had a minor fluctuation with time.

The UGS service is a constant-bitrate service flow with periodic fixed-sized packets, whose throughput remained around 370 Kbps.

The nrtPS service is a non-real time variable-bitrate service flow, whose throughput was smaller but was overall stable.

The throughput of the BE service was higher at the beginning and decreased with time. This was because the BE service did not offer integral reliability. When the network payload was low, the BE service had more polling opportunities, whereas when the network payload became higher it might have had few or even no polling opportunities. Therefore, when all the other service flows started working, the throughput of the BE became low owing to its failure to obtain polling opportunities.

12.4.1.2 Delay

The UGS service was sensitive to delay, therefore it should have had the smallest delay and fluctuation; the rtPS service had a higher service expense and transmission delay than UGS because the SS needed to raise requests periodically; nrtPS was designed for services not sensitive to delay, it didn't have high demand for delay; the BE service didn't provide guarantees for throughput and delay.

From the results of performance testing, the comparison of delay for the four services is listed below, which meets the service requirements in the 802.16 standard.

$$delay_{UGS} < delay_{BE} < delay_{rtPS} < delay_{nrtPS}$$

12.4.1.3 Jitter

As shown by the test results, when the SSs were fixed, there was almost no jitter for all services.

12.4.2 Mobile SSs with Same Speed

12.4.2.1 Throughput

As shown by the results, the movement of SSs did not have any evident impact on throughput.

12.4.2.2 Delay

Except for nrtPS, all delay decreased during the movement of SSs; this might be due to the fact that the SSs approached BS gradually during the movement. For nrtPS the delay increased possibly because the high throughputs of UGS and rtPS had caused network congestion.

12.4.2.3 Jitter

From the test results we found:

The jitter of UGS was almost zero, while rtPS had a higher jitter than UGS, remaining within 0–0.03 ms, which met the service requirements for jitter sensitivity.

The nrtPS service had a higher jitter and a tendency to grow. But since nrtPS is designed for services with a low delay sensitivity, this also met the service requirements.

The jitter of BE was quite random, which was explained by the fact that the BE service supports a non real-time packet data service which has no bit rate or jitter requirement and which does not require a guarantee of throughput and delay.

12.4.3 Mobile SSs with Varying Speed

When the speed of movement increased, the rate of variation of service delay and jitter also increased.

Overall, the performance test results demonstrated that generally the simulation matched the QOS definitions of the four service flows in the WiMAX/802.16 standard and also demonstrated that WiMAX could support the moderate speed of movement of subscriber stations in a cell.

12.5 Enhancement – A New Solution of Cell Selection

12.5.1 System Model

In this section we explain in detail the system model of our proposed WiMAX cell selection method. We can split the system model into four phases based on a chronological sequence, as shown by Figure 12.22:

System Model for an Enhanced Cell Selection method in WiMAX network	
Phase 1	Estimation of BS maximum effective capacity and the idle capacity using the actual traffic throughput statistics
Phase 2	Idle capacity advertisement from BS to MS via MOB_NBR_ADV message
Phase 3	MS triggers a scanning and handover process when the capacity threshold has been reached
Phase 4	Decision algorithm for target cell selection => Stay or Switch?

Figure 12.22   System model for enhance cell selection method in WiMAX network.

12.5.1.1   Phase 1: Effective Capacity Estimation

We refer to 'effective capacity' as the available resources in a WiMAX BS for data transmission to MSs. We examine the 802.16d OFDM PHY layer specifications in order to estimate the effective capacity. Within each OFDM frame, not all of the symbols are used for data transmission; there are overheads such as control and management information ensuring data integrity and synchronization between BS and MSs. Figure 12.23 shows a typical OFDM frame structure [4], where we have marked the overheads in black.

We first calculate the number of symbols N_{symbol} in each OFDM frame:

$$T_{symbol} = (N_{FFT}/f_s)^*(1 + G) \tag{12.1}$$

$$N_{symbol} = T_{frame}/T_{symbol} \tag{12.2}$$

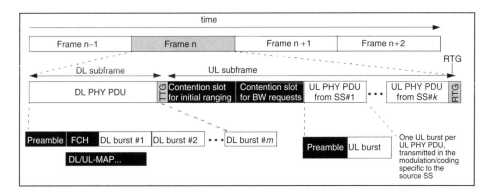

Figure 12.23   802.16 OFDM frame structure.

N_{FFT} is the number of subcarriers. For OFDM PHY, N_{FFT} is equal to 256. f_s is the sampling factor and G is CP ratio. T_{symbol} and T_{frame} are the symbol and frame duration respectively.

Next, from the total number of symbols we exclude the overhead symbols as illustrated by Figure 12.23. 802.16d specifies that in the DL subframe the preamble takes two OFDM symbols followed by a frame control header (FCH) of one symbol. The UL subframe starts with a contention slot for ranging consisting of a long preamble (two symbols), RNG-REQ message (two symbols) and three symbols to span round trip delay, followed by the BW contention requests which are two symbols long. We also need to exclude the guarding gaps TTG and RTG which are in the order of one symbol long. [4], [6], [7].

The size of the DL and UL maps might vary according to the number of active MSs connected to the BS, the size of DL-MAP is $(64 + 32^*n)/N_{bpsymbol}$ and the size of UL-MAP is $(56 + 48^*n)/N_{bpsymbol}$, n stands for the number of active MSs and $N_{bpsymbol}$ is the number of useful bits per symbol. Additionally each UL burst has to start with short preamble so it depends on the number of MSs with uplink transmission, denoted n_{UL}. In total we have the number of overhead symbols $N_{symbol\_overhead}$:

$$N_{symbol\_overhead} = 13 + \frac{120 + 80^*n}{N_{bpsymbol}} + n_{UL} \qquad (12.3)$$

Bear in mind that additional overheads such as MAC PDU packing or periodical DCD/UCD message broadcast could eventually increase the number of overhead symbols. Yet considering that the impact is relatively minor, we have not taken these overheads into account in our estimation.

Finally, we calculate the $N_{bpsymbol}$ and derive the maximum effective capacity of a WiMAX BS:

$$N_{bpsymbol} = (192^*Coding\_rate^*Efficiency) - 8 \qquad (12.4)$$

$$C_{effective} = \frac{(N_{symbol} - N_{symbol\_overhead})^*N_{bpsymbol}}{T_{frame}} \qquad (12.5)$$

If we take a simplified scenario, for a common 5 MHz WiMAX PMP network with a frame duration of 4 ms and CP ration of 1/4, modulated in a 16QAM 3/4 scheme, in which only the BS sends downlink data transmission to a number of MSs (20 in average), we could obtain the following numerical values from equations (12.1)–(12.5) $N_{symbol} = 90$, $N_{symbol\_overhead} = 16$, $N_{bpsymbol} = 578$ and the maximum effective capacity of the BS is:

$$C_{effective} = 10.198 \ Mbps$$

We have validated this estimation in our simulation as a preliminary check and the output has shown a match between the theoretical value and the simulation result. Refer to section 12.4.

12.5.1.2   Phase 2. Idle Capacity Advertisement

In our cell selection solution, the neighbour advertisement capability of the 802.16e specification plays an important role as a carrier to update dynamically the MSs about the effective idle capacity of the serving BS and those of neighbouring BSs.

As presented in the previous subsection, each BS can estimate its maximum effective capacity, on a real-time basis, according to its PHY layer parameters and active MS numbers. Through statistics the BS is also aware of the current data traffic throughput. Therefore, each BS could obtain the effective idle capacity:

$$C_{idle} = C_{effective} - S_{throughput} \qquad (12.6)$$

On a periodic basis, the BS updates its effective idle capacity information to the neighbouring BSs over the backbone and each BS broadcasts the idle capacity information both of itself and of the neighbouring BSs to the connected MSs via the MOB_NBR_ADV messages, together with the DCD/UCD information.

12.5.1.3   Phase 3: Handover Trigger

As a consequence of idle capacity advertisement, the MS will be aware of the serving and neighboring BSs' traffic load. An additional trigger for a handover process based on this capacity information is to be introduced: only on satisfying the following conditions should a MS trigger the scanning and handover process.

Condition 1: $C_{idle}(Serving\ BS) < Requested\ dataflow\ rate(MS)$

Condition 2: $\exists k \in \{i|BS_i\ is\ a\ neighbor\ BS\}$:

$C_{idle}(BS_k) > \alpha^* Requested\ dataflow\ rate(MS)\ (\alpha \geqslant 1)$

Condition 3: $P\{Handover\_Trigger = True\} < p_{switch}\ (0 \leqslant p_{switch} \leqslant 1)$

Condition 1 defines a threshold when the idle capacity of a current serving BS can no longer meet the requested dataflow rate of the MS.

Condition 2 is a stricter condition designed as to avoid inutile scanning when a MS knows upfront that no neighbouring BS offers better idle capacity. α is the assurance factor; by choosing a greater α the MS will be guaranteed to switch only to another WiMAX cell with sustainable idle resources, in order to compensate for the packet loss due to cell switch handover.

Condition 3 is a global filter which we introduced in order to impose an overall control on cell switch frequency. p_{switch} denotes the probability that a MS meeting *Condition 1* and *Condition 2* will trigger the handover process as a random event. A greater p_{switch} implies more frequent cell switch, a cell with heavy traffic load will be relieved quickly but congestion in the target cell might be produced due to massive cell switch; whereas a smaller p_{switch} implies less frequent cell switch therefore less handover cost, but the traffic load will be balanced more slowly. We will analyze the influence of parameter p_{switch} in the next subsection.

Of course, the original WiMAX handover trigger based on signal power attenuation (CINR or RSSI values) is still valid and works in parallel with this additional trigger based on BS idle capacity.

12.5.1.4   Phase 4: Target Cell Decision Algorithm

Once the WiMAX handover process has been triggered, the MS scans the neighbouring BSs and makes a decision to stay or to switch to another cell. The decision factor for each candidate BS depends on two factors: idle capacity and signal strength. We have combined the two factors into a weighted target cell decision function:

$$D_k = \beta_1^* \frac{C_{idle}(BS_k)}{C_{idle}(Serving\ BS)} + \beta_2^* 10^* \log \frac{P_{signal}(BS_k)}{P_{signal}(Serving\ BS)} \qquad (12.7)$$

P_{signal} denotes the signal power.

In equation (12.7), we put $C_{idle}(Serving\ BS)$ and $P_{signal}(Serving\ BS)$ as denominators to normalize the function. β_1 and β_2 are the weights for idle capacity and signal power. Based on the decision function, MS selects the candidate BS_k with the highest D_k as the target cell to switch.

12.5.2   Simulation Result

Our algorithm is implemented in a NS-2 simulator. NIST 802.16 and Mobility modules are installed to add support for WiMAX and handover functionality. The simulation scenario consists of a test area covered by four WiMAX BS, BS_0 is surrounded by three neighbouring BSs in a triangular position with overlapped contiguous areas. Sixty MSs are randomly dispersed in the coverage of BS_0 at the beginning and random movement starts towards the entire test area. We employed a simplified traffic model that each MS requests a constant bit rate UDP video flow at 0.5 Mbps. Table 12.5 lists the main parameters of the simulation scenario.

12.5.2.1   Simulation Result for Effective Capacity Estimation

As a preliminary step, we verify the appropriateness of our estimation of effective capacity. Applying the same OFDM parameters we used for theoretical numeric estimation, we spread an increasing number of MSs in one WiMAX cell and obtain the maximum data traffic throughput.

Table 12.5   Main parameters of simulation scenario

Parameter	Value
Frequency band	5 MHz
Propagation model	Two Ray Ground
Modulation scheme	16 QAM 3/4
BS coverage	1000 m
Reception power threshold	1.27e-13
Frame duration	4 ms
Antenna model	Omni antenna
Contention size	5
Link going down factor	1.2
Simulation duration	50 s
BS number	4
MS number	60
Requested dataflow rate of each MS	0.5 Mbps

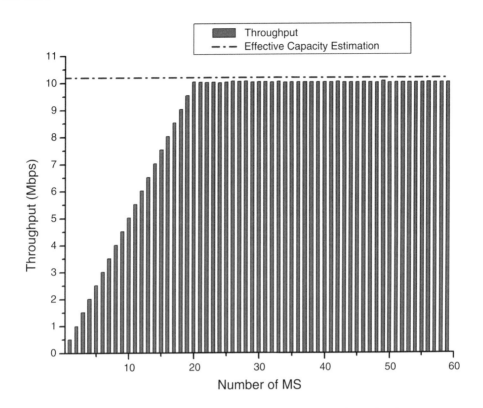

Figure 12.24 Single cell traffic load vs. effective capacity estimation.

Figure 12.24 illustrates that with an increasing number of MSs, the traffic load within a single WiMAX cell reaches its maximum value, which approaches closely, yet remains within the theoretical estimation of the effective capacity value. The slight difference is explained by the existence of some extra overheads in the OFDM frame such as padding space in MAC PDU and the periodical broadcast messages, which are neglected in our theoretical estimation.

12.5.2.2 Performance Evaluation

Parameters β_1 and β_2 are test-scenario-dependent factors. By an evaluation of the capacity, transmission power, MS dataflow rate and the coverage overlaps in our simulation scenario, we set up the parameter setting $\{\beta_1 = 2, \beta_2 = 1\}$ so as to place greater emphasis on the capacity factor. Another parameter α depends on the MS dataflow rate, the cost of BS congestion (i.e. packet loss) and the cost of cell handover. Through incremental attempts, we fine-tuned the parameter $\alpha = 4$. Finally, we select a set of values $p_{switch} = \{0.2, 0.5, 0.8\}$.

In order to evaluate the performance of the proposed method, we analyze the overall data throughput (excluding the management and control messages) in function of time and compare it with the original WiMAX cell selection method based on signal power

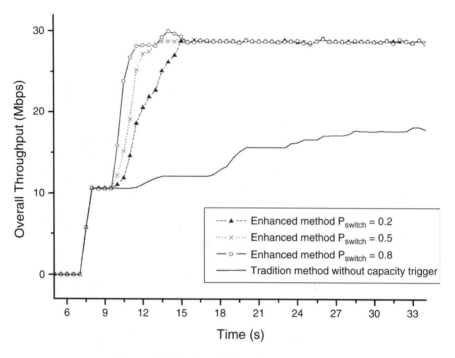

Figure 12.25 Overall throughputs vs. time.

criteria (refer to Figure 12.25). We also present the data packet loss due to the handover process for different p_{switch} values in Figure 12.25.

Figure 12.25 shows a significant improvement in performance: after 15 s the WiMAX network with the enhanced cell selection method has achieved an overall throughput close to 30 Mbps, meaning that BSs are providing almost the full dataflow traffic requested by the 60 MSs. We observe that the period between 9 s and 15 s is the transition period during which MSs connected initially to BS_0 decide to switch to a neighbouring cell due to traffic congestion. As a comparison, in a WiMAX network with the traditional method, without capacity estimation and advertisement, MSs will switch to a neighbouring cell only at low reception signal power; we notice that the total throughput has gradually reached 17 Mbps by the end of 33 seconds.

As we have expected, the influence of different p_{switch} is illustrated in both Figure 12.25 and Figure 12.26. With a greater $p_{switch} = 0.8$ MSs switch between cells more frequently. As a result the entire network reaches its maximum traffic throughput quickly, together with some fluctuations during the transition period (refer to Figure 12.25). When $p_{switch} = 0.5$ and 0.2, the increase of the overall throughput appears to be slower and smoother. However, the sacrifice of applying a greater p_{switch} is the higher loss of packets during more frequent handovers. As shown in Figure 12.26, a network with a greater p_{switch} suffers from a peak packet loss during the transition period; whereas the small p_{switch} brings less abrupt packet loss. To summarize, by defining the p_{switch}, we define the tradeoff between traffic optimization rate and minimal data loss.

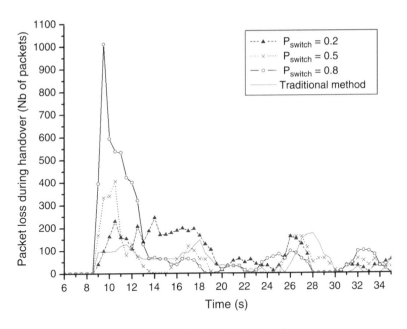

Figure 12.26 Packet loss during handover.

12.6 Summary

In this chapter, we simulated the QoS of a WiMAX network in several scenarios and proposed an enhanced cell selection solution for a WiMAX network. In the QoS simulation, we simulated the scenario of fixed and mobile subscriber stations in a cell and analyzed the performance test results. We also compared the result with the service flow QoS definitions in the 802.16 standard and arrived at a conclusion. In our proposal of the enhanced cell selection solution, we present capacity estimation and advertisement functions that are in line with the IEEE 802.16 specifications and a conditional handover trigger with target cell decision algorithm is designed for mobile stations. A tradeoff parameter p_{switch} designed to balance the traffic optimization rate and the data loss is also presented. The simulation results show that for a multi-cell coverage area, this method could enhance overall system performance significantly and thus improve the user's experience of applications on a mobile WiMAX station.

References

[1] M. Aguado, E. Jacob, P. Saiz, J. Matias, M. Higuero, N. Toledo and M. Berbineau, 'Scanning on Handover Enhancement Issues in Video Application Deployments on WiMAX Mobile Networks', Broadband Multimedia Systems and Broadcasting, 2008 IEEE International Symposium, pp.1–7, March–April 2008.

[2] P. Boone, M. Barbeau and E. Kranakis, 'Strategies for Fast Scanning and Handovers in WiMAX/802.16' *International Journal of Communication Networks and Distributed Systems* **1**(4/5/6): 414–32, 2008.

[3] S. Tabbane, 'Evaluation of Handover Target Cell Determination Algorithms for Heterogeneous Cellular Environments', Communications ICC 97 Montreal, 'Towards the Knowledge Millennium' IEEE International Conference Vol. 1, pp. 396–400, 1997.

[4] IEEE 802.16-2004, IEEE Standard for Local and Metropolitan Area Networks, Air Interface for Fixed Broadband Wireless Access Systems, Oct. 2004.
[5] IEEE 802.16e, IEEE Standard for Local and Metropolitan Area Networks, Air Interface for Fixed Broadband Wireless Access Systems, Amendment 2: Physical and Medium Access Control Layers for Combined Fixed and Mobile Operation in Licensed Bands and Corrigendum 1, Feb. 2006.
[6] A. Belghith and L. Nuaymi, 'WiMAX Capacity Estimation and Simulation Results', Vehicular Technology Conference, 2008. VTC Spring 2008. IEEE, 2008.
[7] P. Mach and R Bestak, 'WiMAX Performance Evaluation', Networking, 2007. ICN '07. p. 17, April 2007.

Appendix List of Standards

WiMAX (Stage 2)	Forum, WiMAX Forum Network Architecture. Stage 2: Architecture, Tenets, Reference Model and Reference Points. V. 1.2, WiMAX Forum Std., 2009.
WiMAX (Stage 3)	Forum Network Architecture. Stage 3: Detailed Protocols and Procedures, WiMAX Forum Std., 2009.
802.1X	IEEE Std 802.1X-2004, "802.1X IEEE Standard for Local and metropolitan area networks, Port-Based Network Access Control", Revision of IEEE Std 802.1X-2001, IEEE, 2004.
802.11	IEEE 802.11-2007 "IEEE Standard for Information technology-Telecommunications and information exchange between systems-Local and metropolitan area networks-Specific requirements – Part 11: Wireless LAN Medium Access Control (MAC) and Physical Layer (PHY) Specifications", IEEE, 2007.
802.11e	IEEE 802.11-2005 "IEEE Standard for Information Technology – Telecommunications and Information Exchange between Local and Metropolitan Area Networks – Specific Requirements – Part 11: Wireless LAN Medium Access Control (MAC) and Physical Layer (PHY) Specifications" Amendment 8: Medium Access Control (MAC) Quality of Service Enhancements.
802.11i	IEEE Std 802.11i-2004, "Amendment to IEEE Std. 802.11, 1999 Edition, Amendment 6: Medium Access Control (MAC) Security Enhancements, Part 11: Wireless LAN Medium Access Control (MAC) and Physical Layer (PHY) specifications", IEEE, 2004.

WiMAX Security and Quality of Service: An End-to-End Perspective Edited by Seok-Yee Tang, Peter Müller and Hamid Sharif
© 2010 John Wiley & Sons, Ltd

802.11n	IEEE 802.11n-2009 "IEEE Draft Standard for Information Technology – Telecommunications and Information Exchange between Local and Metropolitan Area Networks – Specific Requirements – Part 11: Wireless LAN Medium Access Control (MAC) and Physical Layer (PHY) Specifications" Amendment: Enhancements for Higher Throughput.
802.16 Conformance01-2003	IEEE Standard for Conformance to IEEE 802.16, Part 1: Protocol Implementation Conformance Statement (PICS) Proforma for 10–66 GHz WirelessMan-SC air interface.
802.16 Conformance02-2003	IEEE Standard for Conformance to IEEE 802.16, Part 2: Test Suite Structure and Test Purpose for 10–66 GHz WirelessMan-SC air interface.
802.16 Conformance03-2004	IEEE Standard for Conformance to IEEE 802.16, Part 3: Radio Conformance Tests (RCT) for 10–66 GHz WirelessMAN-SC Air interface.
802.16 Conformance04-2006	IEEE Standard for Conformance to IEEE 802.16, Part 4: Protocol Implementation Conformance Statement (PICS) Proforma for Frequencies below 11 GHz.
IEEE 802.16.2-2004	IEEE Recommended Practice for Local and metropolitan area networks, Coexistence of fixed broadband wireless access systems.
802.16-2001	IEEE 802.16-2001, "IEEE Standard for Local and metropolitan area networks, Part 16: Air Interface for Fixed Broadband Wireless Access Systems", Approved 6 December 2001, IEEE Press, 2002.
802.16-2004	IEEE Standard for Local and Metropolitan Area Networks Part 16: Air Interface for Fixed Broadband Wireless Access Systems.
802.16-2004/Cor1-2005	IEEE Standard for local and metropolitan area networks. Part 16: Air interface for fixed and mobile broadband wireless access systems. Amendment 2: Physical and medium access control layers for combined fixed and mobile operation in licensed bands and corrigendum.
802.16a-2003	EEE Standard for Local and metropolitan area networks – Part 16: Air Interface for Fixed Broadband Wireless Access Systems – Amendment 2: Medium Access Control Modifications and Additional Physical Layer Specifications for 2–11 GHz

802.16c-2002	IEEE Standard for Local and metropolitan area networks – Part 16: Air Interface for Fixed Broadband Wireless Access Systems-Amendment 1: Detailed System Profiles for 10–66 GHz.
802.16d	IEEE Standard for Local and metropolitan area networks, Part 16: Air Interface for Fixed Broadband Wireless Access Systems, 2004.
802.16e	IEEE 802.16e-2005 "IEEE Standard for Local and metropolitan area networks Part 16: Air Interface for Fixed and Mobile Broadband Wireless Access Systems Amendment for Physical and Medium Access Control Layers for Combined Fixed and Mobile Operation in Licensed Bands", IEEE, 2006.
802.16f	IEEE Standard for Local and metropolitan area networks Part 16: Air Interface for Fixed Broadband Wireless Access Systems- Amendment 1: Management Information Base.
802.16g	IEEE Standards for Local and metropolitan area networks – Part 16: Air Interface for Fixed and Mobile Broadband Wireless Access Systems – Amendment 3: Management Plane Procedure and Services.
802.16j	IEEE Standard for Local and metropolitan area networks Part 16: Air Interface for Broadband Wireless Access Systems Amendment 1: Multiple Relay Specification.
802.16k	IEEE Standard for Local and Metropolitan Area Networks Media Access Control (MAC) Bridges Amendment 5: Bridging of IEEE 802.16.
802.16m	IEEE 802.16m-2009 "IEEE Standard for Local and metropolitan area networks Part 16: Air Interface for Broadband Wireless Access Systems" Amendment: Advanced Air Interface.
802.21	IEEE 802.21-2008 "IEEE Standard for Local and metropolitan area networks Part 21: Media Independent Handover", IEEE, Jan. 2008.
RFC1213	The Internet Engineering Task Force (IETF); Management Information Base for Network Management of TCP/IP-based internets: MIB-II.
RFC2131	The Internet Engineering Task Force (IETF); Dynamic Host Configuration Protocol (DHCP).

RFC2459   The Internet Engineering Task Force (IETF); Internet X.509
                 Public Key Infrastructure Certificate and CRL Profile.

RFC2560   The Internet Engineering Task Force (IETF); X.509 Internet
                 Public Key Infrastructure Online Certificate Status
                 Protocol – OCSP.

RFC2865   The Internet Engineering Task Force (IETF); Remote
                 Authentication Dial In User Service (RADIUS), Network
                 Working Group Std., 2000.

RFC 3344   The Internet Engineering Task Force (IETF); IP Mobility
                 Support for IPv4 (Mobile IP), Network Working Group Std.,
                 2002.

RFC3588   Internet Engineering Task Force (IETF); Diameter Base
                 Protocol, Network Working Group Std., 2003.

RFC3748   Internet Engineering Task Force (IETF); Extensible
                 Authentication Protocol (EAP), Network Working Group Std.,
                 2004.

RFC 4187   Internet Engineering Task Force (IETF); Extensible
                 Authentication Protocol Method for 3rd Generation
                 Authentication and Key Agreement (EAP-AKA), Network
                 Working Group Std., 2006.

RFC5246   Internet Engineering Task Force (IETF); The Transport Layer
                 Security (TLS) Protocol Version 1.2, Network Working Group
                 Std., 2008.

RFC5281   Internet Engineering Task Force (IETF); Extensible
                 Authentication Protocol Tunneled Transport Layer Security
                 Authenticated Protocol Version 0 (EAP-TTLSv0), Network
                 Working Group Std., 2008.

H.261   ITU-T H.261:1993 "Video Codec for Audio-Visual Services at
                 px64 Kbits/s", ITU-T Recommendation H.261.

H.264   ITU-T H.264: 2009 "Advanced Video Coding for Generic
                 Audio-Visual Services", ITU-T Recommendation H.264.

MPEG-4   ISO/IEC 14496-10:2005 "Information Technology – Coding of
                 Audio Visual Objects – Part 10: Advanced Video Coding".

MPEG-7   ISO/IEC 15938-5:2003 "Information
                 Technology – Multimedia Content Description Interface – Part
                 5: Multimedia Description Definition Schemes".

AES	FIPS 197, "Announcing the Advanced Encryption Standard (AES)", Federal Information Processing Standards Publication 197, Nov. 2001.
DES	FIPS PUB 46-3, "Data Encryption Standard (DES)", Federal Information Processing Standards Publication, Reaffirmed 1999 October 25, U.S. Department of Commerce / National Institute of Standards and Technology, 1999.
RSA	"PKCS #1 v2.1: RSA Cryptography Standard", RSA Laboratories, June 2002.

Index